PROTEIN BIOTECHNOLOGY

PROTEIN BIOTECHNOLOGY

Gary Walsh

Industrial Biochemistry Programme
CECS Department
University of Limerick
Ireland

and

Denis R. Headon

Cell and Molecular Biology Group
Department of Biochemistry
University College Galway
Ireland

JOHN WILEY & SONS

Chichester · New York · Brisbane · Toronto · Singapore

Copyright © 1994 by John Wiley & Sons Ltd,
Baffins Lane, Chichester,
West Sussex PO19 1UD, England
Telephone (+44) (243) 779777

Reprinted September 1996

Other Wiley Editorial Offices

John Wiley & Sons, Inc., 605 Third Avenue,
New York, NY 10158-0012, USA

Jacaranda Wiley Ltd, 33 Park Road, Milton,
Queensland 4064, Australia

John Wiley & Sons (Canada) Ltd, 22 Worcester Road,
Rexdale, Ontario M9W 1L1, Canada

John Wiley & Sons (SEA) Pte Ltd, 37 Jalan Pemimpin #05-04,
Block B, Union Industrial Building, Singapore 2057

Library of Congress Cataloging-in-Publication Data

Walsh, Gary.
    Protein biotechnology / Gary Walsh, Dennis Headon.
        p.   cm.
    Includes bibliographical references and index.
    ISBN 0 471 94396 7—ISBN 0 471 94393 2 (pbk.)
    1. Proteins—Biotechnology. I. Headon, Denis.   II. Title.
    TP248.65.P76W35   1994
    660′.63—dc20   93-50775
                                                        CIP

British Library Cataloging in Publication Data

A catalogue record for this book is available from the British Library

ISBN 0 471 943967 (cloth)
        0 471 943932 (pbk)

Typeset in 10/12pt Times at Alden Press Limited, Oxford and Northampton

Printed and bound in Great Britain by Bookcraft (Bath) Ltd, Midsomer Norton, Somerset

Contents

Preface

Biotechnology has become a buzz word since the development of the techniques of genetic engineering in the 1970s. For millenia, biotechnology has been practised in the production of cheese, wines, yoghurts and other fermented products. Over many hundreds of years efforts have been expended to understand the underlying mechanism of living cells and to apply this knowledge for the common good. This process has progressed enormously over the past two decades with many of the new insights obtained proving to be of applied as well as of academic significance. The development of genetic engineering and techniques of monoclonal antibody production have proven particularly important in the development of modern biotechnology.

Over the past number of years, there has been a dramatic increase in the number of texts dealing with various aspects of the biological sciences, mainly detailing various aspects of biotechnology, such as recombinant DNA technology, monoclonal antibody technology, industrial enzymology and diagnostics.

As the majority of endproducts derived from the biotechnology industry are proteinaceous, it is surprising that few publications have appeared which comprehensively review the production and application of such products.

This textbook aims to provide a comprehensive and an up-to-date review of protein biotechnology by considering the various proteins produced, their sources, methods of production and purification, and their applications. Particular emphasis is placed upon the basic biological functions and characteristics of these proteins, which facilitates a greater appreciation of their potential applications.

The text caters mainly for students studying biotechnology who have prior knowledge of basic biochemical principles and should also be of value to students pursuing degrees in biochemistry, microbiology or any branch of the biomedical sciences. Its scope renders it useful to those currently working in the biotechnological industry.

We also wish to convey our sincere gratitude to the various people, companies, and institutes, whose input was of invaluable assistance in compiling this book. Special thanks is due to Sandy Lawson for her efforts in typing the manuscript. The assistance of Mike Kelly and Gerry Killeen in compiling figures and tables is also gratefully acknowledged. Thanks is also due to Nancy Shanley for proofreading the text and to Deirdre Kerans for her assistance in editing the manuscript.

A special word of gratitude is due to the various organisations, who, by providing relevant product information, contributed greatly to the industrial relevance of this book. While these organisations are listed separately, a special thanks is due to Dr Ken Jones of Affinity Chromatography Ltd, Dr Sally Adams of British Biotechnology, Dr David Bloxham of Celltech Ltd, Ms Judith Shuler of the U.S. Pharmaceutical Manufacturers' Association, Dr Ian Garner of Pharmaceutical Proteins Ltd and Dr G Allen of the Wellcome Research Laboratories, all of whom were especially helpful. We would also like to thank the staff of our publishers, John Wiley & Sons, whose professionalism and dedication were much in evidence during all stages of the project. Finally, we would like to dedicate the book to our families, friends and especially our former teachers. In this regard, G.W. would particularly like to mention John Donlon, Des Mulrooney, the staff of CCM and the late Prof. Michael P. Coughlan, whose sudden and untimely death stunned and saddened us all.

Gary Walsh
Denis R. Headon

Galway
January, 1994

Acknowledgments

We wish to acknowledge the information provided by the following which proved invaluable in writing this textbook.

Affinity Chromatography Ltd, Freeport, Ballasalla, Isle of Man.

Alltech, Inc., 3031 Catnip Hill Pike, Nicholasville, Kentucky 40356, U.S.A.

Amicon Ltd, Upper Mill, Stonehouse, Gloucestershire GL10 2BJ, U.K.

Beckman Instruments, Inc., 200 South Kraemer Boulevard, Brea, Ca., 92621-6209, U.S.A.

Biocon Biochemicals Ltd, Kilnagheary, Carrigaline, Co. Cork, Ireland.

Biopharma Process Systems Ltd, Stratton Rd, Winchester SO23 8JQ, U.K.

Bio Products Laboratory, Dagger Lane, Elstree, Herts WD6 3BX, U.K.

Biozyme Laboratories Ltd, Unit 6, Gilchrist-Thomas Estate, Blaenavon, Gwent NP4 9RL, South Wales, U.K.

Boehringer Mannheim U.K. Ltd, Bell Lane, Lewes, East Sussex BN7 1LG, U.K.

British Biotechnology Ltd, Watlington Rd, Cowley, Oxford OX4 5LY, U.K.

British Patent Office, State House 66–71 High Holborn, London WC1R 4TP, U.K.

Cambridge Biotech Ltd, Blackthorn Rd, Sandyford Industrial Estate, Foxrock, Dublin 18, Ireland.

Celltech Ltd, 216 Bath Rd, Slough SL1 4EN, Berkshire, U.K.

Enzymatix Ltd, Cambridge Science Park, Milton Rd, Cambridge CB4 4WE, U.K.

Flemming GmbH, Tulla, Co. Clare, Ireland.

Hoechst A.G., Postfach 80 03 20, D6230, Frankfurt am Main 80, Germany.

Institute of Biology, 20–22 Queensbury Place, London SW7 2DZ, U.K.

Merck, Postfach 4119, 6100 Darmstadt 1, Frankfurter Straße 250, Germany.

Pall Filtration, Portsmouth, Hants PO1 3PD, U.K.

Pfizer Center Research, Eastern Point Rd, Groton, CT 06340, U.S.A.

Pharmaceutical Manufacturers' Association, 1100 Fifteenth St, N.W., Washington D.C., 20005, U.S.A.

Pharmaceutical Proteins Ltd, Kings Buildings, West Mains Rd, Edinburgh EH9 3JQ, Scotland.

Pharmacia LKB Biotechnology, 5-751 82 Uppsala, Sweden.

Quality Biotech Ltd, 6.04 Kelvin Campus West of Scotland Science Park, Glasgow G20 0SP, Scotland.

Serono Diagnostics Ltd, 21 Woking Business Park, Albert Dr., Woking, Surrey GU21 5JY, U.K.

SmithKline Beecham Pharmaceuticals, Great Burgh, Yew Tree Bottom Rd, Epsom, Surrey KT18 5XQ.

Unipath Ltd, Wade Road, Basingstoke, Hants RG24 0PW, U.K.

Wellcome Research Laboratories, Langley Court, South Eden Park Road, Beckenham, Kent BR3 3BS.

Chapter 1

The scope of protein biotechnology

- What is protein biotechnology?
- The range of industrially significant proteins.
- Proteins used in the health-care industry.
- Additional proteins of industrial significance.
- Protein biotechnology: the future.

WHAT IS PROTEIN BIOTECHNOLOGY?

Protein biotechnology is big business, but what exactly is it? It has been assigned many different definitions by different groups. One of these definitions is that protein biotechnology is concerned with the commercial production and isolation of specific proteins from plant, animal or microbial sources, and/or the subsequent utilization of these proteins in order to achieve a pre-defined biological event.

Many technologies, both old and new, are intrinsically linked with what is now termed protein biotechnology (Table 1.1). Fermentation and allied disciplines are often used in the production of specific proteins. Recombinant DNA technology also plays a significant role in this regard. Indeed this enabling technology which facilitates production of virtually any protein in almost any living system continues to be the driving force behind the development of most novel biotechnologically significant protein products. Immunological techniques, including monoclonal antibody technology, form the backbone of processes concerned with the production of antibodies and related products. Protein purification and stabilizing techniques are also of central importance in this industry.

This book concentrates upon the technical aspects of protein biotechnology but the commercial factors underlying the industry are of equal importance. Biotechnology merely represents the commercial application of scientific discoveries in the biological

Table 1.1. The biological disciplines and enabling techniques that contribute to protein biotechnology

Microbiology	Biochemistry
Chemistry	Genetics
Physiology	Plant/animal sciences
Fermentation technology	Recombinant DNA technology
Monoclonal antibody technology	Protein purification techniques
Protein engineering	Protein immobilization technology

sciences. Many instances may be quoted in which the development of a particular product or process was a resounding technical success, while being a dismal commercial failure. Long-term commercial viability is obviously essential to the success of any biotechnological venture. Failure on commercial grounds may be due to many different contributing factors. Such factors include adverse customer perception of product safety or efficacy, the appearance on the market of an alternative, cheaper or more efficacious competing product and the adverse effect on export markets of currency fluctuations, government price control or regulatory agencies.

Many biotechnological processes have been developed, used and perfected over a number of millennia. Such processes include brewing, wine-making and cheese-making. Although these processes are dependent upon the enzymatic conversion of suitable substrates into desired products, those who first developed the methodologies were ignorant of this fact.

Other protein-dependent processes such as the inclusion of proteolytic enzymes in detergent preparations were first introduced at the beginning of the twentieth century, although they were not widely adopted until the late 1960s. Indeed, it was during the 1940s to 1960s that many large-scale industrial processes dependent upon enzyme-catalysed biotransformations were first developed (Chapter 8).

Many of the more exciting and sophisticated protein products currently entering the marketplace have their roots in discoveries originating in the mid-1970s. Both recombinant DNA technology and monoclonal antibody technology were first developed during this scientifically exciting period. Subsequent chapters testify to the central roles played by both these methodologies in the development of new protein products.

Until the 1970s most significant technological advances had been derived from the study of chemistry, physics or related disciplines. Innovations in the biological arena were few and far between.

During the course of the last decade however, this trend has all but been reversed. The pace of advancement and discovery in the biological sciences has accelerated dramatically during this period. Most such discoveries have obvious commercial application. Although traditional biotechnological processes have benefited from such technological advances, it is within the areas of therapeutics and diagnostics that these new discoveries have found most ready application.

The early 1980s witnessed an explosion in the number of new biotechnology companies formed to take commercial advantage of these exciting discoveries. Most of these companies (Table 1.2) are based in the USA, and they targeted their product development specifically towards the human health-care industry.

Table 1.2. Some biotechnology companies founded in the early 1980s to take commercial advantage of discoveries in the biological sciences

Alpha Beta	Genelabs
Advanced Tissue Sciences	Genzyme
Amgen	Immunex
Affinity Chromatography Ltd	Immune Response
Biogen	Immunogen
British Biotechnology	Imclone
Cal Bio	Pharmaceutical Proteins Ltd
Celltech	Repligen
Genetech	Synergin
Genetics Institute	T-cell Sciences

It is estimated that the total worldwide sales of biotechnological products will exceed US $50 billion by the turn of the century. Worldwide sales for biotechnologically-based therapeutic products alone was in the order of $2.5 billion in 1991. This figure is set to grow substantially over the coming decade. The biotechnology boom is well and truly under way.

THE RANGE OF INDUSTRIALLY SIGNIFICANT PROTEINS

A wide variety of proteins find industrial application. These include enzymes, antibodies, hormones, blood factors, growth factors and regulatory factors. Such proteins are employed as therapeutic and diagnostic agents and in the manufacture of a wide variety of biologically derived industrial commodities. These proteins and their application in the health-care and industrial sectors form the subject matter of this book. These proteins are obtained from a number of different sources including microorganisms, plants and animals (see Chapter 2).

Traditionally, any protein of industrial interest was obtained from whichever producing organism yielded it in highest quantities, with due consideration of relevant safety and economic factors. Recombinant DNA technology has removed such source limitations and many proteins obtained naturally from animal, plant or microbial sources are now produced in appropriate recombinant expression systems. This technology, more than any other, has nurtured and stimulated the rapid expansion of the protein biotechnology industry.

PROTEINS EMPLOYED IN THE HEALTH-CARE INDUSTRY

The vast majority of therapeutic substances produced by the pharmaceutical industry have traditionally been chemical in nature. Until recently, the only biologically based therapeutic agents available were blood products, vaccines and a limited number of hormonal and enzymic preparations. All of these substances were derived from natural sources such as blood donations or animal tissue. Product availability was therefore limited by both tissue availability and the level at which the protein was produced in such tissues. Moreover, all such products bore an associated risk of disease transmission as it was difficult to guarantee that processing would remove or inactivate all potential pathogens which might be derived from infected raw materials.

The introduction of monoclonal antibody technology enabled the production of unlimited quantities of antibodies, exhibiting virtually any desired specificity. Many such monoclonal antibody preparations are now routinely used for therapeutic and diagnostic purposes. Recombinant DNA technology and related techniques allow unlimited production of therapeutically significant proteins such as growth factors, regulatory factors, enzymes and blood proteins. Many such products were previously unavailable for medical use due to the minute quantities in which they are produced by the body.

While some of these exciting new biopharmaceutical products are already available to the medical community, many others await final marketing approval from relevant regulatory authorities. Some such biopharmaceuticals are listed in Table 1.3.

Table 1.3. Some protein-based biopharmaceutical products currently available. Some have gained marketing approval from regulatory bodies such as the American Food and Drug Administration (FDA). Others await final approval or are still undergoing clinical trials. Most are produced as recombinant protein products

Protein	Application
Blood clotting factors	Treatment of haemophilia and related blood disorders
Colony stimulating factors	Treatment of cancer, low blood cell count, adjuvant to chemotherapy, adjuvant to AIDS therapy
Superoxide dismutase	Prevention of oxygen toxicity
Erythropoietins	Treatment of anaemia and related disorders
Epidermal growth factor	Cancer treatment, wound healing, treatment of skin ulcers
Fibroblast growth factor	Treatment of certain ulcer types
Insulin-like growth factor	Treatment of type II diabetes mellitus
Platelet-derived growth factor	Treatment of certain ulcer types
Human growth hormone	Treatment of cancer, AIDS and growth deficiency in children
Interferons (α, β, γ)	Treatment of cancer, allergies, asthma, AIDS, arthritis and infectious diseases
Interleukins (IL-1, 2, 3 and 4)	Treatment of cancer, AIDS and bone marrow failure. Prevention of drug or radiation-related bone marrow suppression
Insulin	Treatment of diabetes mellitus
Monoclonal antibodies	Various uses. Treatment of cancer and rheumatoid arthritis. Employed for both *in vivo* and *in vitro* diagnostic purposes
Tissue plasminogen activator	Treatment of angina, stroke and heart attack
Tumour necrosis factor	Cancer treatment
Vaccines	Recombinant protein antigens have been developed to treat/vaccinate against HIV, hepatitis B, malaria and herpes

As with traditional pharmaceuticals, the development of bio-pharmaceutical products still remains a high-risk, high-reward venture. It may take up to 12 years for any pharmaceutical substance to reach the pharmacy shelves from the time of its inception in the laboratories. The American Pharmaceutical Manufacturers' Association estimated the cost of developing a single new pharmaceutical drug to be in the order of $200–250 million. Against this the ultimate financial rewards, when successful, are usually considerable.

The potential market for products such as recombinant insulin, human growth hormone, tissue plasminogen activator and other such products is estimated to be in the billions of dollars. The worldwide market for interferon alpha alone, used in the treatment of hairy cell leukaemia, herpes and hepatitis, is fast approaching $1 billion.

The number of biopharmaceutical products reaching the market-place is set to mushroom over the coming decade. With more and more conventional pharmaceutical manufacturers investing heavily in this new technology there is little doubt that an ever increasing proportion of newly developed therapeutics will be proteinaceous in nature.

ADDITIONAL PROTEINS OF INDUSTRIAL SIGNIFICANCE

Proteins used for therapeutic purposes are generally produced in relatively small quantities (grams or kilograms), and are subject to extensive downstream processing in order to assure a high degree of protein purity. Production costs are generally of secondary importance relative to product safety and efficacy. The emphasis is, therefore, on low bulk, high value preparations.

Many of the proteins destined for use in the health-care industry have been developed only recently. A host of other proteins have, however, enjoyed widespread industrial application for several decades. The majority of such industrial proteins are enzymes and are normally obtained by fermentation from various microbial sources. Unlike proteins destined for therapeutic application, these enzymes are produced in bulk quantities, often hundreds of tonnes, and are subject to little subsequent downstream processing (Chapter 3). They therefore represent relatively crude preparations, typically exhibiting many activities in addition to the desired enzymatic activity. Bulk enzyme production is a highly competitive industry and many economic factors play an important role in determining the commercial success of such bulk proteinaceous products.

The majority of enzymes used by industry in bulk quantities catalyse depolymerization reactions. These include carbohydrases, proteases and lipases. Such enzymes are extensively used in food

Table 1.4. Some enzymes which are produced in bulk quantities and their industrial applications

Enzyme	Industrial application
Proteases	Inclusion in detergent preparations Cheese-making Brewing/baking industries Meat/leather industries Animal/human digestive aids
Amylases	Starch processing industry Fermentation industries
Cellulases/hemicellulases	Brewing industry Fruit juice production Animal feed industry
Pectinases	Fruit juice/fruit processing industry
Lipases	Dairy industry Vegetable oil industry
Glucose isomerase	Production of high-fructose syrups
Lactase	Hydrolysis of milk lactose
Cyclodextrin glycosyltransferase	Production of cyclodextrins for pharmaceutical and other industries
Penicillin acylase	Production of semisynthetic penicillins

and beverage production and in various processing industries (Table 1.4). Large quantities of proteases are also incorporated into household detergent preparations.

A variety of other enzymes find widespread industrial application, as many are utilized in the production of pharmaceutically important products such as cyclodextrins, steroids and semisynthetic antibiotics.

Most enzymes produced in bulk are currently obtained from microbial sources that have not been altered by recombinant DNA methods. Suitable producer strains are first isolated by extensive screening of candidate microorganisms. The chosen strain may then be subjected to mutational influences in order to generate mutants which hyperproduce the enzyme of interest. In many instances such classical screening procedures have yielded producer strains whose production efficiency is unlikely to be enhanced even by recombinant DNA methodologies. Several industrial-scale proteins, however, are now produced by genetically altered microorganisms. Examples include chymosin (Chapter 8) and phytase (Chapter 9).

Bulk enzyme production is a multi-million dollar industry. In the early 1990s, the total annual sales of such enzymes was estimated to

Table 1.5. Some enzymes produced in bulk, their industrial application and their estimated sales value (1993)

Enzyme	Industrial application	Sales value (US $ million)	Percentage of total enzyme sales
Alkaline proteases	Incorporation into detergent preparations	300	45
Amylases	Starch processing	100	15
Carbohydrases, lipases, acid and neutral proteases	Baking, brewing, confectionery, fruit processing	200	30
Others (cyclodextrin glycosyltransferases, penicillin acylase, β-glucanase, phytase, etc.)	Various	65	10

be approaching US $700 million (Table 1.5). By far the greatest proportion of this figure was generated by sales of alkaline proteases destined for inclusion in detergent preparations. It is predicted that the sales value of such enzymes will continue to increase by a minimum of 5% per annum.

PROTEIN BIOTECHNOLOGY: THE FUTURE

The protein biotechnology industry has grown at a startling rate; however it remains an industry still in its infancy. The potential economic, social and (in particular), health benefits that may accrue from this industry are incalculable. As our scientific knowledge and understanding continue to grow, future development of biotechnological products is limited almost only by the imagination. Substantial investment in research and development has become the hallmark of this fledgling industry. Continued support of both pure and applied research is required if the current impetus within the industry is to be sustained long-term.

FURTHER READING

Books

Bickerstaff, G.F. (1987). *Enzymes in Industry and Medicine*, Edward Arnold, London.

Fogarty, W. & Kelly, C. (eds) (1990). *Microbial Enzymes and Biotechnology*, 2nd edn. Elsevier, New York.

Harris, E. & Angal, S. (eds) (1990). *Protein Purification Applications: A Practical Approach*. IRL Press, Oxford.

Wiseman, A. (ed.) (1985). *Handbook of Enzyme Biotechnology*, Ellis Horwood, Chichester.

Articles

Bienz-Tadmor, B. (1993). Biopharmaceuticals go to market: patterns of worldwide development. *Bio/Technology*, **11,** 168–172.

Bienz-Tadmor, N. *et al.* (1992). Biopharmaceuticals and conventional drugs: clinical success rates. *Bio/Technology*, **10,** 521–525.

Biotech: America's dream machine. Cover story. *Business Week*, March 2, 1992.

Biotechnology Medicines in Development (1991). Report presented by the US Pharmaceutical Manufacturers Association.

Carlsson, R. & Glad, C. (1989). Monoclonal antibodies into the '90s: the all purpose tool. *Bio/Technology*, **7,** 567–573.

Gibbons, A. (1992) Biotech's second generation. *Science*, **256,** 766–768.

Chapter 2

Protein sources

The choice of a protein source represents one of the first steps in the planning of a production process for any protein. Some proteins of commercial interest may be obtained from a variety of sources. For example, many proteases, carbohydrases and other catalytic activities are produced by a range of microorganisms. Traditionally, however, the choice of source has been more limited as the protein of interest was found to be present in only one or at most a few potential biological sources. Until recently insulin was produced exclusively from pancreatic tissue of slaughterhouse animals, while blood products were obtainable only from blood. In many such instances the natural level of protein production was often low. Large quantities of the source material were required to prepare appreciable amounts of the protein product. For example, during the initial purification and characterization studies of the hypothalamic factor thyrotrophin releasing hormone, almost 4 tonnes of hypothalamic tissue yielded only 1 mg of pure releasing hormone.

Many such difficulties have been overcome with the advent of recombinant DNA technology. Today, in principle, the gene or cDNA coding for any protein can be isolated and inserted into an appropriate expression system, thus allowing its production in a foreign organism, usually a microorganism. Large-scale fermentation of such recombinant organisms can yield appreciable quantities of any protein (Table 2.1). Many commercially available proteins are now produced by such methods. Examples include recombinant human insulin, human growth hormone and interferons, produced by transformed *Escherichia coli*. These and many other examples detailing large-scale production of recombinant proteins are discussed throughout this book.

Table 2.1. Some recombinant proteins produced by genetic engineering methods

Protein	Application
Insulin	Treatment of diabetes
Interleukins	Anticancer
Interferons	Antiviral, anticancer
a_1-Antitrypsin	Prevention of life-threatening emphysema
Hepatitis B surface antigen	Hepatitis B vaccine
Growth hormone	Induction of growth
Chymosin	Casein hydrolysis in cheese manufacture
Phytase	Liberation of phosphate groups from phytic acid

MICROORGANISMS AS SOURCES OF PROTEINS

Many proteins of industrial interest are produced from microbial sources. The majority are synthesized by a limited number of microorganisms which are classified as GRAS ("generally recognized as safe"). GRAS microorganisms include bacteria such as *Bacillus subtilis* and *Bacillus amyloliquefaciens*, in addition to various other bacilli, lactobacilli and *Streptomyces* species. GRAS-listed fungi include members of the genera *Aspergillus*, *Penicillium*, *Mucor* and *Rhizopus*. Yeasts such as *Saccharomyces cerevisiae* are also generally recognized as safe. GRAS-listed microbes are non-pathogenic, non-toxic and generally should not produce antibiotics. Microorganisms represent an attractive source of proteins, as they can be cultured in large quantities in a short time by established methods of fermentation (Figure 2.1). Thus they can produce an abundant, regular supply of the desired protein products. Microbial proteins are often more stable than analogous

Figure 2.1. Large scale bioreactor utilized in the production of biotechnological products (photograph courtesy of Boehringer Mannheim, UK)

proteins obtained from plant or animal sources. Furthermore, microbes can be subjected to genetic manipulation more readily than animals and plants.

Many industrially significant proteins obtained by methods of fermentation are secreted by the producing microorganisms directly into the culture medium. Such extracellular protein production greatly simplifies subsequent downstream processing as there is no requirement to disrupt the microbial cells in order to effect protein release; thus there are fewer proteins to separate from the product of interest. Whole cells may be removed from protein-containing extracellular media by methods such as centrifugation or filtration. In many such cases few subsequent purification steps are required to yield a final product of the required purity. Specific examples of industrially important proteins secreted into the extracellular medium during fermentation include various amylolytic and proteolytic enzymes produced by bacilli, in addition to cellulases and other activities produced by fungi such as *Trichoderma veridiae*. Many such examples are detailed in subsequent chapters.

In some instances the protein of interest may be intracellular. In such cases it is necessary to disrupt the cells upon completion of fermentation and cell harvesting. Such an approach releases not only the protein of interest, but also the entire intracellular content of the cell. This, in turn, renders more complicated the subsequent purification procedures required to obtain the final product. Specific examples of intracellular proteins of industrial significance include asparaginase, penicillin acylase and glucose isomerase. However, recombinant DNA technology provides a means of producing such proteins in a host that will secrete the protein directly into the growth medium.

Traditionally, identification of the most suitable microbial protein source involved screening a wide range of candidate microorganisms. Obviously, the existence of a simple, rapid, sensitive assay to identify the protein of interest greatly facilitates such screening activities. Initial screens serve to identify microbial species expressing the protein of interest, with further screens pinpointing microbial species producing the largest quantities of the protein. Frequently, organisms found to produce elevated levels of the protein of interest would then be subjected to mutational studies in an effort to isolate overproducing strains. Advantageous mutations can result in product enhancement in a number of ways. A mutation in the regulatory sequence of the gene encoding the protein can result in increased levels of expression of the gene product. A mutational event occurring in the gene itself can result in an altered amino acid sequence which may render the protein more functionally efficient or enhance its stability.

Microorganisms collected from many different environments are routinely screened in the hope of identifying more attractive sources of various industrially useful proteins. Few environments have as great a variety of microbial populations as fertile soil. Soil bacteria are among the most common group of organisms subjected to routine screening. Soil bacilli (apart from the *Bacillus cereus* group) are suitable enzyme producers as they generally conform to GRAS requirements. They are also easily cultured in simple media and produce a variety of industrially important enzymes extracellularly. Hundreds of different species of fungi also inhabit the soil, especially near the soil surface where aerobic conditions prevail. Such fungi are active in degrading a wide variety of biological materials present in the soil. They thrive on such material by secreting extracellular enzymes capable of degrading large polymeric plant molecules such as cellulose, hemicellulose and pectin, with subsequent assimilation of the liberated nutrients. Many soil fungi thus represent attractive sources of cellulases and related hydrolytic enzymes.

Microbial sources of various industrially important proteins were identified and isolated largely by screening strategies such as those outlined above. Recent advances in the area of recombinant DNA technology have, however, facilitated the development of an entirely new strategy with regard to the production of protein from microbial sources.

Protein production by genetically engineered microorganisms

Genetic manipulation by mutation and selection has played a central role in increasing expression levels of a myriad of microbial proteins. This approach, however, could be at best described as haphazard, with the researcher having little control over the genetic alterations achieved. The advent of recombinant DNA technology has changed this situation. This technology can be utilized in a highly directed manner to achieve specific genetic alterations. Rational improvements in source productivity can thus be achieved—a goal that mutational studies of the past realized only by chance. This is not to suggest that recombinant techniques will replace traditional mutational studies as the only method of strain improvement. Strain improvements by "trial and error" mutational methods will continue to play an important developmental role, as large numbers of such experiments can be carried out conveniently and relatively inexpensively if suitable screening procedures are in place to detect the desired product.

Recombinant DNA technology can be used to increase the level of production of an endogenous microbial protein by a number of methods. These include (a) introduction of additional copies of the relevant gene into the microorganism, and (b) introduction of a copy or copies of the relevant gene into the organism where control of expression has been placed under a more powerful promoter. Such strategies can greatly increase production of the protein of interest. It must be noted, however, that failure will result if attempts to increase the levels of expression of the desired protein in any way compromise normal cellular function.

Heterologous protein production in *E. coli*

Although recombinant DNA techniques can be used to increase levels of expression of specific homologous microbial proteins (i.e. protein normally expressed in the microbe), the main impact of such technology relates to the expression of novel proteins from other species in such microorganisms. As previously mentioned, many proteins of industrial interest are produced naturally in minute quantities. This fact alone often rendered commercial production of such proteins impractical. Many regulatory proteins are produced in very small quantities. Specific examples include hypothalamic regulatory factors, pituitary hormones and immuno-modulatory proteins such as cytokines. Upon their discovery, it was quickly realized that many such proteins could be of immense therapeutic value. Low natural availability limited their clinical assessment and hence their use. The expression of such eukaryotic gene products in microbial species has overcome such difficulties. The expression of recombinant proteins in cells in which they do not naturally occur is often referred to as heterologous protein production. Quite a number of therapeutically important proteins are now produced as heterologous proteins in *E. coli*. Many are in clinical use while others are undergoing clinical evaluation. The first heterologous protein to be employed clinically was human insulin produced in *E. coli*. This product was first approved for use in 1982 by the regulatory authorities in the UK, West Germany, the Netherlands and the USA. Additional examples of heterologous proteins produced in *E. coli* are listed in Table 2.2. Antibody fragments capable of binding antigen with the same affinity as the complete antibodies are now routinely produced in *E. coli* expression systems.

Table 2.2. Some heterologous proteins produced in *E. coli*

Protein	Level of expression achieved (% of total protein)
Somatostatin	<0.05
Insulin A chain	20.00
Insulin B chain	20.00
Bovine growth hormone	5.00
Human growth hormone	5.00
a_1-Antitrypsin	15.00
Interleukin 2	10.00
Human tumour necrosis factor	15.00
Interferon α	ND
Interferon β	15.00
Interferon γ	25.00
Calf prochymosin	8.00

ND, not determined

Probably the bacterial species most commonly employed in the production of heterologous proteins is *E. coli*. This stems almost entirely from the fact that, traditionally, the study of prokaryotic genetics focused on *E. coli* as a model system; hence more was known about the genetic characteristics of *E. coli* than any other microorganism. In some instances, the foreign gene or cDNA coding for the protein of interest may be attached directly to a complete or partial *E. coli* gene resulting in the production of a fusion protein. In other cases, bacterial regulatory elements, promoters and terminators are used to regulate transcription of the foreign gene. This approach is termed *direct expression*. Direct expression yields a protein whose amino acid sequence is identical to that of the protein obtained from its native source. Expression of a heterologous protein fused to a bacterial protein generally necessitates chemical or enzymatic cleavage in order to release the native protein.

The majority of proteins synthesized naturally by *E. coli* are intracellular. Few of the cells' native proteins are exported to the periplasmic space and fewer still are exported into the extracellular medium. Thus the majority of heterologous proteins expressed in *E. coli* accumulate in the cell cytoplasm, where they can represent up to 25% of total cellular protein. Attempts have been made to construct recombinant protein production systems in which the foreign protein is secreted into the periplasmic space or extracellular medium. Such attempts have mainly focused on the fusion of the foreign gene with a signal peptide sequence from an endogenous *E. coli* gene, the protein product of which is normally secreted. So far such attempts have met with only limited success. Extracellular secretion of recombinant protein products is considered desirable as it simplifies subsequent downstream processing. The exact cellular mechanisms by which *E. coli* (and indeed many other cell types) actually regulates the secretion of proteins into the extracellular medium is not fully understood. It therefore remains a difficult task to logically construct an extracellular secretory mechanism for heterologous proteins produced in such microbial systems.

Inclusion body formation. Extremely high levels of diverse heterologous proteins have been produced in *E. coli*. In most such cases, the resultant protein accumulates in the cell cytoplasm in the form of insoluble aggregates termed inclusion bodies or refractile bodies. It is believed that these aggregates are not derived from either native or fully unfolded forms of the protein but are composed of partially folded intermediates. Inclusion bodies may be readily viewed by dark field microscopy and are composed predominantly of the expressed heterologous protein. It is not understood why recombinant proteins form inclusion bodies in *E. coli*. In many cases, it has been shown that a reduction in bacterial growth temperature from 36–37 °C to 30 °C or below can significantly reduce the level of inclusion body formation.

The formation of inclusion bodies should not be regarded as a wholly negative phenomenon. As inclusion bodies are very dense they tend to sediment readily under the influence of a low centrifugal force. This can be exploited to achieve quick and effective partial purification of the aggregated protein, typically by low-speed centrifugation immediately after cell homogenization. Inclusion bodies sediment more rapidly than cellular debris under the influence of a centrifugal force of the order of 500–1000 g. The pellet containing the inclusion bodies can subsequently be resuspended and washed several times to reduce any cosedimenting cellular material. After isolation of the inclusion body, denaturants are used to solubilize the aggregated polypeptides. Denaturants employed include urea, guanidinium chloride, various detergents and organic solvents, in addition to incubation under conditions of alkaline pH. Once solubilization of the inclusion body has been achieved, the denaturant is slowly removed by techniques such as dialysis, dilution or diafiltration. In this way, suitable conditions are induced which promote refolding of the protein into its native, biologically active conformation. Suitable conditions required for such renaturation can vary from protein to protein, and recovery of high yields of activity is not always guaranteed. This is particularly true in the case of large proteins for which correct refolding into their active conformation is a complicated process. For this reason, many high molecular weight proteins such as antibodies and factor VIII may still be produced more successfully in mammalian cell culture systems.

The recombinant protein, if desired, may be subjected to additional purification using a range of chromatographic techniques. Such purification steps may be undertaken after protein renaturation, although in many cases it is feasible to employ such techniques while the solubilized protein is still in the presence of the denaturant.

Other characteristics of heterologous protein production in E. coli. Technical developments currently allow large scale industrial production of a variety of heterologous proteins in *E. coli*. This represents a major step forward for modern protein biotechnology. *E. coli*, in theory, has the potential to become a source of virtually any protein of industrial interest. However, there still remain a number of disadvantages with regard to utilizing genetically modified *E. coli* as the source of such proteins (Table 2.3). As already discussed, one such disadvantage is inclusion body

Table 2.3. Some advantages and disadvantages of heterologous protein production in *E. coli*

Advantages	Disadvantages
Genetic characteristics of *E. coli* are well established	Intracellular accumulation of recombinant protein in inclusion bodies
Suitable fermentation technology is well established	Unable to undertake post-translational modifications of proteins
Can generate potentially unlimited supplies of the recombinant protein	Adverse public perception of products produced by recombinant methodologies
Economically attractive	

formation. Another disadvantage is the inability of *E. coli* to perform post-translational modifications of recombinant eukaryotic proteins. *E. coli* does not have the ability to glycosylate, amidate or acetylate proteins. Eukaryotic proteins normally subjected to such post-translational modifications when produced in their natural source are devoid of such modifications when produced in *E. coli*. Lack of post-translational modification may or may not have a significant effect upon the biological activity of the heterologous protein produced. Interferon beta and interleukin 2 produced in *E. coli* are biologically active despite the fact that they are not glycosylated. However, it should be stressed that glycosylation is a common feature of many eukaryotic proteins. Absence of glycosylation may, in some cases, alter the biological characteristics of the recombinant protein. This is particularly noteworthy with regard to proteins used for therapeutic purposes. Absence of a carbohydrate moiety, or other such modifications, can potentially result in modification of the biological activity of the molecule, change in its circulatory half-life values and altered immunogenicity.

Recombinant proteins expressed in *E. coli* may also contain an extra amino acid residue (methionine) at their N-terminal end. This is due to the fact that translation in *E. coli* is always initiated by an N-formylmethionine residue. Although *E. coli* has the enzymatic capability to deformylate and subsequently remove this additional amino acid, removal is less efficient with regard to processing of heterologous proteins. The presence of an extra N-terminal methionine in such recombinant proteins may alter their biological characteristics.

Many, if not most heterologous proteins currently produced in *E. coli* are destined for therapeutic purposes. *E. coli,* along with other Gram negative bacteria, possess lipopolysaccharide substances in its outer membrane. Such lipopolysaccharides (also known as endotoxin) are pyrogenic. Their presence in injectable products can lead to endotoxic shock and, in severe cases, death of recipient patients. It is therefore vital to ensure that lipopolysaccharides are removed during downstream processing. The clinical significance of lipopolysaccharides in such therapeutic preparations is discussed more fully in Chapter 4.

Heterologous protein production in yeast

Yeast cells have also become important host organisms for production of heterologous proteins. Yeasts are attractive hosts for the following reasons:

- They retain many of the advantages as outlined for *E. coli* expression systems.

- Yeasts such as *Saccharomyces cerevisiae* are GRAS-listed organisms. Many have played a central role in a range of traditional biotechnological processes such as brewing and baking. As a result a wealth of technical data has accumulated with regard to fermentation and manipulation of such yeast.
- The molecular biology of yeasts has been a focus of scientific research over a number of years.
- Yeast cells, unlike bacteria such as *E. coli*, possess subcellular organelles and are thus capable of carrying out post-translational modifications of proteins.

Initially the majority of heterologous proteins produced in yeast were produced in *Saccharomyces cerevisiae*. Owing to its traditional industrial importance, this yeast is among the best studied of all organisms. Specific examples of heterologous proteins produced in *Saccharomyces cerevisiae* include interferon, epidermal growth factor and calf prochymosin. Indeed, the first vaccine produced by recombinant DNA methods to be administered to humans (hepatitis B surface antigen) was produced in *Saccharomyces cerevisiae*. Prior to this preparation becoming commercially available, human hepatitis B vaccines were prepared by purification of hepatitis B surface antigen (HBsAg) from the plasma of human carriers. This method of preparation was unsatisfactory, firstly because vaccine production was restricted by the supply of infected human plasma, and secondly because stringent controls were required during downstream processing in order to ensure that the final product was free of infectious hepatitis B or other infectious agents that might be present in the donor blood.

Although many heterologous proteins have been successfully produced in *Saccharomyces cerevisiae* this system is also subject to a number of drawbacks (Table 2.4). Expression levels of heterologous proteins are often low, typically representing less than 5% of

Table 2.4. Some advantages and disadvantages of heterologous protein production in yeasts

Advantages	Disadvantages
Most are GRAS listed	Recombinant proteins usually expressed at low levels
Proven history of use in many biotechnological processes	Retention of many exported proteins in the periplasmic space
Fermentation technology well established	Adverse public perception of products produced by recombinant methodologies
Ability to carry out post-translational modifications of recombinant proteins	Some post-translational modifications differ significantly from those in animal cells

total cellular protein. Such values compare unfavourably with heterologous protein production in *E. coli*, where production levels of over 25% of total cellular protein have been achieved. Many heterologous proteins produced and secreted by *Saccharomyces cerevisiae* are not released into the culture medium, but are retained in the periplasmic space. This is especially true in the case of heterologous proteins of high molecular weight. In such cases downstream processing is rendered more complicated. Although yeast systems have the ability to carry out post-translational protein modifications, it has been noted that the level and type of modifications do not entirely resemble the modifications observed when the protein is produced in its natural source. Such variations in post-translational modifications may or may not have a significant effect on the biological characteristics of the heterologous protein produced.

While *Saccharomyces cerevisiae* remains the most popular yeast host for expression of heterologous proteins, several other yeasts are utilized in the production of heterologous proteins of industrial interest. Popular alternatives to *Saccharomyces cerevisiae* include *Hansenula polymorpha*, *Kluyveromyces lactis* and *Picha pastoris*. Such yeasts have served in the production of proteins such as invertase, glucoamylase and human epidermal growth factor. In some specific cases, use of alternative expression systems to *Saccharomyces cerevisiae* has resulted in improved glycosylation characteristics. This is, however, by no means true in all cases, and the question of which yeast expression system is most appropriate to a given situation remains an open one.

Heterologous protein production in fungi

Many of the initial studies in which specific proteins were produced as heterologous products employed *E. coli* or *Saccharomyces cerevisiae* as host systems. Filamentous fungi represent attractive hosts for heterologous gene expression for a number of reasons:

- They possess the enzymatic capability to carry out post-translational modifications.
- They have been extensively employed on an industrial scale for many decades in the production of a variety of enzymes.
- They are capable of synthesizing and secreting large quantities of certain proteins into the extracellular medium, in marked contrast to *E. coli* or species of *Saccharomyces*.

Extracellular production of heterologous products is desirable as it simplifies subsequent product purification. Some industrial strains of *Aspergillus niger* are known to produce naturally up to 20 g of the enzyme glucoamylase per litre of fermentation medium. Owing to their industrial significance, large-scale fermentation systems for these fungi have been long since developed and optimized. A wide variety of industrially important proteins have now been produced as heterologous products in a number of fungal species, most notably in *Aspergillus* species (Table 2.5). In many instances, the genetic information of interest has been coupled to a strong fungal promoter, for example the fungal α-amylase promoter, in an attempt to maximize protein production. While the resultant protein products with suitable signal sequences have invariably been secreted into the medium in biologically active form, the levels of expression have been somewhat disappointing. Levels ranging from a few hundred milligrams to over a gram of heterologous protein per litre have been reported.

Table 2.5. Some proteins of industrial significance which have been expressed in recombinant fungal systems

Protein	Organism
Human interferon	*Aspergillus niger, A. nidulans*
Bovine chymosin	*A. niger, A. nidulans*
Aspartic proteinase (from *Rhizomucor michei*)	*A. oryzae*
Triglyceride lipase (from *Rhizomucor michei*)	*A. oryzae*
Human lactoferrin	*A. oryzae, A. niger*

Bovine chymosin, also known as rennin, is an aspartyl protease which is used in cheese production. The enzyme cleaves a specific bond present in milk κ-casein, thus promoting curd formation. Traditionally, this enzyme has been obtained from the fourth stomach of unweaned calves. It has been produced both in *E. coli* and yeast systems, though with limited success. The enzyme has also been successfully expressed in various species of *Aspergillus*. Such systems are capable of secreting moderate amounts of this enzyme directly into the medium and hence are commercially more attractive than the *E. coli* or yeast systems. Production of chymosin from recombinant fungi overcomes problems such as variability of supply often encountered when the sole source of this enzyme was calf stomach.

Human lactoferrin (hLF) is another protein of potential industrial significance which has been successfully expressed in *Aspergillus oryzae* and *A. niger*; it is an iron-binding glycoprotein of molecular weight, 78 kDa. Originally discovered in milk, it is also present in a variety of other external fluids such as tears and saliva. The exact biological role of hLF is the subject of debate. Proposed functions include promotion of increased intestinal iron absorption and protection against microbial infection, as removal of iron would inhibit growth of iron-requiring microorganisms. Lactoferrin levels of the order of up to 1 g per litre are normally associated with human milk. Bovine milk contains much lower levels of this protein. This may be one reason why breast-fed infants perform better than their counterparts fed bovine milk. Inclusion of recombinant lactoferrin in the latter diet might redress this nutritional situation.

Limitations on the production of heterologous proteins in microbial species can also be due to codon usage where specific codons for amino acids are used differentially in one species compared with the species from which the genetic information was obtained.

PLANTS AS SOURCES OF INDUSTRIALLY IMPORTANT PROTEINS

Plants represent a traditional source of a wide range of biologically active molecules. Narcotics such as opium are perhaps the best-known of such products. Crude opium consists of the dried milky exudate obtained from unripe capsules of certain species of plants mainly found in parts of Asia, India and China. The most important medical constituents of opium are the alkaloids—the best-known of which is morphine. Morphine is extracted and purified from crude opium preparations, generally by ion-exchange chromatography.

For a number of reasons higher plants are not regarded as prolific producers of many commercially important proteins.

- Many industrially important proteins synthesized in plants are also found in other biological sources. In most cases, the alternative source becomes the source of choice for both technical and economic reasons.
- Plant growth is seasonal and hence a constant source of material is not always obtainable.
- Higher plants tend to accumulate waste substances in structures termed vacuoles. Upon cell disruption these wastes, which include a number of powerful precipitating and denaturing agents, often irreversibly inactivate many plant proteins.

Despite such drawbacks, a number of industrially important proteins are obtained from plants. Two plant proteins—monellin and thaumatin—are recognized as the sweetest naturally occurring substances known. Such proteins have exciting potential uses in the food industry and are reviewed in Chapter 9. Beta-amylases are also produced by many higher plants; the most widely used are obtained from barley. These enzymes play an important role in the starch processing industry, as discussed later. However, perhaps the best-known plant-derived protein produced on an industrial scale is the proteolytic enzyme, papain.

Papain, also known as vegetable pepsin, is obtained from the latex of the green fruit and leaves of *Carica papaya*. It was first isolated and characterized in 1937 in the USA by A. K. Balls and colleagues. Papain is a cysteine protease. Its active site contains an essential cysteine residue which must remain in the reduced state if proteolytic activity is to be maintained. The purified enzyme exhibits broad proteolytic activity against a wide range of substrates. It consists of a single polypeptide chain containing 212 amino acid residues, with a molecular weight of 23 400 Da. The term "papain" is applied not only to the purified enzyme but also to the crude dried latex. Papain has a variety of industrial applications, the best known of which is its use as a meat tenderizing agent. During the tenderization process, the proteolytic activity of papain is directed at collagen, which is the major structural protein in animals, representing up to one-third of all vertebrate protein. It is the collagen present in connective tissue and blood vessels that renders meat tough. Hydrolysis of the peptide bonds of collagen is achieved by boiling or by proteolytic action. Such treatment results in the degradation of tough, insoluble collagenous fibrils, yielding soluble polypeptides of lower molecular weight. Papain is often injected directly into the blood stream of animals immediately prior to slaughter. This achieves an even body distribution of the enzyme. Papain has a relatively high optimum temperature (65 °C) and retains activity at temperatures up to 90 °C. Because of its thermal stability, papain maintains its proteolytic activity during the initial stages of cooking. This enzyme has also been used in other industrially important processes (Table 2.6).

Table 2.6. Some industrial applications of papain

Meat tenderization

Bating of animal skins

Clarification of beverages

Digestive aid (high protein diet)

Debriding agent (cleaning of wounds)

Ficin is another commercially available protease derived from plant sources. It is generally extracted from the latex of certain tropical trees and, like papain, is a cysteine protease. It exhibits considerably higher proteolytic activity than papain and has similar industrial applications. Purified ficin has a molecular weight of about 25 000 Da, though the term "ficin" is applied not only to the purified enzyme but also to the crude latex extract. Most large-scale industrial applications of papain and ficin do not require highly purified enzyme preparations. Plant enzymes, in particular those destined for application in the food processing industry, must be obtained only from non-toxic, edible plants.

Production of heterologous proteins in plants

Advances in recombinant DNA technology facilitate genetic manipulation not only of microorganisms but also of eukaryotic cells. A number of different heterologous proteins and peptides are now produced in a variety of plants. Genetic manipulation of plant systems may be undertaken for a number of reasons. Introduction of foreign genes or cDNAs may be undertaken in order to confer a new function or ability on the resultant plant tissue. Novel DNA sequences may be introduced into plant cells by several means. These include use of *Agrobacterium* as a carrier, and direct injection of the DNA into certain plant cells. Using such techniques, plants can be engineered to produce insecticides, which when expressed may play a protective role. Plants may also be altered genetically to produce heterologous proteins of industrial interest. Expression of some such foreign proteins in plants has been reported.

Attempts to produce antibodies in a variety of heterologous systems have not usually been successful. This is most probably due to the complex structural nature of the mature antibody. Antibodies consist of four polypeptide chains, two light chains (identical) and two heavy chains (identical). Correct intrachain folding and interchain association is required to form a functional antibody. Such interactions are complex and are both covalent and non-covalent in nature. Functional antibodies have been produced in plants with limited success. Plant expression systems have the ability to carry out a number of post-translational modifications and can successfully glycosylate a range of heterologous proteins. However, recombinant glycoproteins produced by transgenic plant cells normally contain a glycosylation pattern different to the pattern associated with the protein produced in its natural source. Certain oligosaccharide epitopes commonly found on plant glycoproteins are highly immunogenic in mammals. This suggests that mammalian proteins intended for therapeutic application, if expressed in plant cells, might be highly immunogenic.

Heterologous peptide production in plant seeds

It is now possible to produce a range of heterologous peptides of commercial interest in plant expression systems. In recent years, a wide range of peptides of considerable commercial value have been identified. Many such peptides are of therapeutic significance. These occur naturally in the body and perform a variety of biological functions. Examples include thyrotrophin releasing factor (TRF), a 3 amino acid peptide produced in the hypothalamus which stimulates synthesis and release of the hormone thyrotrophin from the anterior pituitary gland. Oxytocin is a 9 amino acid peptide hormone secreted by the posterior pituitary. It stimulates uterine muscle contraction. Luteinizing hormone releasing hormone (LHRH) is a decapeptide produced by the hypothalamus which stimulates the release of luteinizing hormone (LH) and follicle stimulating hormone (FSH) from the pituitary gland. Other examples of peptides of clinical significance include bradykinin, a 9 amino acid hormone which inhibits inflammation of tissue, and the endorphins—a group of neuropeptides often referred to as the body's own opiates. Endorphins are endogenous ligands of the opiate receptors and hence exhibit a biological activity similar to morphine. Several endorphin peptides have been characterized, the most important of which are known as α, β, and γ endorphins, in addition to met-enkephalin and leu-enkephalin.

Most such peptides are synthesized in minute quantities in the body. As a result, purification from their natural source is fraught with technical difficulties. Many such peptides may be synthesized chemically. The cost of such chemical synthesis increases enormously with increasing peptide length. Production of a peptide containing modified amino acid residues by chemical synthesis may also present technical difficulties. However, despite such potential drawbacks, a number of peptides available commercially are synthesized chemically. Many have also been produced as heterologous peptides in fermentation systems utilizing pro-karyotes or yeast expression systems. Certain peptides are also successfully produced in plant seeds. Leu-enkephalin, for example, has been produced in this manner.

The seeds of higher plants contain large quantities of storage proteins. Some such storage proteins may constitute in excess of 50% of total seed protein. Production of leu-enkephalin was achieved by inserting its DNA coding sequence into the gene coding for a seed storage protein termed 2S albumin. The family of 2S albumins are among the smallest seed storage proteins, having a molecular weight of the order of 12 kDa. This family of proteins is derived from a group of structurally related genes—all of which exhibit both conserved and variable sequences. The variable regions vary not only in sequence but also in length. The strategy employed to produce leu-enkephalin involved substituting part of this variable sequence with a DNA sequence coding for the 5 amino acid neurohormone. The DNA construct was flanked on both sides by nucleotides encoding amino acid sequences recognised by the proteolytic enzyme trypsin. Expression of the altered 2S albumin gene resulted in production of a hybrid storage protein containing the leu-enkephalin sequence. The enkephalin was subsequently released from the altered protein by tryptic cleavage and purified by high-performance liquid chromatography (HPLC). Because of the incorporation of the tryptic cleavage sites, the purified product contained an extra lysine residue which was subsequently removed by treatment with carboxypeptidase C—a proteolytic enzyme which hydrolyses only the peptide bond at the carboxyl terminus of a peptide or polypeptide.

Although production of heterologous proteins and peptides in plant seeds has been shown to be technically feasible, it is still unclear to what extent such production methods will be adopted by industry. Incorporation of significantly larger peptides into storage proteins may have adverse effects on the synthesis and stability of such hybrid proteins and thus may not be feasible. Economic considerations will constitute the important deciding factor. As yet, it is not clear if such methods of production would be economically more attractive when compared to chemical synthesis or microbial fermentation.

ANIMAL TISSUE AS A PROTEIN SOURCE

A wide variety of commercially available proteins are obtained from animal sources. This is particularly true with regard to numerous therapeutic proteins such as insulin and blood factors. Many such examples are discussed in later chapters and thus are only briefly reviewed here. The existence of slaughterhouses facilitates the collection of significant quantities of the particular tissue required as protein source.

Perhaps the best-known protein obtained from animal sources is insulin. This polypeptide hormone is produced in the pancreas by the beta cells of the islets of Langerhans. It is employed therapeutically in the treatment of insulin-dependent diabetes. Until the early 1980s insulin was obtained exclusively from pancreatic tissue derived from slaughterhouse cattle and pigs. Insulin containing an identical amino acid sequence to that of human insulin may be produced by enzymatic modification of porcine insulin. The amount of insulin obtained from the pancreatic tissue of three pigs satisfies the requirements of one diabetic patient for approximately 10 days. The increasing worldwide incidence of diabetes raised fears that one day demand for insulin supplies could exceed supply from slaughterhouse sources. This is no longer of concern however, as potentially unlimited supplies of insulin can now be produced by microbial cells carrying the genetic information for production of human insulin. Additional examples of commercially available hormones obtained from animal sources are listed in Table 2.7.

Table 2.7. Some proteins of industrial or medical significance which have traditionally been obtained from animal sources. Many of the listed examples may now be obtained from microbial sources by the application of recombinant DNA procedures

Protein	Source	Application
Insulin	Porcine/bovine pancreatic tissue	Treatment of insulin-dependent diabetes
Glucagon	Porcine/bovine pancreatic tissue	Reversal of insulin-induced hypoglycaemia in diabetic patients
Follicle stimulating hormone	Porcine pituitary glands	Induction of superovulation in animals
	Urine of post-menopausal women	Treatment of reproductive dysfunction (human)
Human chorionic gonadotrophin	Urine of pregnant women	Treatment of reproductive dysfunction
Erythropoietin	Urine	Treatment of some blood disorders
Blood factors	Human plasma	Treatment of a variety of blood disorders such as haemophilia
Polyclonal antibodies	Human/animal blood	Various diagnostic and therapeutic applications
Chymosin (rennin)	The fourth stomach of unweaned calves	Cheese manufacture (hydrolysis of casein)

Most industrially significant proteins obtained from human and other animal sources are destined for therapeutic use. One disadvantage with regard to the therapeutic utilization of such proteins from animal sources, especially human sources, is the potential presence of pathogens in the raw material. The large number of haemophiliac patients who contracted acquired immune deficiency syndrome (AIDS) from blood transfusions infected with human immunodeficiency virus (HIV) stand as testament to this fact. Outbreaks of bovine spongiform encephalitis (BSE or "mad cow" disease) in cattle herds from various countries serve as another example. A number of precautions must thus be taken when animal tissue is used as a protein source. The most obvious precaution involves the use of tissue obtained only from disease-free animals. Downstream processing procedures used in purifying the protein of interest must also be validated, thereby showing that the purification steps employed can eliminate pathogens which may be present in the starting material.

Many pathogens, in particular viral pathogens, exhibit marked species specificity. Thus therapeutically used proteins obtained from a particular animal species are not usually administered to other animals of that same species. For example, purified follicle stimulating hormone used to superovulate cattle is usually sourced from porcine and not bovine pituitary glands. Not all proteins obtained from animal tissue are used for therapeutic purposes. Rennet, which is used on a large scale during the manufacture of cheese, is a dried extract obtained from the true stomach of young calves. Rennet contains rennin, a proteolytic enzyme which serves as the predominant milk clotting enzyme in such animals. A number of other commercially available proteases used in industry and in research are also obtained from animal tissue.

Heterologous protein production in animals

Exciting advances have been recorded in the development of transgenic animals. Initial experiments in this area concentrated on attempts to improve various animal characteristics. It was hoped that, for example, animal growth rates could be dramatically improved by genomic integration of extra copies of the growth hormone gene. Generation of such transgenic animals is normally achieved by direct microinjection of DNA into ova, although the success rate of this method is still somewhat low. One goal of such "molecular farming" is the introduction of specific functional genes into animals, thereby conferring on them desirable characteristics such as more efficient feed utilization, improved growth characteristics or generation of leaner meat.

Another goal of such transgenic technology is to confer on the transgenic animal the ability to produce large quantities of industrially important proteins. This has been achieved with varying degrees of success in mice, goats, sheep and cattle. Initial successes in expressing high levels of growth hormone in transgenic animals highlighted some potential problems associated with this technique. In many cases, it was found that chronically high circulatory levels of growth hormone significantly outside the normal physiological range resulted in many adverse physiological effects. Elevated circulatory levels of many proteins of potential therapeutic value would also almost certainly promote similar adverse effects on normal transgenic animal metabolism. Specific animal tissues were also targeted as heterologous protein expression sites. One such tissue target is the mammary gland. By targeting expression of the foreign gene to the mammary gland the heterologous protein may be secreted directly into the milk. As such it is physically removed from the animal's circulatory system. Expression of the genetic information of interest can be targeted to this tissue by fusing the gene to the signal sequence of a milk protein and using mammary gland specific promoters. If this foreign DNA construct is successfully microinjected into an egg, the egg is subsequently fertilized and implanted into a surrogate mother and the embryo is then brought to term, and if the foreign DNA has been successfully incorporated into the newborn's chromosomal DNA and is expressed, the resultant transgenic animal is capable of producing the protein of interest and secreting that protein selectively into its milk.

The earliest success involving the production of a heterologous protein of considerable therapeutic potential was recorded in the mid-1980s when the production of human tissue plasminogen activator (tPA) in the milk of transgenic mice was reported. This was achieved by injecting into mice embryos a DNA construct consisting of the promoter and upstream regulatory sequence from the mouse whey acidic protein gene to the gene coding for human tPA. Whey acidic protein is the most abundant whey protein found in mouse milk. Biologically active tPA was recovered from the milk of the resultant transgenic mice.

Tissue plasminogen activator is a serine protease, with a molecular weight of the order of 70 000 Da. It catalyses the enzymatic conversion of plasminogen to plasmin by hydrolysing a single Arg-Val bond. Plasmin also exhibits proteolytic activity, cleaving fibrin and thus promoting lysis of blood clots. Tissue plaminogen activator thus normally plays an important role in the wound healing process and is of clinical value in the treatment of coronary thrombosis, where it has been shown to be effective in dissolving the fibrin clots responsible for coronary occlusions. Recombinant human tPA produced by a mammalian cell line has been used clinically to treat heart attack victims. It is now also available from the milk of transgenic goats.

High production levels of active human α_1-antitrypsin in the milk of transgenic sheep has also been achieved (Figure 2.2). Human α_1-antitrypsin is a glycoprotein normally synthesized in the liver. Its molecular weight is approximately 52 000 Da and it consists of a single polypeptide chain of 394 amino acid residues with three carbohydrate side chains N-linked to asparagine residues. It is the major serine protease inhibitor found in mammalian plasma and is normally present at a concentration of up to 2 g per litre. Large numbers of people suffer from a genetic defect in the α_1-antitrypsin gene and as such are at high risk of developing life-threatening emphysema. In order to reduce this risk, α_1-antitrypsin may be administered by injection to affected individuals. Current demand for such purposes is met by purification of the protein from

Figure 2.2. Photograph of "Tracy", a transgenic sheep. Tracy expresses high levels of active human α_1-antitrypsin in her milk. One of the two lambs shown above is also transgenic (photograph courtesy of Pharmaceutical Proteins Ltd)

plasma. As each affected individual requires therapeutic doses of approximately 200 g per year, large quantities of source plasma are required. Production of α_1-antitrypsin in the milk of transgenic animals may well prove to be an attractive alternative source of this protein product.

Although this technology is still in its infancy, production of pharmaceutically important proteins in the milk of suitable transgenic animals holds considerable potential. Apart from tPA and α_1-antitrypsin, various other heterologous proteins have been successfully produced in milk. Examples of such successes include human antihaemophilic factor IX from sheep milk, human interleukin 2 and human urokinase. Such production systems have a number of potential advantages over alternative production methods such as animal cell culture. Desirable features include:

- High production capacities. During a typical 5 month lactation period, one sheep can produce 2–3 litres of milk per day. If the recombinant protein is expressed at a level of 1 g per litre, a single sheep could produce in excess of 20 g of product per week.
- Ease of collection of source material. This only requires the animal to be milked. Commercial automated milking systems are readily available. Such systems require only moderate design alterations as they are already designed to maximize hygienic standards during the milking process.

- Low capital investment requirements and low operational costs. Traditional production methods yielding recombinant proteins require considerable expenditure on fermentation equipment. Using this technology such costs are reduced to raising and maintaining the transgenic herds.
- Producer animal numbers can be expanded by breeding programmes.

A number of technical details relating to the production of therapeutically important proteins in the milk of transgenic animals remain to be optimized. Yields of heterologous proteins are often found to be extremely variable. In some cases, values of less than 1 g per litre have been recorded, although much higher values are generally obtained. Mammary tissue is capable of carrying out a broad range of post-translational modifications of the proteins synthesized. Detailed characterization of the nature of such modifications, in particular in relation to glycosylation patterns, remains to be carried out. Any significant alterations in glycosylation patterns relative to the patterns observed when the protein of interest is produced in its natural source, may alter its biological characteristics.

PROTEIN PURIFICATION FROM CELL CULTURE SYSTEMS

Cell culture also represents an important source of several medically important proteins, virtually all of which are destined for therapeutic or diagnostic applications. Monoclonal antibodies as well as various vaccines and interferons are among the best-known examples of proteins produced by such methods. Increased interest in the production of proteins by cell culture has prompted an upsurge in research and development into animal cell manipulation and culturing methods. This has resulted in a greater understanding of the basic processes involved, although certain technical difficulties remain to be overcome.

Animal and microbial cells exhibit many basic differences in their cellular physiology and structure. Microbial cell fermentation technology has been adapted in order to promote successful culture of animal cells. Animal cells do not possess a cell wall, and thus are more susceptible to physical damage when compared with their microbial counterparts. Fermentation tanks in which animal cells are cultured usually contain agitation blades of modified design in order to reduce the damaging shear forces generated by such rotating blades. The potential physical damage caused to animal cells can be reduced still further if fermentation is conducted in air-lift reactors, in which liquid culture motion is promoted within the vessel by sparging a mixture of air and carbon dioxide into the reactor at its base. Several other differences between animal and microbial cells also influence animal cell culture and production reactor design:

- The nutritional requirements of animal cells are more complex than those of microbial cells.
- Animal cells tend to have lower oxygen requirements and grow more slowly than their microbial counterparts.
- Far greater numbers of animal cells are required to seed the reactors effectively.

Animal cells can be differentiated into anchorage-dependent and anchorage-independent cell types, based upon their mode of growth. Anchorage-dependent cells will grow only on a solid substratum. Such cells grow in monolayer fashion and exhibit contact inhibition. Anchorage-independent cells, on the other hand, do not require a solid support for growth. Most established cell lines in addition to transformed cell lines fall into this latter category. Although both cell types may be grown in submerged cell culture fermenters, anchorage-dependent cells must be attached to suitable carrier beads.

Monoclonal antibodies are among the most notable proteins produced in animal cell culture. Owing to their biological attributes, monoclonal antibodies have numerous diagnostic, therapeutic and preparative uses. Monoclonal antibodies are widely used as analytical tools in many areas of research. Their large-scale production is now a well-established industrial process. Culture of hybridomas is still the most common method of monoclonal antibody production. However, many other mammalian cell lines also have the ability to assemble and secrete antibodies, and expression of antibody genes in non-lymphoid mammalian cell lines, such as Chinese hamster ovary (CHO) cells, has been successfully achieved. Highly efficient mammalian cell expression systems can thus be used to produce numerous heterologous proteins, including complex antibodies.

One obstacle which may limit the widespread use of monoclonal antibodies as therapeutic agents is that such antibodies, usually derived from a mouse hybridoma cell line, are themselves antigenic when administered to humans. One possible method of circumventing this problem involves the development of human monoclonal antibodies produced by virally transformed lymphoblastoid cell lines, or the development of human hybridoma systems. Another strategy employed to overcome the problem of immunogenicity involves the production of "humanized" antibodies by recombinant DNA technology. Production of such antibodies entails the construction of a hybrid gene, coding for the required mouse variable region (antigen binding domain), fused to a human gene coding for the constant antibody domains. Expression of the resultant hybrid gene in a mammalian cell line such as a Chinese hamster ovary cell line or a myeloma cell line results in the production of a hybrid antibody. Such hybrid humanized antibodies can now be produced on an industrial scale in various cell lines. This technology may well overcome the human–antimouse IgG response which threatens to limit the therapeutic usefulness of entirely mouse-derived monoclonal antibodies. Production of humanized antibodies is discussed more fully in Chapter 5.

Various interferons have also been produced on an industrial scale by cell culture methods. The potential therapeutic application of interferons as antiviral and anticancer agents was appreciated shortly after their initial discovery in the late 1950s. Their clinical application was limited for many years due to inadequate supplies. Until the early 1980s all the interferon employed clinically was obtained directly from human leukocytes. Such supplies contained approximately 1% interferon and were extremely limited. By the early 1980s, developments in cell culture techniques facilitated the production of large quantities of essentially pure interferon. Interferon alpha, for example, is produced industrially by mass cell culture techniques in 8000 litre bioreactors using a human lymphoblastoid cell line termed Namalwa. Recombinant interferon is also produced by fermentation of bacterial cells. A number of other proteins are produced by animal cell culture techniques. Baby hamster kidney (BHK) cell lines have been used for many years in the commercial production of foot and mouth disease vaccines. Production of other industrially important proteins, such as human growth hormone, human colony stimulating factor 1 and tissue plasminogen activator by animal cell culture, has also been reported, at least on a pilot scale.

As mentioned earlier, many animal cell lines exhibit relatively complex nutritional requirements. Basal tissue culture media typically contain a carbohydrate source, as well as various vitamins, mineral salts and amino acids. Traditionally, various other supplements have also been added to the culture media. Such supplements include antibiotics, to prevent growth of any contaminant bacteria, and serum. Serum is added as a source of non-defined essential nutrients. Fetal calf serum was often found to be the most efficient serum type. It is usually added to the culture medium to a final concentration range of 0.5–25%. The use of fetal calf serum in large-scale animal cell culture systems would be prohibitively expensive. It was often replaced in the past by serum obtained from newborn calves. More recent trends, however, involve the substitution of the serum by more defined substances such as a mixture of bovine serum albumin, insulin and transferrin.

The addition of serum or substances derived from serum to animal cell culture media is undesirable for a number of reasons. The addition of such a complex mixture of proteins and other biomolecules can only render the subsequent purification protocol more complex. Addition of serum also risks contamination of the culture medium with blood-borne pathogens. Increased research focusing on animal cell culture now allows the formulation of cell culture media low in protein and devoid of serum.

Insect cell culture systems

A number of recombinant proteins have been produced by insect cell lines in culture, although most such systems do not as yet enjoy widespread industrial use. Expression of heterologous proteins in insect cell lines often involves the use of a baculovirus expression system. This virus has the ability to infect and replicate in a number of established insect cell lines with the resultant production of high levels of a virally encoded structural protein termed polyhedrin. Polyhedrin synthesis can constitute up to 50% of total proteins produced by the infected host insect cell. The strategy most often employed in producing heterologous proteins using this system involves placing the gene coding for the protein of interest under the influence of the polyhedrin promoter. Thus far, expression levels of heterologous proteins achieved using this system have been highly variable, generally falling within the range of 1–600 mg of protein per litre of culture medium.

PROTEIN ENGINEERING

Recent advances in molecular biology and related techniques now allow scientists to engineer proteins. This entails the introduction of specific, predefined alterations in the amino acid sequence. This is best achieved by the process of site-directed mutagenesis. Site directed mutagenesis involves introducing a specific change in the nucleotide sequence of a particular gene or cDNA. A protein of the required altered amino acid sequence can then be obtained by expression of the mutated DNA in an appropriate host system. Oligonucleotides of appropriate sequence can be synthesized to achieve the alterations in the DNA.

This technique should allow researchers to logically introduce changes in the amino acid sequence of proteins in order to achieve a predefined goal. As discussed in Chapter 5, this technology has been used to produce humanized monoclonal antibodies. These antibodies contain antigen binding sites derived from mouse antibodies, with the remainder of the antibody being human in origin. Such humanized antibodies exhibit several advantages over mouse-derived monoclonal antibodies when administered for therapeutic purposes (Chapter 5).

All proteins attain their biological function by virtue of their molecular structure. An understanding of the relationship between structure and function is essential if protein engineering is to be employed logically to enhance the physicochemical properties of many proteins. In recent times much effort has been directed towards engineering enzymes in order to confer on them enhanced catalytic activity, and/or temperature or pH stability. Although X-ray crystallography allows determination and study of the three-dimensional structure of proteins it is still not possible to predict from the amino acid sequence the conformation that will be adopted by a polypeptide. The effect of substituting amino acids by site-directed mutagenesis on protein conformation and hence on its physicochemical properties cannot as yet be accurately predicted. Several studies have shown that replacing a specific lysine residue with arginine, or replacing glycine with alanine, can result in increased thermal stability. For example, replacement of lysine 253 with arginine in glucose isomerase resulted in enhanced thermal stability of the resultant engineered enzyme. Replacement of the lysine with arginine seemed to alter the enzyme's conformation in such a way as to enhance the overall level of hydrogen bonding; this of course exerts a positive influence on protein stability. Glucose isomerase is utilized industrially in the production of high-fructose syrup from glucose syrups.

Introduction of cysteine residues which allow formation of disulphide bonds can also enhance protein stability. This has been illustrated in the case of lysozyme which has a cysteine residue at position 97. Conversion of an isoleucine residue at position 3 of the molecule to cysteine was shown to result in disulphide bond formation with cysteine 97. This in turn enhances the stability of the enzyme. Protein engineering holds great promise for the future. The full potential of this technology, however, will not be attained until more is known about the forces that govern protein folding.

Human tissue plasminogen activator is an important therapeutic agent which has been subject to modification by methods of protein engineering; it is a serine protease which plays an important role in promoting fibrinolysis, the degradation of blood clots (Chapter 5). Removal of a tripeptide sequence (Tyr-Phe-Ser) from a specific domain within tPA was shown to decrease significantly the plasma clearance rate of this molecule. Native tPA exhibits a very short plasma half-life. This necessitates its administration by prolonged intravenous (IV) infusion. An engineered protein exhibiting an extended plasma half-life could be administered as a single IV injection. Further modifications have also been made to tPA which not only increase its plasma half-life but also render it more potent in terms of promoting lysis of blood clots.

CONCLUSION

Proteins of industrial interest have been traditionally obtained from a wide range of sources. The advent of recombinant DNA technology has rendered it technically feasible to produce virtually any protein of interest in a number of different organisms. However it must be borne in mind that technical feasibility does not guarantee economic success.

Many microbial strains have been identified which are quite efficient at naturally producing a variety of industrially important proteins. These strains have been identified by employing standard screening selection and mutational techniques. This is particularly true with regard to organisms producing enzymes such as amylases and proteases in bulk.

Increased understanding of the factors governing gene expression and protein synthesis confers upon the scientific community the ability to render even more efficient protein production from most sources. Perhaps the greatest industrial impact of such genetic manipulation relates to the production of high-value proteins for therapeutic and diagnostic purposes in novel hosts. Recombinant DNA technology now allows large quantities of such proteins to be produced, most notably in microbial production systems.

FURTHER READING

Books

McNeil, B. & Harvey, L.M. (1990). *Fermentation, A Practical Approach.* IRL Press, Oxford.

Watson, J. *et al.* (1992). *Recombinant DNA*, 2nd edn. Scientific American Books, New York.

Articles

Allen, G. (1982). Structure and properties of human interferon-α from Namalwa lymphoblastoid cells. *Biochem. J.*, **297**, 397–408.

Allen, G. & Fantes, K.H. (1980). A family of structural genes for human lymphoblastoid (leukocyte-type) interferon. *Nature*, **287**, 408–411.

Balls, A.K. *et al.* (1937). Crystalline papain. *Science*, **86**, 379.

Browne, M.J. *et al.* (1993). Improving therapeutic agents through protein engineering. *J. Chem. Technol. Biotechnol.* **57**, 286–287.

Buckholz, R. & Gleeson, M. (1991). Yeast systems for the commercial production of heterologous proteins. *Bio/Technology*, **9**, 1067–1071.

Cullen, D. *et al.* (1987). Controlled expression and secretion of bovine chymosin in *Aspergillus nidulans*. *Bio/Technology*, **5**, 369–375.

Datar, R. *et al.* (1993). Process economics of animal cell and bacterial fermentations: a case study analysis of tissue plasminogen activator. *Bio/Technology*, **11**, 340–357.

Denman, J. *et al.* (1991). Transgenic expression of a variant of human tissue-type plasminogen activator in goats milk: purification and characterization of the recombinant enzyme. *Bio/Technology*, **9**, 839–842.

Ebert, K.M. *et al.* (1991). Transgenic production of a variant of human tissue-type plasminogen activator in goats milk: generation of transgenic goats and analysis of expression. *Bio/Technology*, **9**, 835–838.

Finter, N.B. *et al.* (1984). Mass human cell culture as a source of interferons. *Lab. Technology*, March–April, 57.

Fish, N.M. & Hoare, M. (1988). Recovery of protein inclusion bodies. *Biochem. Soc. Trans.*, **16**, 102–104.

Fox, B. (1991). Killer plants. *New Scientist*, 26 October, 28.

Fradd, R.B. (1993). Searching for miracle biopharmaceuticals. *Bio/Technology*, **11**, 870–871.

Goodenough, P. (1993). Protein engineering—the next step in evolution. *Biologist*, **40**, 67–72.

Gordon, K. *et al.* (1987). Production of human tissue plasminogen activator in transgenic mouse milk. *Bio/Technology*, **5**, 1183–1187.

Hartley, D.L. & Kane, J.F. (1988). Properties of inclusion bodies from recombinant *Escherichia coli*. *Biochem. Soc. Trans.*, **16**, 101–102.

Hu, W.S. & Peshwa, M.V. (1993). Mammalian cells for pharmaceutical manufacturing. *Am. Soc. Microbiol. News*, **59**, 65–68.

Kim, S.-H. *et al.* (1988). Crystal structures of two intensely sweet proteins. *Trends Biochem. Sci.*, **13**, 13–15.

Kingsman, S. *et al.* (1985). Heterologous gene expression in *Saccharomyces cerevisiae*. *In* Russell, G.E. (ed.) *Biotechnology and Genetic Engineering Reviews*, Vol 3, 377–416. Intercept.

Krebbers, E. & Vandekerckhove, J. (1990). Production of peptides in plant seeds. *TIBTECH*, **8**, 1–3.

LaVallie, E. *et al.* (1993). A thioredoxin gene fusion expression system that circumvents inclusion body formation in the *E. coli* cytoplasm. *Bio/Technology*, **11**, 187–193.

Maiorella, B. *et al.* (1988). Large-scale insect cell-culture for recombinant protein production. *Bio/Technology*, **6**, 1406–1410.

Marston, F. (1986). The purification of eukaryotic polypeptides synthesized in *Escherichia coli*. *Biochem. J.*, **240**, 1–12.

McAleer, W. *et al.* (1984). Human hepatitis B vaccine from recombinant yeast. *Nature*, **307**, 178–180.

Moffat, A. (1992). High-tech plants promise a bumper crop of new products. *Science*, **256**, 770–771.

Parekh, R.B. *et al.* (1989). N-glycosylation and the production of recombinant glycoproteins. *Trends in Biotechnol.*, **7**, 117–121.

Pen, J. *et al.* (1993). Phytase-containing transgenic seeds as a novel feed additive for improved phosphorus utilization. *Bio/Technology*, **11**, 811–814.

Rhodes, M. & Birch, J. (1988). Large-scale production of proteins from mammalian cells. *Bio/Technology*, **6**, 518–523.

Schein, C.H. (1989). Production of soluble recombinant proteins in bacteria. *Bio/Technology*, **7**, 1141–1147.

Smith, R.A. *et al.* (1985). Heterologous protein secretion from yeast. Science, **229**, 1219–1224.

Svensson, B. & Sogaard, M. (1992). Protein engineering of amylases. *Biochem. Soc. Trans.* **20**, 34–42.

Swain, W. F. (1991). Antibodies in plants. *TIBTECH*, **9** 107–109.

Uhlen, M. *et al.* (1988). Protein engineering to optimize recombinant protein purification. *Biochem. Soc. Trans.*, **16**, 111–112.

Van Brunt, J. (1986). Fungi: the perfect hosts? *Bio/Technology*, **4**, 1057–1062.

Van Brunt, J. (1988). Molecular farming: transgenic animals as bioreactors. *Bio/Technology*, **6**, 1149–1154.

Ward, P. *et al.* (1992). Production of biologically active recombinant human lactoferrin in *Aspergillus oryzae*. *Bio/Technology*, **10**, 784–789.

Weaver, J.F. *et al.* (1988). Production of recombinant human CSF-1 in an indusible mammalian expression system. *Bio/Technology*, **6**, 287–290.

Wiech, H. *et al.* (1991). Protein export in procaryotes and eucaryotes. *FEBS Lett.*, **285**, 2, 182–188.

Wright, G. *et al.* (1991). High level expression of active human alpha-1-antitrypsin in the milk of transgenic sheep. *Bio/Technology*, **9**, 830–834.

Chapter 3

Downstream processing of protein products

- Scale-up of protein purification.
  - Extent of scale-up required.
- Source availability.
- Initial recovery of protein.
- Cell disruption.
  - Microbial cell disruption.
- Removal of whole cells and cell debris.
  - Centrifugation.
  - Filtration.
  - Aqueous two-phase partitioning.
- Nucleic acid removal.
- Concentration and primary purification.
  - Concentration by precipitation.
  - Concentration by ion-exchange.
  - Concentration by ultrafiltration.
  - Diafiltration.
- Column chromatography.
  - Size exclusion chromatography (molecular sieving).
  - Ion-exchange chromatography.
  - Hydrophobic interaction chromatography.
  - Affinity chromatography.
  - Immunoaffinity purifications.
  - Chromatography on hydroxyapatite.
  - Chromatofocusing.
  - Protein chromatography based on aqueous two-phase separation.
  - HPLC of proteins.
- Engineering proteins for subsequent purification.
- Bulk enzyme preparation.
- Purification of proteins for therapeutic and diagnostic use.
- Scale-up of protein purification systems.
- Protein stabilization and finished product formats.

- Stabilization of the finished product.
  - Lyophilization.
- Labelling and packing of finished product.
- Further reading.

Once a suitable protein source has been identified it becomes necessary to design an appropriate downstream processing procedure to isolate the desired protein. Downstream processing encompasses not only purification of the protein from the starting material, but also many other manufacturing activities such as quality control evaluation, end product stabilization, adjustment of product potency to within specified limits, filling of product into suitable final product containers, labelling of containers, etc. The extent of purification required depends upon a number of factors, the most important of which include the degree of protein purity required as well as the nature of the starting material. As discussed in Chapter 2, the application of recombinant DNA technology now facilitates the production of any protein of interest in many host organisms.

The majority of proteins currently available commercially are produced by microbial cell fermentation or by animal cell culture. The protein may be intracellular or may be secreted into the medium. Generally, if the protein is intracellular, its subsequent purification requires more exhaustive downstream processing procedures than those required for a protein secreted directly into the culture medium. However, if the protein of interest accumulates intracellularly in the form of inclusion bodies, a single low-speed centrifugation step subsequent to cell disruption can result in considerable purification. If the required protein is found in the periplasmic space of Gram negative bacteria, a considerable level of purification may be achieved simply by harvesting the cells, followed by disruption of the outer wall, with the resultant release of the contents of the periplasmic space.

Disruption of the cell results in the generation of a much more complex mixture of biological molecules, from which the protein of interest must then be purified. Extracellular production of a protein of interest is usually regarded as being advantageous in terms of the subsequent downstream processing requirements; in such instances disruption of intact cells is not required to liberate the protein. The downstream processing of many proteins secreted into the extracellular medium is, however, often far from simple. Although cells secrete relatively few proteins into the extracellular medium, other molecules which are components of the medium may complicate the purification procedures. This is particularly true with regard to animal cell culture where various serum and/or non-serum derived proteins are added to the initial culture medium in order to promote cell growth. This addition to the culture medium of a complex cocktail of proteins heightens the demands made upon subsequent purification procedures.

The quantity of the desired protein present in the initial preparation will affect the downstream processing. The higher the level, the greater the ratio of desired protein to other protein impurities. High levels will also allow a more exhaustive protein purification scheme to be undertaken, as even if the yield of product obtained is relatively low, the finite quantity of finished product obtained may still be acceptable. As a general rule, the greater the level of purification, the lower the yield of final product obtained.

The degree of purity required is an overriding factor in the design of any downstream processing procedure. This is largely dependent on the intended application of the final product. As a general principle, the extent of downstream processing of any protein is maintained at the minimum required to produce an acceptable final product. Unnecessary steps in the purification of a protein product can be expected to reduce the biological stability of the final product, and the yield of product decreases as the number of purification steps increases. As outlined in Chapter 1, most proteins of industrial interest may be grouped into one of two broad categories:

- Proteins produced in bulk as relatively crude preparations. Such proteins usually are enzymes which have a wide variety of applications in the food and beverage industries.
- Proteins destined for therapeutic and/or diagnostic applications. These proteins are generally produced in quantities orders of magnitude lower than bulk enzyme preparations, and to a very high degree of purity.

Proteins used for therapeutic or *in vivo* diagnostic purposes are generally subjected to the most stringent purification procedures, as the presence of molecular species other than the intended product may have an adverse clinical impact (Chapter 4). Proteins used for *in vitro* diagnostic and analytical purposes are also usually highly purified. In such cases, however, the level of purification required is not as high as for those proteins intended for *in vivo* administration. In some instances, design of a downstream purification procedure which removes specific contaminating proteins is more important than purification of the protein to homogeneity. Despite ever-increasing numbers of publications detailing various aspects of protein biotechnology, few reveal detailed protocols employed by manufacturers in the downstream processing of specific protein products. Production of industrially important proteins, especially the production of bulk industrial enzymes, is highly competitive and many companies are understandably reluctant for commercial

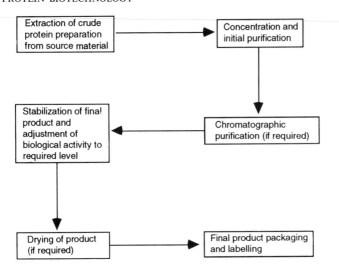

Figure 3.1 Flow diagram outlining the major steps in the downstream processing of protein products

reasons to reveal specific details of their production methods. The major steps generally undertaken during downstream processing are outlined in Figure 3.1.

SCALE-UP OF PROTEIN PURIFICATION

Protein purification protocols are initially designed at laboratory level. Scale-up studies are then undertaken in order to produce sufficient quantities of the protein to meet market demands as economically as possible. Generally speaking, the costs associated with the production of a unit quantity of any protein decline with increasing scale of production (Table 3.1).

Table 3.1. Reasons why production cost per unit quantity of product declines with increased scale of production

Most chemicals and other raw materials required may be purchased more cheaply in bulk

Many overhead costs remain independent of the scale of production

Labour costs (per unit of product produced) decrease sharply with increased scale of production

Many of the techniques and procedures used in laboratory-scale purification of proteins are not amenable to scale-up. For example, many methods employed to disrupt bacterial cells in the laboratory are not suitable for large-scale application, for either technical or economic reasons. Disruption of the bacterial cell wall by sonication, while feasible on a small scale, is inefficient when applied to large-scale procedures. Treatment with lysozyme is sometimes employed on a laboratory-scale to degrade the bacterial cell wall; however, the use of lysozyme on a large scale would be uneconomic. Other techniques routinely used in laboratory-scale purification procedures must be modified before they can be successfully employed on a large scale. Laboratory-scale centrifugation, for example, is generally carried out in a batch (fixed volume) centrifuge, whereas continuous-flow centrifuges are employed in industrial-scale purification systems. Problems associated with scale-up of specific techniques should always be borne in mind when designing a purification procedure at laboratory level.

Rational design of a protein purification procedure, with all stages being amenable to direct scale-up, is especially desirable when working with a protein of therapeutic interest. Such proteins are initially produced in small quantities which are then subjected to animal trials. If encouraging results are obtained, limited clinical studies may then be initiated. Such trials, while requiring relatively little protein, could take several years to complete and are expensive to carry out. If successful, however, scale-up of production is normally initiated in line with projected market demands. Any but the most minor of changes made to the original purification process during such scale-up would invalidate the earlier clinical studies. In such cases, therefore, it is critically important to ensure that the purification system initially developed in the laboratory can be scaled up without difficulty. The three stages involved in the scale-up of protein purification are outlined in Figure 3.2.

Figure 3.2. Sequence of steps undertaken during design and scale-up of a protein purification protocol. Initial studies are undertaken at laboratory level. These identify the minimum steps required to yield a protein product of desired purity and format. Pilot-scale studies are then undertaken which serve to pinpoint and resolve difficulties associated with process scale up. Finally, the process-scale operation is undertaken

A prior detailed knowledge of the physicochemical properties of a protein is of great benefit in designing a rational purification procedure. Such a procedure should contain the minimum number of steps required to yield a protein product within the designated product specifications. Many techniques, in particular chromatographic steps, are equally effective no matter what the source material of the protein is (e.g. microbial, animal or plant).

Extent of scale-up required

It is difficult to define what exactly constitutes a small-scale or large-scale protein purification process. Laboratory-scale purification procedures generally yield microgram to milligram quantities of the protein product. Pilot-scale production often yields milligram to gram quantities, while large-scale purification yields quantities in the order of grams to kilograms. Initially it is often difficult to decide what level of scale-up is desirable for any given protein product. It may not always be clear what the level of market demand for the protein in question will be. Although there are several hundred proteinaceous products currently on the market, most are produced in relatively small quantities. There are probably no more than about a hundred proteins whose annual level of production is in the range of kilograms, while a very small number of proteins are marketed whose annual production levels exceed 100–1000 kg.

Large-scale production of proteins destined for therapeutic or diagnostic applications yields gram to kilogram quantities of the protein in question. Market demands for monoclonal antibodies used in diagnostic kits is in the range of grams per year. Thus, if a single cell line produces up to 50 mg of monoclonal antibody per litre per day, a single laboratory-scale bioreactor can produce more than adequate quantities of such a monoclonal antibody. Monoclonal antibodies used for therapeutic purposes (for passive immunization, for example) are required in larger quantities—in the order of kilograms per year. Larger bioreactors are required to satisfy such market demands. Several other therapeutic proteins such as cytokines will probably be required in quantities of the order of kilograms per year. Many such proteins are extremely potent and hence low dosages are required. On the other hand, some such products may find numerous clinical applications and their market demand will increase.

At the other end of the scale, bulk industrial enzymes are produced in large quantities. Examples include α-amylase used in starch processing, proteases used in the detergent industry, and pectinases used in clarification of fruit juices. Product quantities are often many thousands of kilograms of protein per year. Such bulk enzymes are relatively impure preparations with little downstream processing required because of the nature of their application.

Although the scale of production and level of purification required can vary enormously from product to product, most purification systems share some common attributes. The first step in most purification protocols is an extraction step. This is normally followed by preliminary treatments of the crude extract to clarify and concentrate the crude material and, in some cases, to achieve a very limited level of enrichment. In the case of proteins produced in large quantities, i.e. most bulk enzymes, little further processing is required. In the case of most other proteins, however, a number of additional purification steps are required. The majority of high-resolution purification steps used are based on chromatographic procedures, including ion-exchange, gel filtration, affinity, pseudoaffinity and hydrophobic interaction chromotography, all of which are routinely used on an industrial scale. Once the desired level of purity has been achieved, stabilizers and/or preservatives are added to the product and the level of biological activity is adjusted to within the desired limits. The final product is subsequently packed into suitable containers. While in some cases this final product is a liquid, many proteins are marketed in powdered (dry) form.

SOURCE AVAILABILITY

The availability of sufficient supplies of source material is one important non-technical consideration which invariably influences the success of any protein production venture. This is particularly true in cases where the protein of interest is obtained from animal or plant tissues. Availability of plant material is generally seasonal. Furthermore, the biochemical composition of the plant material may also vary from growing season to growing season. Availability of sufficient quantities of animal tissue also influences protein production from animal sources. Production of follicle stimulating hormone (FSH) from pituitary glands may require the collection of millions of pituitary glands per year. Such large numbers could only be obtained from very large slaughterhouse facilities as collection of such material from a large number of smaller slaughtering facilities, could render the whole process uneconomical. Generally speaking, such availability problems are rare if the protein of interest is produced by fermentation methods, as fermentation scale may be adjusted upward to meet demand.

INITIAL RECOVERY OF PROTEIN

As outlined in Chapter 2, proteins of industrial interest are obtained from a wide range of sources. Despite such a diversity of origin, proteins derived from animal, plant and microbial species are generally purified using very similar techniques. The initial step of any purification procedure involves recovery of the protein from the source from which it was produced. The complexity of this step depends largely upon whether the protein of interest is intracellular or extracellular. Many proteins produced by fermentation of microorganisms or by animal cell culture are secreted into the medium. Initial product recovery in such cases involves the separation of the whole cells from the fermentation medium by filtration or centrifugation. The protein of interest is present in the cell-free medium, often in very dilute form.

In the case of intracellular microbial proteins, cell harvesting from the culture medium is followed by resuspension of the cells in buffer or water with subsequent cell disruption. Resuspension of microbial cell pastes can often be achieved by simple mechanical stirring, although in some cases a more vigorous approach is required. A number of mechanical mixers and related devices are commercially available which can efficiently resuspend large quantities of cell paste. Such cell pastes are resuspended in much smaller volumes than the original volume of fermentation broth from which they were prepared. Such reduced volumes facilitate more efficient handling during subsequent processing steps.

Most proteins obtained from animal or plant tissues are intracellular in nature. The initial step in processing such material involves collection of the appropriate tissue required. Specific examples include the collection of pituitary glands, from which hormones such as FSH and luteinizing hormone (LH) may be purified; collection of blood from which various blood products are obtained; collection of internal organs such as liver and kidneys, from which various enzymes and other proteins of interest may be obtained; and collection of pancreatic tissue, which is the traditional source of insulin.

The material collected is transported from the site of collection, usually a slaughterhouse, to the processing facility. The tissue may be transported in a refrigerated or deep-frozen state. Upon arrival, solid tissue is normally cut up into smaller pieces which are more suitable for subsequent homogenization. Liquid products such as blood require different handling initially. Blood is often collected in containers containing sodium citrate to prevent clotting; in such cases the first step in the recovery of the desired plasma protein involves centrifugation, which separates the suspended cellular elements from the supernatant, termed plasma. If the blood is allowed to clot, the clot—which consists of various cellular elements embedded in a network of the protein, fibrin—may be removed by centrifugation. The resultant clear supernatant fluid is termed serum. (Blood plasma therefore consists of blood serum plus the protein fibrinogen, which is the precursor of the clotting protein, fibrin.) The protein of interest can subsequently be isolated from the plasma or serum by various methods, including fractionation by precipitation and column chromatographic procedures.

CELL DISRUPTION

If the protein required is intracellular, collection of the source cells or tissue is followed by disruption of the cell. Mammalian cell tissues are relatively easily disrupted, because animal cells (unlike their bacterial or plant counterparts) are devoid of a protective cell wall. Many techniques used in laboratory-scale homogenization of animal tissue can readily be scaled up. Most such techniques rely on physical disruption of the animal cells. A well-known example is that of the Potter homogenizer. Cell disruption is achieved in this case by shear forces generated between a rotating pestle and the inside wall of a test tube-like container. Efficient homogenization of plant tissue is more difficult to achieve, mainly because of the presence of the outer cell wall. Disruption of plant tissue is often achieved by physical means, in which the plant material is subjected to homogenization by rapidly rotating blades, as in the Waring blender.

Microbial cell disruption

Disruption of microbial cells is difficult owing to their cell walls. Despite this, a number of efficient systems exist which are capable of disrupting large quantities of microbial biomass (Table 3.2). Disruption techniques such as sonication or treatment with the enzyme lysozyme are rarely if ever employed on an industrial scale, either because of inadequate equipment or on economic grounds. Large-scale cell disruption by chemical means has been used successfully in some instances. Chemicals utilized include a variety of detergents and antibiotics, solvents such as toluene or acetone, and treatment under alkaline conditions or with chaotropic agents such as urea or guanidine.

Table 3.2. Some chemical, physical and enzymatic techniques that may be employed to achieve microbial cell disruption

Treatment with chemicals:
  detergents
  antibiotics
  solvents (toluene, acetone)
  chaotropic agents (urea, guanidine)

Exposure to alkaline conditions

Sonication

Homogenization
Agitation in the presence of glass beads

Treatment with lysozyme

Protein extraction procedures utilising detergents are effective in many instances but suffer from a number of drawbacks. The mode of detergent action primarily involves solubilization of the cell membranes. Ionic detergents such as sodium lauryl sulphate are more efficient than non-ionic detergents such as polysorbates. The major disadvantage associated with the detergent system is that such detergents often cause protein denaturation and precipitation, which limits their usefulness for such purposes. Many other chemicals, including various solvents, or methods such as incubation under alkaline conditions also suffer from this disadvantage. Furthermore, if the chemicals used do not adversely affect the protein, their presence may adversely affect a subsequent downstream processing step. In addition the presence of such materials in the final preparation, even in trace quantities, may be unacceptable for a number of reasons. Detergent-based cell disruption systems have been successfully employed in a number of specific cases. Triton has been used to render *Nocardia* cells permeable in the large-scale extraction and purification of cholesterol oxidase; this enzyme finds use in the assay of blood cholesterol levels.

Industrial-scale disruption of microbial cells is most often achieved by mechanical methods such as homogenization, or by vigorous agitation with abrasives. Homogenization is probably the most popular technique used to achieve large-scale microbial cell disruption. During the homogenization process a cell suspension is forced through an orifice of very narrow internal diameter at extremely high pressures. This generates extremely high shear forces. As the microbial suspension passes through the outlet point it experiences an almost instantaneous drop in pressure to normal atmospheric pressure. The high shear forces and subsequent rapid pressure drop act as very effective cellular disruption forces, and result in the rupture of most microbial cell types (Figure 3.3). In most cases a single pass through the cell homogenizer results in adequate cell breakage, but it is also possible to recirculate the material through the system for a second or third pass.

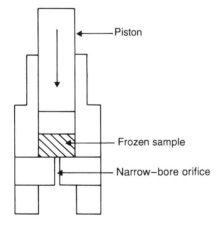

— Piston

— Frozen sample

— Narrow–bore orifice

Figure 3.3. Diagrammatic representation of a cell homogenizer. This represents one of a number of instruments routinely employed to rupture microbial cells

Although prototype homogenizers were developed and used as far back as the 1950s, many design improvements have since been incorporated. An efficient cooling system minimizes protein denaturation, which would otherwise occur due to the considerable amount of heat generated during the homogenization process.

Homogenizers capable of handling large quantities of cellular suspensions are now available, many of which can efficiently process several thousand litres per hour.

An additional method often used to achieve microbial cell disruption, both at laboratory level and on an industrial scale, involves cellular agitation in the presence of glass beads. In such bead mills, the microorganisms are placed in a chamber together with a quantity of glass beads 0.2–0.3 mm in diameter. This mixture is then agitated vigorously resulting in numerous collisions between the microbial cells and the glass beads, thus grinding the cells between the rotating beads. These forces result in efficient disruption of most microbial cells. Operational parameters such as the ratio of microbial cells to beads, in addition to the rate and duration of agitation, may be adjusted to achieve optimum disruption of the particular cells in question. Commercially available industrial-scale bead milling systems can process over a thousand litres of cell suspension per hour. Cooling systems minimize protein inactivation by dissipating the considerable heat generated during this process.

REMOVAL OF WHOLE CELLS AND CELL DEBRIS

Upon completion of the homogenization step, cellular debris and any remaining intact cells can be removed by centrifugation, or in some cases by filtration. As previously mentioned, these techniques are also used to remove whole cells from the medium during the initial stages of extracellular protein purification.

Centrifugation

Batch (fixed volume) centrifuges are capable of processing, at most, a few litres during any one spin cycle. Although such centrifuges can attain high centrifugal forces, low processing capacity limits their use on an industrial scale where often batches of several hundreds or thousands of litres of crude extract must be processed.

Industrial-scale centrifugation is normally achieved using continuous flow centrifuges, through which homogenate is continuously pumped and the clarified solution continually collected. The deposited solids can be removed from the centrifuge bowl by periodically stopping the centrifuge and manually removing the pelleted material. Most modern continuous-flow centrifuges, however, are designed to allow continuous discharge of collected solids through a peripheral nozzle, or alternatively facilitate intermittent discharge of pelleted material via a suitable discharge valve. A number of different continuous-flow centrifuge designs are commercially available. The three basic types are the disc centrifuge, the hollow-bowl centrifuge and the basket centrifuge. Most microbial cells are sedimented by batch centrifugation by applying a centrifugal force of approximately $5000\,g$ for 15 minutes or less. Efficient removal of cell debris requires the application of higher centrifugal forces for longer periods, typically $10\,000\,g$ for 45 minutes.

Although higher centrifugal forces are attained by batch centrifuges, modern continuous-flow centrifuges generate sufficient gravitational force to allow effective processing at high flow rates. Typically this may be of the order of several hundred litres per hour. Centrifugation remains the method of choice to effect cell and cellular debris separation both at laboratory level and on an industrial scale. However, a number of factors (most notably the high capital and running costs of such centrifugation equipment) have led many process scientists to investigate alternative means of seperating cell or cellular debris from the required liquid. The most popular of these is filtration.

Filtration

Both whole cells and cell debris may be removed from solution by filtration. Either depth filters or membrane filters may be used. Depth filters consist of randomly oriented fibres (usually manufactured from glass-fibre or cellulose) which form an irregular network of channels or mesh-like structures. Such filters retain particles not only on their surface but also within the depth of the filter. Depth filtration is often used to remove whole cells from fermentation media, as discussed later in this chapter. Such filters are also used to remove or reduce levels of cellular debris, denatured protein aggregates or other precipitates from solution. Depth filtration is also often used to clarify fruit juices and other beverages.

Membrane filtration, also termed microfiltration, is achieved using thin, membrane-like sheets of polymeric substances such as cellulose acetate or cellulose nitrate, nylon or polytetrafluoroethylene (PTFE), in which very small pores have been generated. Pore sizes available generally range between $10\,\mu m$ and $0.02\,\mu m$. Membrane filters of pore diameter $0.2–0.45\,\mu m$ will retain all microbial cells. The retention of particles or microbial cells occurs only on or in the surface layer of the filter. The material to be filtered is applied to the filtration system under pressure in order to achieve satisfactory flow rates. On an industrial scale this is generally attained by use of a suitable pumping system. Membrane filtration enjoys increasing popularity as a system of choice to remove cell and cellular debris from solutions. The method is efficient and requires relatively simple equipment. Filter configuration may be of the flat disc type, though more often than not the filter is shaped into a cartridge (Figure 3.4). This is achieved by placing the rectangular membrane filter sheet on a supporting mesh of the same size and folding it into a pleated structure. The two ends are then sealed to form a cylinder which is placed between a plastic core and outer structure which physically protects and supports the filter material itself. Pleating allows a large filtration surface area to be accommodated in a compact area. Such filters are normally housed in stainless steel filter housing systems.

Figure 3.4. Photographs illustrating (a) a range of cartridge filters and (b) a range of filters and their stainless steel housings. In each case, the pleated filter is protected by an outer plastic supporting mesh (photographs courtesy of Pall Process Filtration Ltd)

(a)

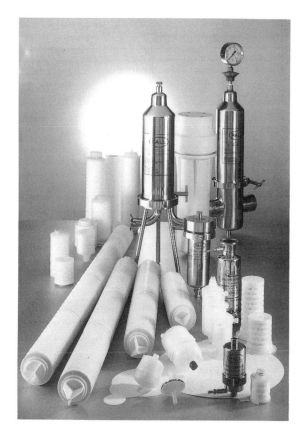

(b)

One of the main shortcomings of membrane filtration systems is their tendency to clog easily. This results in a sharp decrease in flow rate, and the blockages can also lead to a pressure build-up in the filtration system which can potentially destroy filter integrity. Most modern filters, however, withstand the application of relatively high pressures.

Incorporation of a suitable prefiltration system invariably increases effective filter life span and sustains higher flow rates through the main filter system. Thus, a fermentation broth or a homogenate may first be passed through a depth filter, the eluate from which is then passed through a membrane filter. Another approach is to use two or three membrane filters of decreasing pore size, connected in series. Such systems are popular when micro-filtration is used to yield a sterile flow of liquid.

Proteins are thermolabile, as are many other biomolecules, and may not be subjected to terminal product sterilization by autoclaving. Under such circumstances, filtration is most often employed to yield a final sterile product. Removal of all microbial cells is achieved by filtration through an "absolute" $0.2\,\mu$m filter. Membrane filters may be classified as "absolute" or "nominal". Absolute filters are guaranteed to remove all particles larger than the indicated filter pore size, i.e. they are 100% effective. Nominal filters, on the other hand, may not be 100% effective, so although they are considerably cheaper than absolute filters, nominal filters should not be used for critical operations such as sterilization. Sterile filtration is often achieved by passing a solution or suspension through a filtration series consisting of a $1\,\mu$m filter followed by a $0.45\,\mu$m filter, followed by a $0.2\,\mu$m filter. Reduction in microbial levels (the bioburden) may be achieved by using a $5\,\mu$m and a $1\,\mu$m filter connected in series, often followed by a $0.45\,\mu$m filter.

Many modern filtration systems incorporate several filters of different pore sizes into a single cartridge system (Figure 3.5). For example, sheets of a prefilter, two membrane filters of differing pore size and a supporting mesh may be placed on top of each other in series, folded into pleats and formed into a cylindrical cartridge by joining both ends. One such filter can thus effectively replace a series of two or three filters of monopore sizes in a filtration system. Due to the gradation of filtration achieved, the resulting filter has a greatly extended life-span. Most modern cartridge filters can be repeatedly sterilized by autoclaving or on-line steaming, and they may also be operated for long periods at elevated temper-atures ($70-75\,°$C).

To ensure their effectiveness, filters are often integrity tested, both before and after use. Membrane integrity tests most commonly employed include the bubble point test and the pressure hold method. Both tests rely on monitoring the effect of pressurized air on the filter system.

Although most filtration media are relatively inert, the possibility exists that certain proteins or other biomolecules may adhere to the filter matrix due to electrostatic, hydrophobic or other interactions. It is often considered prudent to pretest the proposed filter media for filtration on a laboratory-scale before employing it on a large scale, to ensure that no such protein–filter interactions occur.

In summary, therefore, filtration techniques may be used at a number of stages during a protein purification process. Filtration may be used to achieve separation of whole cells from fermentation

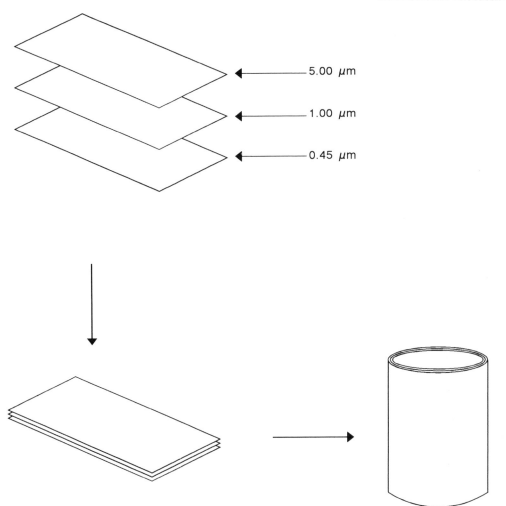

5.00 μm

1.00 μm

0.45 μm

media or to remove whole cells and cell debris after cell disruption. It may also be used at later stages in a purification process to reduce or totally eliminate microbial cells from the product stream. In many instances, such separations may also be achieved by centrifugation.

Although centrifugation represents the traditional method used to achieve many such separations there is increasing industrial utilization of filtration. The increasing popularity of filtration is largely due to reduced capital and operating costs compared with centrifugation, as well as reduced inactivation of particularly labile proteins; such proteins may be damaged by shear forces or frothing generated during continuous-flow centrifugation processes. Most filtration systems also offer a reduction in actual processing time when compared with centrifugation. As discussed later in this

Figure 3.5. Schematic representation of modern cartridge filter in which filter sheets of three different pore sizes are housed

chapter, another form of filtration, termed ultrafiltration, can be used to concentrate protein solutions. Such filtration relies on the use of a membrane filter of extremely small pore size, usually with a cut-off point in the range of 10 000–100 000 Da.

Aqueous two-phase partitioning

The technique of aqueous two-phase partitioning may be used to separate whole cells or cell debris from soluble protein. It can also be used to achieve a limited degree of protein purification and concentration. Although studied at laboratory and pilot plant scale for many years, aqueous two-phase partitioning does not enjoy widespread industrial use. This is mainly due to a lack of understanding as to how the technique works at a molecular level.

Aqueous two-phase partitioning is based on the fact that many water-soluble (aqueous) polymers are incompatible with each other, or with salt solutions of high ionic strength. Thus, if one such polymer is mixed with a salt solution or with a second incompatible polymer, upon standing two phases are formed. Such partitioning systems may be employed to separate proteins from cell debris or other impurities. The debris partitions to the lower, more polar and denser phase, while soluble proteins tend to partition into the top, less polar and less dense phase. Subsequent separation of the two phases achieves effective separation of cellular debris from soluble protein.

The most commonly employed polymers are polyethylene glycol (PEG, a polymer consisting of ethylene molecules linked by ether bonds), and dextran (a polymer consisting of repeating glucose residues, linked by $\alpha 1 \rightarrow 6$ bonds). The most common polymer–salt system is based on polyethylene glycol and a phosphate salt (normally sodium or potassium phosphate). If solutions of PEG and dextran are mixed with a cell homogenate under appropriate conditions, the protein components partition into the upper, PEG phase, whereas cellular debris accumulates in the lower, dextran phase. In this way, effective separations can be achieved upon prolonged standing. If the mixture is one of PEG and phosphate, the proteins tend to accumulate in the upper, PEG-rich phase, whereas the debris accumulates in the lower, phosphate-rich phase (Figure 3.6).

When two such incompatible water-soluble moieties are mixed, subsequent phase separation occurs at a slow rate. However, the rate of phase separation is increased by applying a centrifugal force. Thus, in practice phase separation is accelerated by employing a subsequent centrifugation step. Once the phase separation has been completed, the phase containing the protein of interest (normally the PEG-containing upper phase) is subjected to further processing. On a laboratory scale, batch centrifuges are normally used to accelerate phase separation. Continuous flow centrifuges are more

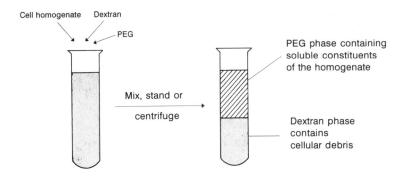

appropriate to industrial-scale operations. Application of the phase containing the protein of interest to an ultrafiltration system (see next section) results in an immediate concentration of the protein product. Alternatively, further purification may be achieved by employing a second phase extraction step, in which the partition coefficient of the (new) system is altered. This may be achieved by, for example, altering the molecular weight of the polymers used (various PEG preparations may be obtained with molecular weights ranging from 500 Da to 20 000 Da), or by changing the ionic strength of the solutions. During separation of this second phase system, some of the protein species remain in the PEG phase while others will partition to the lower phase, depending upon the conditions used. In this way, a limited purification of the protein of interest may be attained.

A high degree of protein localization in one phase may be obtained by attaching a ligand which specifically binds the protein of interest to the least polar polymer of the two phase system (normally PEG). This technique is termed affinity partitioning. In this case conditions are employed that promote partitioning of most contaminant proteins, in addition to nucleic acids and cell debris, to the more polar phase. The protein of interest will, however, partition into the less polar phase due to the presence of the ligand in this phase. Affinity ligands used can be specific, e.g. a substrate or inhibitor of an enzyme of interest, or non-specific, e.g. various dyes which bind a number of protein types. Affinity ligands are discussed in more detail later in this chapter.

Aqueous two-phase partitioning is a potentially useful downstream processing technique for a number of reasons:

- It is a gentle method, having little or no adverse effect on the biological activity of most proteins.
- Many of the polymers used exhibit protein stabilizing properties.
- The yield of protein activity recovered is generally high.
- Little if any technical difficulties arise during process scale-up.

The major disadvantage associated with this technique relates to the lack of understanding of the molecular mechanisms involved in the partitioning process. Development of a two-phase partitioning process to achieve effective partitioning is wholly empirical. Without an intimate understanding of the underlying principles involved, a rational approach to designing such systems cannot be undertaken. Also, technical grade dextran is expensive; this may be overcome by employing crude dextran preparations, or substituting the dextran with other polymers such as polyvinyl alcohol (PVA) or polyvinyl pyrrolidone (PVP).

The disadvantages outlined above, coupled with the availability of alternative, established separation techniques such as centrifugation or filtration, has thus far limited the utilization of this technique on an industrial scale.

NUCLEIC ACID REMOVAL

In many cases during downstream processing it may be necessary to remove or destroy the nucleic acid content of a cell homogenate. Liberation of large amounts of nucleic acids upon homogenization often significantly increases the viscosity of the homogenate. This generally renders the homogenate more difficult to process. Significant increases in viscosity place additional demands upon the method of cell debris removal employed. Increased centrifugal forces for longer periods may be required to pellet cell debris efficiently in a solution of higher viscosity. If a filtration system is used to remove cellular debris, increased viscosity will also adversely affect flow rate and filter performance. Increased viscosity due to liberation of nucleic acids during homogenization is often most noticeable when prokaryotic organisms are used, as the DNA in such organisms is not bounded by an intracellular protective membrane, the nuclear membrane, as is the case in eukaryotic cells.

The inclusion of a specific nucleic acid removal step is not required in all downstream processing procedures. Inclusion or exclusion of such a step depends upon the extent to which liberated nucleic acids affect the viscosity of particular suspensions and on the intended use of the final protein product. Effective nucleic acid removal is particularly important when purifying any protein destined for therapeutic use. Many regulatory authorities insist that the nucleic acid content present in the final preparation be, at most, a few picograms per dose (see Chapter 4).

Effective removal of nucleic acids during downstream processing may be achieved by precipitation or by treatment with nucleases. A number of cationic (positively charged) molecules have been shown to be effective precipitants of DNA and RNA. These positively charged molecules complex with, and precipitate, the negatively charged nucleic acids. Perhaps the most commonly employed nucleic acid precipitant is polyethyleneimine, a long-chain cationic polymer. Low concentrations of polyethyleneimine in homogenates lead to effective precipitation of nucleic acids, with the precipitate being subsequently removed together with cellular debris by centrifugation or filtration. The use of polyethyleneimine during downstream processing of proteins destined for therapeutic applications is often discouraged, as small quantities of unreacted monomer may be present in the polyethyleneimine preparation. Such monomeric species have been shown to be carcinogenic. If polyethyleneimine is utilized in such cases the subsequent processing steps must be shown to be capable of effectively removing any of the polymer or its monomeric units that may remain in solution.

Nucleic acids may also be removed by treatment with nucleases, which catalyse the enzymatic degradation of these biomolecules. Nuclease treatment is quickly becoming the most popular method of nucleic acid removal during downstream processing. This treatment is efficient and inexpensive, and (unlike many of the chemical precipitants used) nuclease preparations themselves are innocuous and thus do not compromise the final protein product.

CONCENTRATION AND PRIMARY PURIFICATION

During the initial stages of many protein purification procedures the protein of interest is often present in dilute solutions, thus large volumes of process liquid must be handled. This is particularly true

Table 3.3. Methods most commonly used in the laboratory to concentrate protein solutions

Precipitation (salt, solvent, etc.)

Ion-exchange chromatography

Ultrafiltration

Vacuum dialysis

Freeze drying

Addition of dry Sephadex G-25*

*Addition of dry Sephadex G-25 (or analogous beads) to a dilute protein solution results in bead swelling. During hydration of the beads, water enters the internal bead structure. As protein molecules are too large to enter the gel matrix, they remain in the decreased volume of liquid surrounding the beads, and hence are effectively concentrated. Unlike most of the other techniques mentioned, this method of concentration is not amenable to scale-up, mainly on economic grounds.

in the case of proteins secreted into the medium during microbial fermentation or during animal cell culture. It thus becomes necessary to concentrate such solutions in order to render the extracts more manageable for subsequent downstream processing steps. Smaller volumes are generally processed considerably faster than large volumes. Methods used to achieve concentration on a laboratory scale are listed in Table 3.3.

For large-scale applications, concentration of extracts is normally achieved by precipitation, ion-exchange chromatography or ultrafiltration. Each of these techniques has its own inherent advantages and disadvantages. All can effectively concentrate protein solutions and may result in an increase in the purity of the protein of interest.

Concentration by precipitation

Protein precipitation can be promoted by agents such as neutral salts, organic solvents and high molecular weight polymers, or by appropriate pH adjustments. Concentration by precipitation is one of the oldest concentration methods known. Ammonium sulphate is perhaps the most common protein precipitant utilized. This neutral salt is particularly popular due to its high solubility, cheapness, lack of denaturing properties towards most proteins, and its stabilizing effect on many proteins. The addition of small quantities of neutral salts such as ammonium sulphate often increases protein solubility— the "salting in" effect; however, increasing salt concentration leads to destabilization of proteins in solution and eventually promotes their precipitation from solution, known as "salting out" (Figure 3.7). At high concentrations, such salts effectively compete with the protein molecules for water of hydration. This promotes increased protein–protein interactions, predominantly interactions between hydrophobic regions on the surface of adjacent protein molecules.

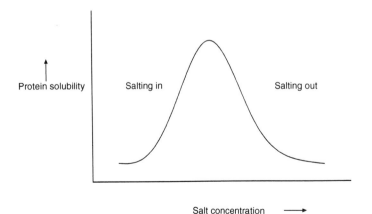

Figure 3.7. Effect of salt concentration on protein solubility. Increases in salt concentration from low initial values often increase protein solubility (salting in). Further increase above an optimal value will result in destabilization of the protein and eventually its precipitation from solution (salting out)

Such increased protein–protein interaction eventually results in protein precipitation.

Addition of various organic solvents to a protein solution may also promote protein precipitation. Added organic solvents lower the dielectric constant of an aqueous solution. This in turn promotes increased electrostatic attraction between bodies of opposite charge in the solution, in this case proteins. Such increasing interactions between proteins of opposite charges eventually lead to protein precipitation. Organic solvents frequently used to promote protein precipitation include ethanol, isopropanol, acetone and diethyl ether. Protein precipitation using organic solvents must be carried out at temperatures at or below $0\,^\circ C$ in order to prevent protein denaturation. As such solvents depress the freezing point of aqueous solutions, it is quite feasible to maintain the solution temperature several degrees below $0\,^\circ C$. Precipitation may also be promoted by the addition of organic polymers such as polyethylene glycol. The addition of such polymers, however, often dramatically increases the viscosity of the resultant solution, making recovery of the precipitate more difficult. Recovery is normally achieved by centrifugation or by filtration. The precipitate can subsequently be redissolved in a smaller volume of liquid and in this way is effectively concentrated.

Concentration by precipitation has played an important role in the downstream processing of many industrially important proteins. The technique is relatively straightforward to perform and requires only a limited amount of equipment. In many cases, precipitation also achieves some degree of protein purification, and high recoveries of biological activity are usually recorded. There are, however, a number of disadvantages associated with this technique. Many of the precipitants used are highly corrosive, in particular towards stainless steel equipment such as centrifuge rotors. Precipitation is often inefficient if used in circumstances where the protein concentration of the extract is low; in such cases, low recoveries of protein may be recorded. Some precipitants, such as acetone and diethyl ether, are highly inflammable and hence are hazardous to work with, while others such as ethanol are expensive. In many cases all traces of the precipitant present in the precipitate must be removed before further processing. Ammonium sulphate must be removed from resuspended precipitates before the resultant protein solution can be applied to an ion-exchange column.

Purification of serum proteins is one example of an industrially important process which traditionally included several protein precipitation steps. Precipitants normally used include both ethanol and ammonium sulphate. Most modern processes, however, rely on methods other than precipitation to achieve initial concentration of the proteins. The decreasing popularity of precipitation as a method of concentration is due not only to the various disadvantages associated with this technique, but also to the availability of efficient alternative methods of concentration and concern about the environmental impact of waste solvents and solutes.

Concentration by ion-exchange

Concentration of protein extracts by ion-exchange chromatography is enjoying increased industrial popularity. At any given pH value, proteins display a positive, negative or zero net charge. Using this parameter of molecular distinction, different protein molecules can be separated from one another by judicious choice of pH, ionic strength and ion-exchange materials. Positively charged proteins will bind to cation exchangers whereas negatively charged proteins will bind to anion exchangers. Elution of such bound protein may be easily achieved by irrigation with a solution of high ionic strength. Cellulose-based ion-exchangers offer an effective and relatively inexpensive method of achieving initial concentration. This may also result in a limited degree of protein purification.

Batch adsorption of proteins in dilute solutions is easily achieved by the direct addition of the ion-exchanger to the solution in question. Examples of such "dilute" protein solutions include fermentation broth or cell culture medium containing extracellular proteins from which the whole cells have been removed, or cell homogenates from which cell debris has been removed. A short period after its addition, the ion-exchange material may be recovered by centrifugation, and is then generally placed in a filter funnel or stainless steel vessel. Such holding vessels have an outlet covered by a mesh which ensures retention of the ion-exchange material. Bound proteins are then eluted from the ion-exchange material by washing with a suitable solution at high ionic strength. The protein-containing eluate is collected and subjected to further processing. The ion-exchange material can be regenerated and reused for the next purification cycle.

If the protein of interest is negatively charged, anion exchangers such as aminoethyl cellulose (AM-cellulose) or diethylaminoethyl cellulose (DEAE-cellulose) may be employed. In the case of positively charged proteins, cation exchangers such as carboxymethyl cellulose (CM-cellulose) or phospho cellulose (P-cellulose) are used. The charge of the proteins depends upon the pH of the solution.

Ion-exchange techniques have become popular on an industrial scale as the cellulose-based ion-exchangers are relatively inexpensive, robust and can easily be regenerated. Very high levels of protein recovery are generally recorded. Batch adsorption and elution also result in considerable clarification of the resultant protein solution. Many impurities which have adverse effects on solution characteristics either do not bind to the cellulose-based ion-exchangers, or do not subsequently elute from such exchangers under the conditions employed. Such undesirable impurities include particulate material which previous clarification steps failed to remove, various lipid and carbohydrate molecules, in addition to partially denatured or aggregated protein. Effective clarification of such protein solutions is desirable as it prevents fouling of columns during the subsequent purification steps. Such initial ion-exchange treatments also lead to some degree of protein purification, as only other molecules with similar charge characteristics will bind and subsequently coelute with the protein of interest.

Concentration by ultrafiltration

Protein solutions may be quickly and conveniently concentrated by ultrafiltration (Figure 3.8), and this method is widely applied in the biotechnology industry. As previously discussed, the technique of microfiltration is effectively utilized to remove whole cells or cell debris from solution. Membrane filters employed in the micro-filtration process generally have pore sizes ranging from 0.1 μm to 10 μm. Such pores, while retaining whole cells and large particulate matter, fail to retain most macromolecular components such as proteins. In the case of ultrafiltration membranes, pore diameters normally range from 1 nm to 20 nm. These pores are sufficiently small to retain proteins of low molecular weight. Ultrafiltration membranes with molecular weight cut-off points ranging from 3000 Da to 100 000 Da are commercially available. Membranes with

Figure 3.8. Ultrafilter used for laboratory or analytical scale molecular separations based upon molecular size (photograph courtesy of Amicon Ltd)

molecular weight cut-off points of 3000, 10 000, 30 000, 50 000 and 100 000 Da are most commonly used.

Traditionally ultrafilters have been manufactured from cellulose acetate or cellulose nitrate. Several other materials such as polyvinyl chloride and polycarbonate are now increasingly used in membrane manufacture. Such plastic-type membranes exhibit enhanced chemical and physical stability when compared with cellulose-based ultrafiltration membranes. An important prerequisite in manufacturing ultrafilters is that the material utilized exhibits low protein adsorptive properties.

No matter which material is used in the manufacture of ultrafilters the pore size obtained is not uniform, a point that must be emphasized. A range of pore sizes of wide deviation from the mean pore size is generally observed. The molecular weight cut-off point quoted for any such filter is thus best regarded as being a nominal figure. It is advisable to use a membrane whose stated molecular weight cut-off point is considerably lower than the molecular weight of the protein of interest. It is also important to realize that the molecular weight cut-off point quoted applies to globular proteins. The overall shape of the protein of interest therefore affects its ultrafiltration characteristics. If the protein is somewhat elongated, it may not be retained by an ultrafilter whose cut-off point is significantly lower than the molecular weight of the protein. Extensive post-translational modification of proteins, in particular glycosylation, may also affect its ultrafiltration behaviour.

Ultrafiltration is generally carried out on a laboratory scale using a stirred cell system. The flat membrane is placed on a supporting mesh at the bottom of the cell chamber, and the material to be concentrated is then transferred into the cell. Application of pressure, usually nitrogen gas, ensures adequate flow through the ultrafilter. Molecules of lower molecular weight than the filter cut-off pore size, water, salt and low molecular weight compounds all pass through the ultrafilter, thus concentrating the molecular species present in the solution whose molecular weight is significantly greater than the nominal molecular weight cut-off point. Concentration polarization, i.e. the build-up of a concentrated layer of molecules directly over the membrane surface which are unable to pass through the membrane, is minimized by a stirring mechanism operating close to the membrane surface. If unchecked, concentration polarization would result in a lowering of the flow rate.

As described for microfiltration, large-scale ultrafiltration systems generally employ cartridge filters within which the ultrafiltration membrane is present in a highly folded format (Figure 3.9). This allows a large filtration surface area to be accommodated in a compact area. Concentration polarization is avoided by allowing

Figure 3.9. Ultrafiltration system used on an industrial scale. This system (SPM 180), which comprises 180 ft^2 (16.7 m^2) of spiral-wound membrane, is routinely utilized for protein concentration and diafiltration (photograph courtesy Amicon Ltd)

the incoming liquid to flow across the membrane surface at right angles (tangential flow). The ultrafiltration membrane may be pleated, with subsequent joining of the two ends to form a cylindrical cartridge. Alternatively, the membrane may be laid on a spacer mesh and this may then be wrapped spirally around a central collection tube, into which the filtrate can flow.

Another widely used membrane configuration is that of hollow fibres. In this case, the hollow cylindrical cartridge casing is loaded with bundles of hollow fibres, which have an outward appearance somewhat similar to a drinking straw, although their internal diameters may be considerably smaller. In this configuration, the liquid to be filtered is pumped through the central core of the hollow fibres. Molecules of lower molecular weight than the membrane rated cut-off point pass through the walls of the hollow fibre. The permeate, which emerges from the hollow fibres all along their length, is drained from the cartridge via a valve. The concentrate emerges from the other end of the hollow fibre and is collected by an outlet pipe; it is referred to as the retentate.

The permeate is normally discarded while the retentate, containing the protein of interest, is processed further. The retentate may be recycled through the system if further concentration is required. Many industrial-scale ultrafiltration systems are automated and computer-controlled.

Ultrafiltration as a method of protein concentration is becoming increasingly popular, for a variety of reasons. The method is regarded as being very gentle, having little adverse effect on the bioactivity of the protein molecules. High recovery rates are usually recorded—some manufacturers claim recoveries of up to 99.9%. Processing times are rapid when compared with alternative methods of concentration, and little ancillary equipment is required. Ultrafiltration may also achieve some degree of protein purification. This, however, is often minimal due to factors such as wide deviation of membrane pore sizes. One drawback relating to this filtration technique is its susceptibility to rapid clogging of the membrane. Viscous solutions also lead to decreases in flow rates and prolonged processing times.

Ultrafiltration may also be used to achieve a number of other objectives besides concentration. As discussed above, it may yield a limited degree of protein purification and may also be effective in depyrogenating solutions; this is discussed further in Chapter 4. The technique is also widely used in industry to remove low molecular weight molecules from protein solutions by a process termed *diafiltration*.

Diafiltration

Diafiltration is a process whereby an ultrafiltration system is utilized to reduce or eliminate low molecular weight molecules from a solution. At a practical industrial level, this normally entails the removal of low molecular weight molecules, such as salts, ethanol and other solvents, buffer components, amino acids, peptides, added protein stabilizers or other molecules, from a partially processed protein solution. Diafiltration is generally preceded by an ultrafiltration step to reduce process volumes. The actual diafiltration process is identical to that of ultrafiltration except for the fact that the level of reservoir is maintained at a constant volume. This is achieved by the continual addition of solvent lacking the low molecular weight molecules which are to be removed. By recycling the concentrated material and adding sufficient fresh solvent to the system such that five times the original volume emerges from the system as permeate, over 99% of all molecules in the original solution that freely cross the membrane will be flushed out. Removal of low molecular weight contaminants from protein solutions may also be achieved by dialysis or by gel filtration chromatography. Diafiltration however, is emerging as the method of choice in many cases as it is quick, efficient and utilizes the same equipment used in ultrafiltration.

COLUMN CHROMATOGRAPHY

The initial steps in any purification process are designed (a) to liberate and concentrate the protein of interest, and (b) to remove particularly undesirable contaminants such as particulate matter, lipids or other substances which may result in fouling or clogging of high-precision chromatographic separatory systems during later stages of the purification process. The degree of purification achieved by such preliminary techniques is often marginal. It is considered essential that such initial steps yield a high percentage recovery of the desired protein product. In most instances, it then becomes necessary to purify the protein of interest further. This is generally achieved by column chromatography. Individual protein types possess a variety of characteristics which distinguish them from other protein molecules. Such characteristics include size and shape, overall charge, the presence of surface hydrophobic groups and the ability to bind various ligands.

Quite a number of protein molecules may be similar to one another if compared on the basis of any one such characteristic. All protein types, however, present their own unique combination of characteristics, a protein chromatographic "fingerprint". Various chromatographic techniques have been developed which separate proteins from each other on the basis of differences in such characteristics. Ion-exchange chromatography separates proteins on the basis of charge, gel filtration separates on the basis of size and shape, hydrophobic interaction chromatography separates on the degree of protein hydrophobicity, while affinity chromatography separates on the basis of the ability of a given protein to bind (or be bound by) specific ligands. Utilization of any one of these methods to exploit the molecular distinctiveness usually results in a dramatic increase in the purity of the protein of interest. A combination of methods may be employed to yield highly purified protein preparations.

While chromatographic techniques can effectively separate various proteins from one another, the percentage recovery of the desired protein may be significantly less than 100%. Typical recovery values might range from 75% to 95%. It is thus necessary to design a purification scheme which yields the desired level of protein purity, using as few chromatographic steps as possible. Optimization of each step employed is therefore critical if overall process yields are to be maximized. In most cases, downstream processing accounts for more than 50% of total production costs, with chromatographic steps accounting for most of this. Improved and more efficient chromatographic systems therefore hold the key to reducing such production costs. Most purification systems employ three to five high-resolution chromatographic steps. The various individual chromatographic techniques which may be employed to purify proteins on an industrial scale are discussed below.

Size exclusion chromatography (molecular sieving)

Size exclusion chromatography, also termed gel permeation or gel filtration chromatography, separates proteins on the basis of their size and shape. As most proteins fractionated by this technique are considered to have approximately similar molecular shapes, separation is often described as being on the basis of molecular weight, although such a description is somewhat simplistic.

Fractionation of proteins by size exclusion chromatography is achieved by percolating the protein-containing solution through a column packed with a porous gel matrix in bead form (Figure 3.10). As the sample travels down the column, large proteins cannot enter the gel beads and hence are quickly eluted. The progress of smaller proteins through the column is retarded as such molecules are capable of entering the gel beads. The internal structure of the matrix beads could be visualized as a maze, through which proteins

Figure 3.10. Chromatographic column. The superformance glass column illustrated in (a) is manufactured by Merck. A range of columns of varying dimensions are available from this and other manufacturers. (b) Process-scale chromatographic system. This particular system is utilized by British Biotechnology in the production of material for clinical trials. The actual chromatographic column is positioned to the left of the picture. Much larger columns may also be employed in conjunction with this system

(a)

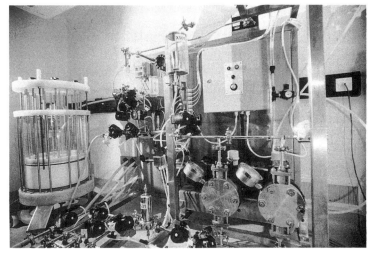

(b)

small enough to enter the gel must pass. Various possible routes through this maze are of different lengths. All proteins capable of entering the gel are thus not retained within the gel matrix for equal time periods. The smaller the protein, the more potential internal routes are open to it and thus, generally, the longer it is retained within the bead structure. Protein molecules are usually eluted from a gel filtration column in order of decreasing molecular size.

In most cases gel matrices utilized are prepared by chemically cross-linking polymeric molecules such as dextran, agarose, acrylamide and vinyl polymers. The degree of cross-linking controls the average pore size of the gel prepared. Most gels synthesized from any one polymer type are thus available in a variety of pore sizes. The higher the degree of cross-linking introduced, the smaller the average pore size, and the more rigid the resultant gel bead. Various highly cross-linked gel matrices such as Sephadex G-25 or Bio-Gel P2 have pore sizes that exclude all protein polymers from entering the gel matrix. Such gels may be used to separate proteins from other molecules which are orders of magnitude smaller, and are often used to remove low molecular weight buffer components and salts from protein solutions.

Gels of larger pore size are used to separate proteins from each other. The gel of most appropriate pore size for chromatographic application in any given circumstance depends largely on (a) the molecular weight of the protein of interest, and (b) the molecular weight of the major proteinaceous contaminants. All companies that produce gel filtration media publish the most effective fractionation range of each gel type. Most gel filtration media are also available in a number of different particle sizes, typically coarse, medium, fine and superfine grades. Although smaller beads (fine or superfine) yield higher resolution, flow rates are lower; thus beads of larger particle size are usually preferred for large-scale industrial applications.

The Sephadex range of fractionation gels (Pharmacia, Uppsala, Sweden) were among the first to be developed. These gels are prepared by cross-linking dextran with epichlorohydrin. Dextran is a polysaccharide composed of glucose monomers, linked predominantly by $\alpha 1 \rightarrow 6$ glycosidic bonds. Some Sephadex gels such as Sephadex G-25 or G-50 are quite rigid due to a high degree of cross-linking. Dextran-based gels of larger pore size, Sephadex G-100 or G-200, are not so rigid, and tend to compress readily. For this reason, the latter types of Sephadex are unsuitable for many industrial applications.

The Sephacryl range of gel filtration media are more rigid and physically stable than the Sephadex range, hence they are more applicable to industrial procedures. The Sephacryl gels are prepared by cross-linking allyl dextran with N,N-methylene bisacrylamide.

The Sepharose range of gels are prepared from the polysaccharide agarose. While these gels are particularly suited to fractionation of proteins of high molecular weight, their industrial usefulness is limited, mainly due to lack of physical stability. This type of gel exhibits a very open pore structure which is stabilized not by covalent linkages, but by hydrogen bonding between adjacent agarose molecules. The Sepharose gel structure disintegrates at temperatures above 40 °C and thus, unlike the Sephadex or Sephacryl gels, may not be sterilized by autoclaving. The chemical and physical stability of Sepharose gels may be enhanced by covalently cross-linking the agarose moieties in the gel structure. Such cross-linking is achieved by incubation of the Sepharose with the cross-linking agent 2,3-dibromopropanol under alkaline conditions.

The Bio-Gel P range of media are prepared by cross-linking acrylamide with N,N'-methylene bisacrylamide. Bio-Gel P is normally prepared in beaded form, and is especially useful when fractionating samples containing enzymes which degrade gels prepared from biological substances such as dextran.

Fractogel, produced by Merck, represents a more recently developed gel material. This gel is a copolymer of oligoethylene glycol, glycidyl methacrylate and pentacrythrol dimethacrylate. The internal pore walls of Fractogel are formed from intertwined polymer agglomerates which result in a very high degree of mechanical stability. Such mechanical strength facilitates use of these gels for large-scale preparative purposes.

Protein fractionation by size exclusion chromatography requires the use of long chromatographic columns. Generally, columns are 25–40 times greater in length than in width. Such dimensions are required in order to achieve adequate resolution of protein mixtures which separate into discrete protein bands on the basis of molecular size and shape, as they migrate down through the column. Industrial-scale gel filtration columns may be several metres in length. Gel utilized for preparative purposes must be mechanically rigid so as to avoid gel compression and reduced flow rates. An alternative approach to industrial-scale column design involves the use of stacking gel systems, which entail packing the gel matrix into a number of identical short, wide columns with the subsequent vertical connection of a number of such stacking columns in series. The connecting distance between each column is kept to a minimum. The overall stack system behaves surprisingly like a single chromatographic column of similar dimensions. The chromatographic gel in such systems, however, experiences lower differential pressure. Furthermore, if one particular column, usually the first in series, becomes fouled, it may be easily disconnected and replaced by a fresh column.

Size exclusion chromatography is rarely used during the initial stages of protein purification. Small sample volumes must be applied to the column in order to achieve effective resolution.

Application volumes are usually in the range of 2–5% of the column volume. Furthermore, columns are easily fouled by a variety of sample impurities. Size exclusion chromatography is thus often employed towards the end of a purification protocol, when the protein of interest is already relatively pure and is present in a small, concentrated volume. After sample application, the protein components are progressively eluted from the column by flushing with an appropriate buffer. In many cases, the eluate from the column passes through a detector. This facilitates immediate detection of protein-containing bands as they elute from the column. The eluate is normally collected as a series of fractions. On a preparative scale each fraction may be several litres in volume. While size exclusion chromatography is an effective fractionation technique, it generally results in a significant dilution of the protein solution relative to the starting concentration. Column flow rates are also often considerably lower than flow rates employed with other chromatographic media. This results in long processing times which has adverse process cost implications.

Ion-exchange chromatography

Several of the twenty amino acids which constitute the building blocks of proteins exhibit charged side chains. At pH 7.0, aspartic and glutamic acid have overall negatively-charged acidic side groups, while lysine, arginine and histidine have positively-charged basic side groups (Figure 3.11). Protein molecules therefore possess both positive and negative charges—largely due to the presence of varying amounts of these five amino acids (N-terminal amino groups and the C-terminal carboxyl groups also contribute to overall protein charge characteristics). The nett charge exhibited by any protein depends on the relative quantities of these amino acids present in the protein, and on the pH of the protein solution. The pH value at which a protein molecule possesses zero overall charge is termed its isoelectric point (pI). At pH values above its pI, a protein will exhibit a nett negative-charge whereas at pH values below the pI, proteins will exhibit a nett positive-charge.

Ion-exchange chromatography is based upon the principle of reversible electrostatic attraction of a charged molecule to a solid matrix which contains covalently attached side groups of opposite charge (Figure 3.12). Proteins may subsequently be eluted by altering the pH, or by increasing the salt concentration of the irrigating buffer. Ion-exchange matrices which contain covalently attached positive groups are termed anion exchangers. These will adsorb anionic proteins, i.e. proteins with a nett negative-charge. Matrices to which negatively-charged groups are covalently attached are termed cation exchangers, adsorbing cationic proteins, i.e. positively-charged proteins. Positively charged functional

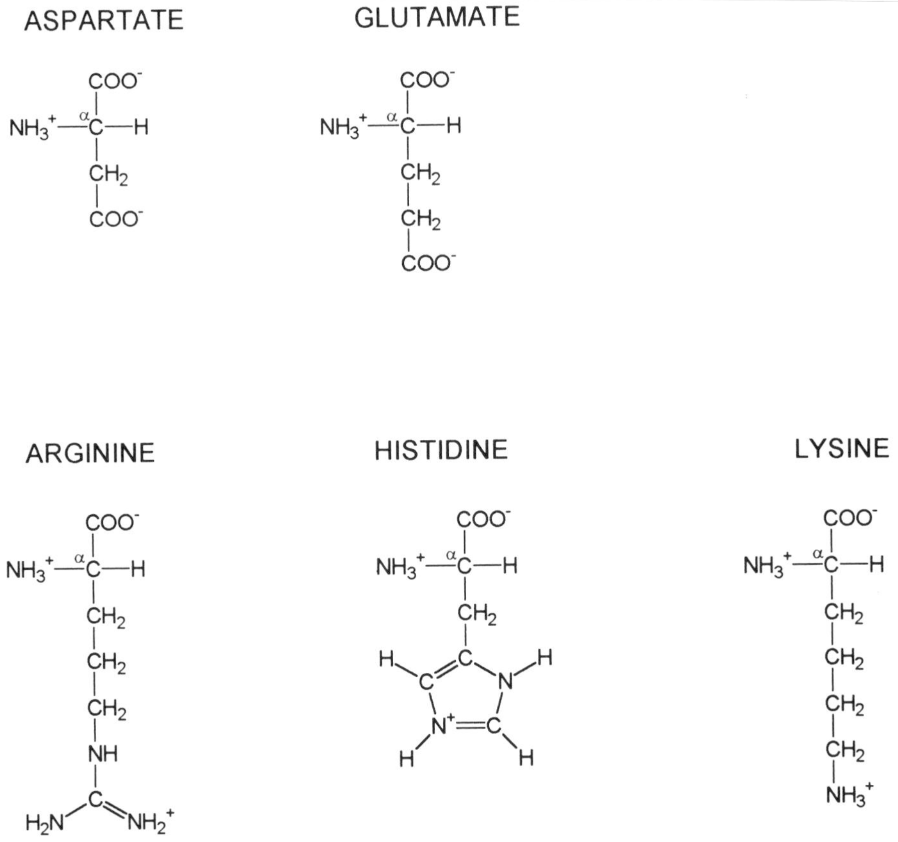

ASPARTATE

GLUTAMATE

ARGININE

HISTIDINE

LYSINE

groups (anion exchangers) include aminoethyl ($C_2H_4N^+H_3$) and diethylaminoethyl ($C_2H_4{}^+NH(C_2H_5)_2$) groups. Negatively-charged groups attached to suitable matrices forming cation exchangers include sulpho ($SO_3{}^-$) and carboxymethyl (CH_2COO^-) groups. Ion-exchangers may also be described as "strong" or "weak". Strong ion-exchange resins remain ionized over a wide pH range, whereas weak ion-exchange resins are ionized within a narrow pH range.

During the cation-exchange process, positively charged proteins bind to the negatively-charged ion exchange matrix by displacing the counter ion (often H^+) which is initially bound to the resin by electrostatic attraction. Elution may be achieved using a salt-containing irrigation buffer. The salt cation, often Na^+ of NaCl, in turn displaces the protein from the ion-exchange matrix. In the case of negatively-charged proteins, an anion exchanger is used, with the protein adsorbing to the column by replacing a negatively charged counterion.

The vast majority of protein purification procedures employ at least one ion exchange step. The popularity of this technique is

Figure 3.11. Structures of amino acids having overall nett charges at pH 7.0. In protein molecules the charges associated with the α-amino and α-carboxyl groups in all but the terminal amino acids are not present as these groups are involved in formation of the peptide bonds

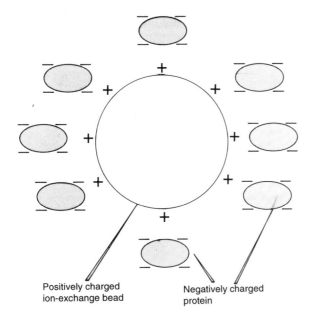

Figure 3.12. Principle of ion-exchange chromatography, in this case anion exchange chromatography. The chromatographic beads exhibit an overall positive charge. Proteins displaying an overall negative charge at the pH selected for the chromatography will bind to the beads due to electrostatic interactions

Positively charged
ion-exchange bead

Negatively charged
protein

based upon the high level of resolution achievable, its straightforward scale-up, together with its ease of use and ease of column regeneration. In addition, it leads to a concentration of the protein of interest. It is also one of the most inexpensive chromatographic methods available. At physiological pH values most proteins exhibit a nett negative charge. Anion exchange chromatography is therefore most commonly used.

A wide range of support matrices are utilized in the manufacture of ion-exchange media. The ideal matrix should be inert, structurally rigid and highly porous. Rigidity is required in order to facilitate scale-up and operation at high flow rates as soft matrices often compact, even under their own weight, when packed in large, industrial scale columns. Gels exhibiting pore sizes large enough to allow free entry of proteins into the internal matrix cavities are desirable, as this dramatically increases binding capacity.

Cellulose-based ion-exchangers were among the first to be widely used in protein purification systems. Cellulose-based cation and anion exchangers are both readily available. Due to ease of compression, such exchangers are not widely utilized in industrial-scale column chromatography, though they are used to concentrate proteins by batch adsorption. Improved cellulose-based ion-exchangers such as DEAE Sephacel (Pharmacia) have been developed, although these are also somewhat confined to laboratory-scale applications. Sephadex ion-exchangers are prepared by covalent attachment of charged groups to Sephadex G-25 or G-50 (see previous section). These ion-exchangers have not been widely adopted by industry, however, as the G-50 types are relatively easily compressed. G-25 exchangers, although quite sturdy, have low binding capacities as the bead pore diameters are too small to allow penetration by proteins.

A number of other ion-exchange resins based upon polymers such as agarose are also available. Such ion-exchange media are generally considered suitable for large-scale applications as they combine high capacity with rigidity, and hence high flow rates. The high capacity of such exchangers reflects their large open pore nature, which allows proteins free access to binding sites within the gel beads. Sepharose-based ion-exchangers typify this group—such ion-exchangers consist of a cross-linked agarose-based matrix. In some cases (e.g. Sepharose fast flow from Pharmacia), a high level of chemical cross-linking renders the resultant beads extremely stable, both chemically and physically. Such physical stability allows faster flow rates, while retaining high resolving power. This, in turn, allows for quick processing turnaround times which are so vital to maximizing downstream processing efficiency.

Fractogel (Merck) represents yet another ion-exchange material suitable for use in large-scale industrial separations. Its suitability derives from its mechanical and chemical stability. To facilitate its use as an ion-exchange medium, the basic gel structure is chemically modified by the covalent introduction of appropriate charged groups, both on external and internal bead surfaces. The gel has a high protein binding capacity, as the pore size is sufficiently large to allow most proteins free entry into the gel matrix.

An alternative ion-exchange resin design, termed "tentacle type" (Merck), has been developed recently. This is suitable for both analytical and preparative applications. Traditional ion-exchangers consist of charged (ionic) groups directly attached to the ion exchange matrix; this obviously presents a very rigid array of binding sites to which the protein is adsorbed, and in turn implies that binding may distort the conformational structure of the protein in question. It is not clear to what extent such distortion may lead to protein denaturation and inactivation. It must be stated, however, that the high recovery of biologically active proteins recorded when such traditional exchangers are employed suggest that such effects are minimal. Tentacle-type exchangers consist of a bead support matrix, to which linear charged polymers, the "tentacles", are attached (Figure 3.13A). Such tentacle structures are flexible and may automatically adopt a configuration that allows maximum binding contact with the adsorbed proteins. This flexibility minimizes or eliminates any induced distortion of bound proteins. The binding capacity of such ion-exchangers may be controlled by altering the tentacle length. Suitable support materials must carry primary or secondary hydroxyl groups on their surface. So far, matrices utilized are based on silica gel or organic polymers.

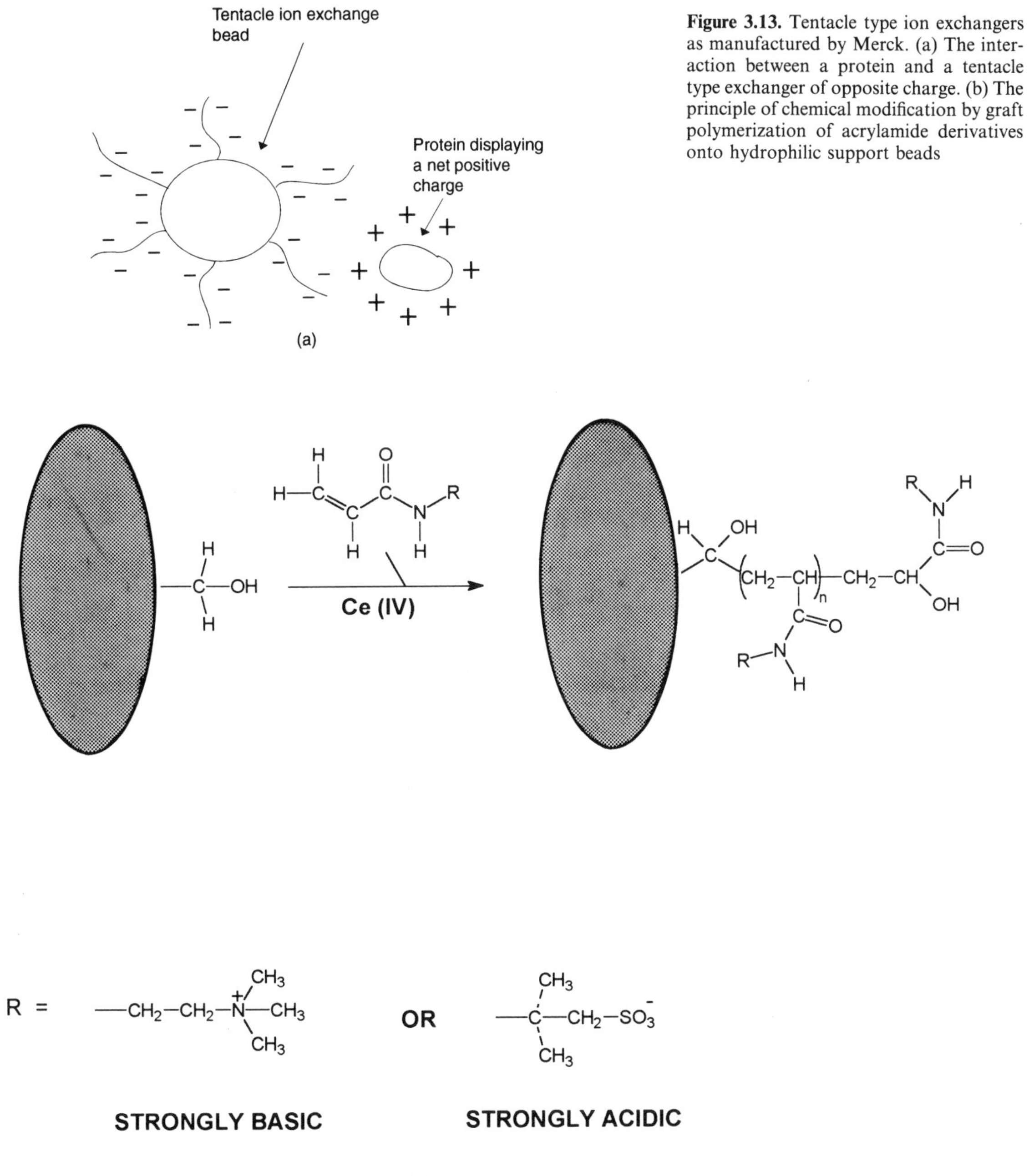

Tentacle ion exchange
bead

Protein displaying
a net positive
charge

(a)

Figure 3.13. Tentacle type ion exchangers
as manufactured by Merck. (a) The interaction between a protein and a tentacle
type exchanger of opposite charge. (b) The
principle of chemical modification by graft
polymerization of acrylamide derivatives
onto hydrophilic support beads

Ce (IV)

$$R = \quad -CH_2-CH_2-\overset{CH_3}{\underset{CH_3}{\overset{+}{N}}}-CH_3 \qquad OR \qquad -\overset{CH_3}{\underset{CH_3}{C}}-CH_2-SO_3^-$$

STRONGLY BASIC   **STRONGLY ACIDIC**

(b)

The tentacle polymer usually consists of repeating acrylamide
groups, which have been substituted with suitable charged groups.
Suitable groups include diethylaminoethyl groups in the case of
anion exchangers or sulpho groups in the case of cation exchangers.
These are covalently attached to the support matrix by a process

known as graft polymerization. In this process, the matrix support is incubated with the substituted acrylamide monomers and cerium (IV) ions (Figure 3.13B). The cerium ions bind the hydroxyl groups on the matrix surface and promote radical formation. This initiates a chemical chain reaction resulting in the covalent attachment of a substituted acrylamide group to the support matrix, with subsequent polyacrylamide chain growth. A linear polyacrylamide chain, usually consisting of 15 to 25 acrylamide monomers, is then synthesized which contains charged groups spaced along its length.

Hydrophobic interaction chromatography

Of the twenty amino acids commonly found in proteins, eight are classified as hydrophobic, due to the non-polar nature of their side chains (R groups, Figure 3.14). Most proteins are folded in such a way that the majority of their hydrophobic amino acid residues are buried internally in the molecule, and hence are shielded from the

Figure 3.14. Structural formulae of the eight commonly occurring amino acids found in proteins which display hydrophobic characteristics

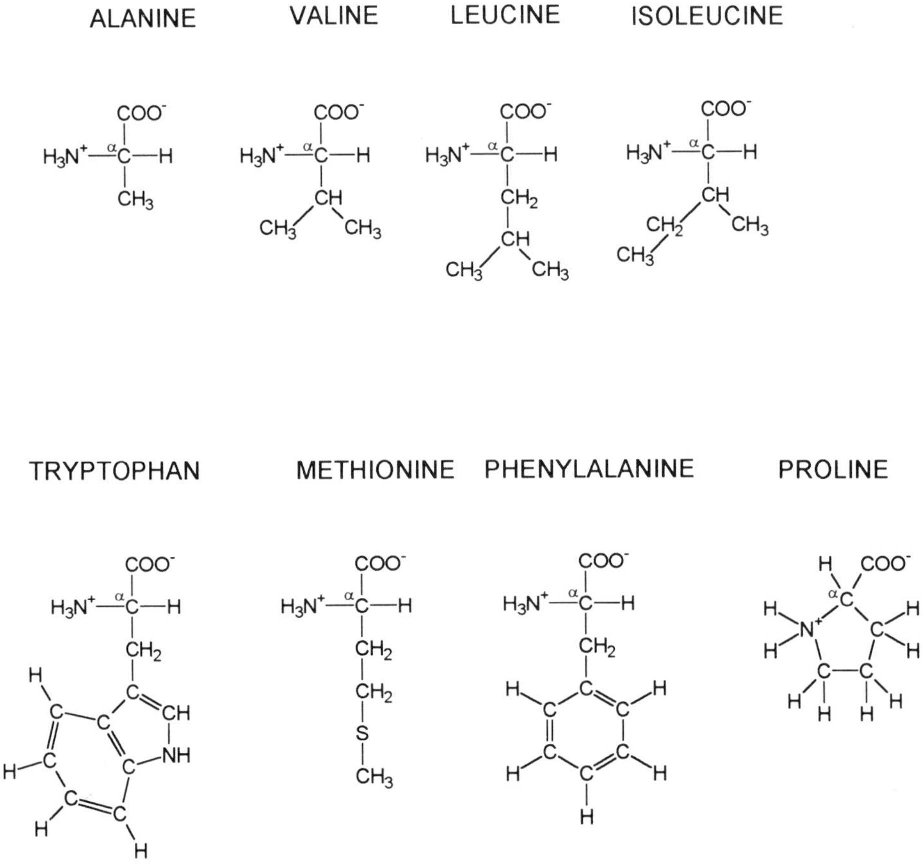

ALANINE   VALINE   LEUCINE   ISOLEUCINE

TRYPTOPHAN   METHIONINE   PHENYLALANINE   PROLINE

surrounding aqueous environment. Internalized hydrophobic groups normally associate with adjacent hydrophobic groups. Such hydrophobic interactions stabilize the protein conformation. A minority of hydrophobic amino acids are, however, present on the protein surface, and hence are exposed to the outer aqueous environment. Such surface hydrophobicity may play a role in dictating the overall stability of the protein conformational structure. It may also play a functional role in protein interactions with other molecules. Different protein molecules differ in the number and type of hydrophobic amino acids on their surface, and hence in their degree of surface hydrophobicity. Hydrophobic amino acids tend to be arranged in clusters or patches on the protein surface.

Hydrophobic interaction chromatography fractionates protein molecules by exploiting differing degrees of such surface hydrophobicity. It depends on the occurrence of hydrophobic interactions between the hydrophobic patches on the protein surface and hydrophobic groups covalently attached to a suitable matrix.

The most popular hydrophobic interaction chromatographic resins are cross-linked agarose gels to which hydrophobic groups have been covalently attached. Specific examples include phenyl and octyl Sepharose gels from Pharmacia, which contain phenyl and octyl hydrophobic groups, respectively (Figure 3.15). Attachment of octyl groups to a suitable matrix yields chromatographic beads which are very hydrophobic in character. Chromatography of highly hydrophobic proteins using octyl Sepharose should be avoided as the protein may bind so strongly to the gel matrix that elution becomes impossible without inducing protein denaturation. A variety of hydrophobic interaction chromatography resins are commercially available from companies such as Bio-Rad and Merck; these resins all consist of hydrophobic moieties attached to specific gel types.

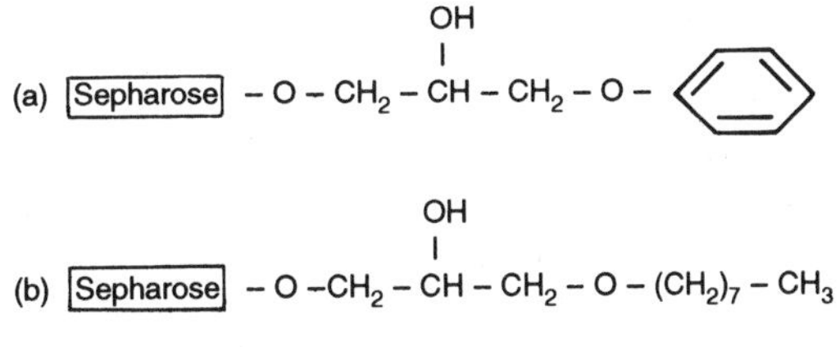

(a) $\boxed{\text{Sepharose}} - O - CH_2 - \overset{\overset{\displaystyle OH}{|}}{CH} - CH_2 - O - \langle \rangle$

(b) $\boxed{\text{Sepharose}} - O - CH_2 - \overset{\overset{\displaystyle OH}{|}}{CH} - CH_2 - O - (CH_2)_7 - CH_3$

Figure 3.15. Chemical structure of (a) phenyl and (b) octyl Sepharose, widely used in hydrophobic interaction chromatography

Protein separation by hydrophobic interaction chromatography is dependent upon interactions between the protein itself, the gel matrix and the surrounding solvent—which is usually aqueous. Increasing the ionic strength of a solution by the addition of a salt, such as ammonium sulphate or sodium chloride, increases the hydrophobicity of protein molecules. This increased hydrophobicity may be explained somewhat simplistically on the basis that the hydration of salt ions in solution results in an ordered shell of water molecules forming around each ion. This attracts water molecules away from protein molecules, which in turn helps to unmask hydrophobic domains on the surface of the protein. As such increases in ionic strength enhance the hydrophobicity of protein molecules, hydrophobic interaction chromatography is often performed following a salt precipitation step, or after ion exchange chromatography.

Not all ionic solutions are equally effective in promoting hydrophobic interactions. Anions may be arranged in order of increasing salting-out effect, that is in the order of increased ability to promote hydrophobic interactions.

Anions:

$$\text{increasing salting-out effect} \rightarrow$$

$$SCN^-, I^-, ClO_4^-, NO_3^-, Br^-, Cl^-, CH_3COO^-, SO_4^{2-}, PO_4^{3-}$$

Cations are often arranged in order of increasing chaotropic effect, the increasing tendency to disrupt the structure of water. Increase of such chaotropic effects leads to a decrease in the strength of hydrophobic interactions.

$$\text{increasing chaotropic effect} \rightarrow$$

Cations: NH_4^+, Rb^+, K^+, Na^+, Cs^+, Li^+, Mg^{2+}, Ca^{2+}, Ba^{2+}

Protein samples are therefore best applied to hydrophobic interaction columns under conditions of high ionic strength. As they percolate through the column proteins may be retained on the column via hydrophobic interaction. The more hydrophobic the protein, the tighter the binding. After a washing step, bound protein may be eluted utilizing conditions that promote a decrease in hydrophobic interactions; this may be achieved by irrigation with a buffer of decreased ionic strength, inclusion of a suitable detergent or lowering the polarity of the buffer by including agents such as ethanol or ethylene glycol. Subsequent to protein elution, hydrophobic interaction chromatographic resins must be washed extensively to remove any tightly bound proteins, which would otherwise quickly decrease their binding capacity. An initial washing step with a detergent such as sodium dodecyl sulphate (SDS) may be required to remove very tightly bound protein. Subsequent washing steps often involve the use of ethanol, butanol and water.

Reverse-phase chromatography may also be used to separate proteins on the basis of differential hydrophobicity. This technique involves applying the protein sample to a highly hydrophobic column to which most proteins will bind. Elution is promoted by decreasing the polarity of the mobile phase. This is normally achieved by the introduction of an organic solvent. Elution conditions are harsh, and generally result in denaturation of many proteins.

Affinity chromatography

Affinity chromatography is often described as the most powerful and highly selective method of protein purification available. This technique relies on the ability of most proteins to bind specifically and reversibly to other compounds, often termed "ligands". A wide variety of ligands may be covalently attached to an inert support matrix, and subsequently packed into a chromatographic column. In such a system, only the protein molecules that selectively bind to the immobilized ligand will be retained on the column. Washing the column with a suitable buffer will flush out all unbound molecules. An appropriate change in buffer composition, such as inclusion of a competing ligand, will result in desorption of the retained proteins.

A wide variety of ligands have been employed to selectively purify proteins by affinity chromatography. Various immobilized substrates, substrate analogues or cofactors may adsorb specific enzymes. Immobilized lectins have been used to purify various glycoproteins, hormone receptors may be used to purify hormones and indeed, immobilized hormones have been used to purify their receptors. Antibodies may be used to purify the antigen that stimulated their production in immunoaffinity chromatography. The use of most biologically derived molecules as affinity adsorbents is costly, and although technically feasible is often commercially unrealistic on economic grounds. There are many exceptions to the above statement, particularly in cases where the final product is of high commercial value. Examples of such exceptions include the use of protein A from *Staphylococcus aureus* as an affinity adsorbent in antibody purification, or the use of monoclonal antibody adsorbents to immunopurify particularly valuable products such as factor VIII.

Most species of *Staphylococcus aureus* produce a protein known simply as protein A. This protein consists of a single polypeptide chain of molecular weight 42 000 Da. Protein A binds the Fc region, i.e. the constant region, of immunoglobulin G (IgG) obtained from human and many other mammalian species with high specificity and affinity. Immobilization of protein A on Sephadex or agarose beads provides a powerful affinity system which may be used to purify IgG. There is considerable variation, however, in the binding affinity of protein A for various IgG subclasses obtained from different mammalian sources. In some cases another protein (protein G) may be used if the binding affinity of protein A for the particular IgG species required is low. Most immunoglobulin molecules that bind to immobilized protein A do so under alkaline conditions, and may subsequently be eluted at acidic pH values. Protein A affinity chromatography has significant industrial potential in terms of the purification of high-value antibodies destined for diagnostic or therapeutic use.

Immobilized gelatin has also been employed to purify fibronectin, a cell adhesion factor required in the formulation of many defined cell culture media preparations. Immobilized heparin has been used to purify various proteins, in particular blood coagulation factors such as antithrombin III, factor VII and factor IX. Heparin, a glycoaminoglycan, is a mixture of variably sulphated polysaccharides with molecular weights in the range 6000–30000 Da. It exhibits anticoagulant properties. Commercially available heparin is normally obtained from pig intestinal mucosa or beef lung.

Lectin affinity chromatography may also be used to purify a range of glycoproteins, some of which are of considerable therapeutic value. Initially, lectins were studied because of their ability to promote agglutination of erythrocytes and a number of other cell types. Lectins are a group of proteins synthesized by plants, vertebrates and a number of invertebrate species. Especially high levels of lectin are obtained from a variety of plant seeds. Plant lectins are often termed phytohaemagglutinins. All lectins have the ability to bind certain carbohydrate-containing molecules, such as α-D-mannose, α-D-glucose and D-N-acetylgalactosamine. Among the best known and most widely used lectins are concanavalin A (Con A), soybean lectin (SBL) and wheatgerm agglutinin (WGA).

Lectin affinity chromatography may be used to purify a variety of commercially important proteins, including various hormones, growth factors and cytokines. Glycoproteins generally bind to lectin affinity columns at pH values close to neutrality. Desorption may be achieved in some cases by alteration of the pH of the eluting buffer; however, the most common method of desorption involves inclusion of free sugar molecules, for which the lectin exhibits a high affinity, in the elution buffer, i.e. the inclusion of a competing ligand.

Although lectin affinity chromatography may be utilized to purify a variety of glycoproteins, it has not been widely adopted by industry, largely because of economic factors. Certain technical factors also militate against use of this technique. Leakage of the lectin affinity ligand into a protein product destined for therapeutic application would be particularly serious, in view of the erythrocyte agglutinating ability of lectins.

Two general approaches may be adopted when employing biospecific affinity chromatography. The first approach involves immobilization of a ligand to which several proteins may bind. This is termed the general ligand approach. Examples include immobilization of adenosine triphosphate (ATP) or cofactors such as nicotinamide adenine dinucleotide (NAD$^+$) in affinity systems designed to purify various ATP or cofactor-binding enzymes. One advantage of such systems is that the same affinity gel may be used to purify a number of different enzymes.

The second approach to biospecific affinity chromatography involves the use of a specific ligand to which only the protein of interest will bind. Ligands used often include enzyme substrates, substrate analogues or inhibitors, in addition to antibodies raised specifically against the protein of interest. Such an approach achieves a much higher degree of separation when compared with the general ligand approach. However, the design and synthesis of a specific immobilized ligand for each protein molecule is required.

Synthesis of any affinity gel may be discussed under three main headings: choice of affinity ligand, choice of support matrix and choice of chemical coupling technology. The chosen ligand should ideally exhibit high specificity towards the protein of interest and binding to it should be reversible, such that desorption may be achieved under relatively mild conditions. The ligand should also be stable, as conditions required for the chemical coupling to the support matrix are often harsh. As already mentioned, many biological macromolecules have been successfully employed as affinity ligands. Most such ligands, however, suffer from a number of limitations, including high cost and low stability. The development of synthetic dye ligands which can bind a variety of proteins has overcome many such disadvantages.

The matrix to be used should be physically and chemically stable, and should be sufficiently rigid to allow high working flow rates. The matrix itself should be inert in order not to contribute to non-specific binding of proteins. It should also exhibit a high degree of porosity, with pores sufficiently large to allow free entry of protein molecules to internally bound ligands. This greatly increases the binding capacity of the column. The matrix used must also be derivatizable and should be inexpensive and reusable. Support matrices employed include agarose, cellulose, silica and various organic polymers. Cross-linked agarose is probably the most popular matrix employed.

A wide variety of chemical immobilization techniques may be used to couple the ligand to the matrix. In many cases, the ligand is not coupled directly to the support, but is attached via a spacer arm (Figure 3.16). A variety of spacer arms may be used, the most popular of which consists simply of a number of methylene groups. Spacer arms are required in many instances in order to overcome steric difficulties associated with the binding of large protein molecules. Matrices may be activated by a number of chemical means which generate a chemically unstable and therefore reactive support matrix. This is subsequently derivatized with the chosen ligand, or initially by a spacer arm followed by coupling of the

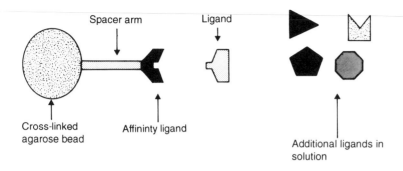

Figure 3.16. Schematic representation of the principles of biospecific affinity chromatography. The chosen affinity ligand is chemically attached to the support matrix (agarose bead) via a suitable spacer arm. Only those ligands in solution that exhibit biospecific affinity for the immobilized species will be retained

ligand. Activation with cyanogen bromide was one of the first matrix activation methods developed. Cyanogen bromide activated agarose is still widely used in the immobilization of a variety of ligands containing primary amino groups. The resultant ligand, however, exhibits a positive charge at neutral pH which confers ion-exchange properties on the gel. The coupling bond is also susceptible to hydrolysis under certain conditions, which may result in leakage of ligand from the column. Many other methods of activation have subsequently been developed, some of which yield extremely stable derivatized gels.

Elution of bound protein from an affinity column is achieved by altering the composition of the elution buffer, such that the affinity of the protein for the immobilized ligand is greatly reduced. A variety of non-covalent interactions contribute to protein–ligand interaction. In many cases, changes in buffer pH or ionic strength, or inclusion of a detergent or agents such as ethylene glycol which reduce solution polarity, may suffice to elute the protein. In other cases, inclusion of a competing ligand promotes desorption. Competing ligands often employed include free substrates, substrate analogues or cofactors (in the case of enzymes) or free carbohydrates (in the case of lectin affinity chromatography). Use of a competing ligand generally results in more selective protein desorption than does the generalized approach such as alteration of buffer pH or ionic strength. In some cases, a combination of such elution conditions may be required. Identification of optimal desorption conditions often requires considerable empirical study.

One particularly elegant example illustrating the concept of the specific ligand approach involves the purification of lactate dehydrogenase (LDH) using an immobilized oxamate affinity column (Figure 3.17). Oxamate is a structural analogue of pyruvate, a substrate for LDH. The reaction catalysed by LDH may be represented as

$$\text{Pyruvate} + \text{NADH} + \text{H}^+ \xrightleftharpoons{\text{LDH}} \text{lactate} + \text{NAD}^+$$

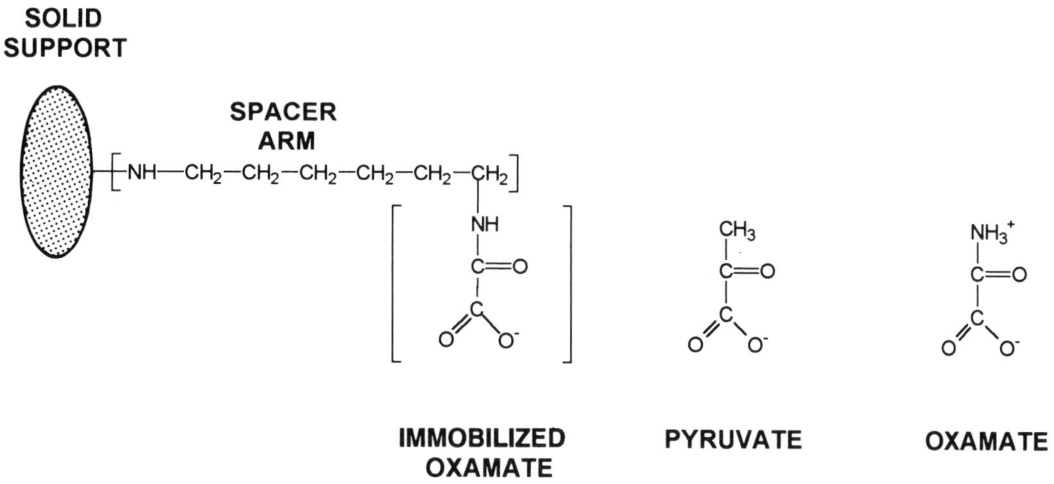

Figure 3.17. Structural relationship of pyruvate, oxamate and the immobilized oxamate employed in the affinity purification of lactate dehydrogenase

Lactate dehydrogenase exhibits an ordered kinetic mechanism in which the enzyme must first bind reduced NAD (NADH) before binding pyruvate or alternatively oxamate. If NADH is removed the enzyme can no longer bind the substrate.

Inclusion of NADH in a crude enzyme preparation containing LDH followed by percolation through an oxamate column promotes binding of LDH to the immobilized oxamate. All other protein molecules pass through unbound. The column may subsequently be washed with buffer, to ensure that all unbound molecules are washed out. The LDH molecule remains adsorbed to the column as long as this washing buffer contains NADH. Omission of the NADH in the elution buffer results in spontaneous elution of the enzyme. With the use of such an affinity column, LDH from crude extracts of human placenta was purified to homogeneity in a single step, with a yield in excess of 98%. While such columns offer resolution superior to any other known single chromatographic method, they are not widely utilized on a preparative scale for a number of technical and economic reasons to be discussed later.

Affinity chromatography offers many advantages over conventional chromatographic techniques. The specificity and selectivity of biospecific affinity chromatography cannot be matched by other chromatographic procedures. Increases in purity of over a thousand fold with almost 100% yields are often reported, at least on a laboratory scale. Incorporation of an affinity step could thus drastically reduce the number of subsequent steps required to achieve protein purification. This in turn could result in dramatic time and cost savings during downstream processing. Despite such promise, biospecific affinity chromatography does not enjoy widespread industrial use. This fact reflects a number of disadvantages associated with this approach. As mentioned earlier, many biospecific ligands are extremely expensive and often exhibit poor stability. Many of the ligand coupling techniques are chemically complex, hazardous, time-consuming and costly. Any leaching of coupled ligands from the matrix also gives cause for concern for two reasons: (a) it effectively reduces the capacity of the system, and (b) leaching of what are often noxious chemicals into high-value protein products, in particular those destined for therapeutic use, is undesirable.

It is not considered prudent to use biospecific affinity chromatography as an initial purification step, as various enzymatic activities present in the crude fractions may modify or degrade the expensive gels. It should, however, be utilized as early as possible in the purification procedure in order to accrue the full benefit afforded by its high specificity.

For all the above reasons, classical affinity systems, despite their great promise, have found limited use outside research laboratories. This situation, however, is changing because of (a) the discovery and use of dye affinity ligands, and (b) improvements in chemical coupling techniques, (Figure 3.18).

The development of dye affinity chromatography, often termed quasiaffinity chromatography, may be attributed to the observation

Figure 3.18. Large-scale manufacture of affinity chromatography medium under clean-room conditions (see Chapter 4) (photograph courtesy of Affinity Chromatography Ltd)

that some proteins exhibited anomalous elution characteristics when fractionated on gel filtration columns in the presence of blue dextran. Blue dextran consists of a triazine dye (cibacron blue F3G-A) covalently linked to a high molecular weight dextran. It is often used to determine void volumes of gel filtration columns. The discovery that some proteins bind the triazine dye soon led to its use as an affinity adsorbent by immobilization on an agarose matrix. A variety of other triazine dyes also bind certain proteins and hence have also been used as affinity adsorbents. Dye affinity chromatography has become popular for a number of reasons. The dyes are readily available in bulk and are relatively inexpensive. Chemical coupling of the dyes to the matrix is usually straightforward, often requiring no more than incubation under alkaline conditions at elevated temperatures for a number of hours. The use of noxious chemicals such as cyanogen bromide is avoided and in many cases incorporation of spacer arms is not required. Furthermore the resultant dye matrix is resistant to chemical, physical and enzymatic degradation. In this way ligand leakage from the column is minimized and is easily recognizable if it does occur, owing to the dye colour. The protein binding capacity of immobilized dye adsorbents is also high and far exceeds the binding capacity normally exhibited by natural biospecific adsorption ligands. Elution of bound protein is also relatively easily achieved. One problem is that textile dyes are produced in bulk and preparations often contain varying amounts of impurities, which may adversely affect the column's chromatographic properties.

It is not possible as yet to predict accurately if a specific protein will be retained on a dye affinity column, or what conditions will allow optimum binding and elution. Such information must be derived by empirical study. Many companies now supply kits containing a variety of different dye affinity ligands which may be used in initial empirical studies. An understanding of the specific interactions which allow many apparently unrelated proteins to bind to dye affinity ligands, while other proteins are not retained, is lacking. Dye affinity chromatography nevertheless represents a powerful chromatographic technique which is set to play an increasingly important role in downstream processing of protein products.

Recent advances in this area include chemical modification of crude textile dyes to yield second-generation dye ligands which exhibit a higher degree of selectivity—an example is the MIMETIC range of dye ligands manufactured by Affinity Chromatography Ltd. Many of these modified dye affinity columns exhibit increased fractional resolution, as well as greatly enhanced chemical stability. Extremes of pH, particularly alkaline conditions, result in significant ligand leakage from first-generation dye affinity columns. Such columns are thus unsuitable for some chromatographic applications, such as purification of therapeutically useful proteins, as the preferred method of column depyrogenation (see Chapter 4) involves prolonged flushing with 1 M NaOH. Most second-generation dye affinity chromatographic media are resistant to such conditions and thus find wider use.

Immunoaffinity purifications

Immobilized antibodies may be used as affinity adsorbents for the antigens that stimulated their production (Figure 3.19). Antibodies, like many other biomolecules may be immobilized on a suitable support matrix by a variety of chemical coupling procedures. Many of the initial immunoaffinity columns utilized mammalian poly-clonal antibodies raised against purified preparations of antigen. Thus a prerequisite for designing an immunoaffinity column was that an effective purification scheme for the antigen in question must already exist. Many immunoaffinity columns employing polyclonal antibody preparations exhibited low binding capacity for the protein of interest. In addition they also adsorbed some other proteins, as the antibodies specific for the protein of interest constituted only a small proportion of the antibodies in the polyclonal antibody preparation.

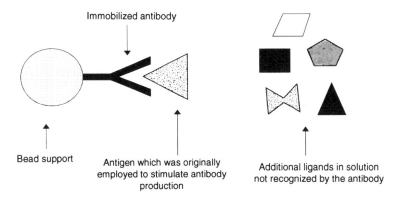

Immobilized antibody

Bead support

Antigen which was originally employed to stimulate antibody production

Additional ligands in solution not recognized by the antibody

Figure 3.19. Principle of immunoaffinity chromatography. Only antigen that is specifically recognized by the immobilized antibody will be retained on the column

Monoclonal antibody technology has resulted in an increased application of this technique, as it makes possible the production of large quantities of antibody which is monospecific towards the antigen of interest. Furthermore, it is not necessary to purify fully the antigen of interest in order to generate monoclonal antibodies.

Although immunoaffinity chromatography is probably the most highly specific of all forms of biospecific chromatography, its general applicability as a protein purification method has been delayed for a number of reasons. Like other forms of affinity chromatography, it is regarded as being a relatively high-cost technique.

Generation and immobilization of appropriate monoclonal antibodies is certainly not an inexpensive exercise. Ligand (antibody) leakage from the column may occur, and is undesirable. In many cases, elution of the bound protein from the immobilized antibody is not readily achieved. Desorption generally requires conditions that result in partial denaturation of the bound protein. This is often achieved by alteration of buffer pH or by employing agents such as urea or guanidine. One of the most popular elution methods employed involves irrigation with a glycine–HCl buffer at pH 2.2–2.8. In some cases, elution is more readily attainable at alkaline pH values. Specific examples have been documented in which protein elution was performed under relatively mild conditions, such as a change of buffer system or an increase in ionic strength; however, such examples are exceptional.

Despite such drawbacks, immunoaffinity chromatography offers extreme specificity and may well find application in the purification of high-value products. Proteins such as interferon and interleukin 2 have been purified by such techniques, at least on a laboratory scale. The technique may also be usefully employed in removal of low levels of specific protein contaminants from highly purified protein preparations.

Chromatography on hydroxyapatite

Hydroxyapatite occurs naturally as a mineral in phosphate rock and in addition constitutes the mineral portion of bone. It also may be used to fractionate protein by batch or column chromatography.

Hydroxyapatite is prepared by mixing a solution of sodium phosphate (Na_2HPO_4) with calcium chloride ($CaCl_2$). A white precipitate known as brushite is formed. Brushite is then converted to hydroxyapatite by heating to 100 °C in the presence of ammonia. Although hydroxyapatite may be obtained commercially, it is relatively easily prepared in-house.

$$Ca_2HPO_4 \cdot 2H_2O \xrightarrow{\text{heat } 100\,°C} Ca_{10}(PO_4)_6(OH)_2$$

$$\underset{\text{brushite}}{} \qquad \underset{\text{hydroxyapatite}}{\phantom{Ca_{10}(PO_4)_6(OH)_2}}$$

The underlying mechanism by which this substance binds and fractionates proteins is poorly understood. Protein adsorption is believed to involve interaction with both the calcium and phosphate moieties of the hydroxyapatite matrix. Elution of bound molecules from such columns is normally achieved by irrigation with a potassium phosphate gradient.

Calcium phosphate–cellulose gels represent a variation of hydroxyapatite chromatography. These gels are prepared by incubation of a calcium phosphate slurry with a slurry of cellulose powder, which seems to result in coating of the cellulose matrix with the calcium phosphate.

Chromatography on hydroxyapatite is not often encountered at an industrial level, most probably owing to the availability of other chromatographic methods whose mode of fractionation is readily understood. Preparation of hydroxyapatite, although straightforward, represents an additional inconvenience. In some cases variation in chromatographic characteristics from batch to batch has been observed. However, chromatography on hydroxyapatite does represent an additional technique by which proteins may be fractionated.

Chromatofocusing

Protein fractionation by chromatofocusing represents a relatively new chromatographic technique, which separates proteins on the basis of their isoelectric points. This technique involves percolating a buffer of one pH through an ion-exchange column which is pre-equilibrated at a different pH. Due to the natural buffering capacity of the exchanger, a continuous pH gradient may be set up along the length of the column. In order to achieve maximum resolution a linear pH gradient must be constructed. This necessitates the use of an eluent buffer and exchanger which exhibit even buffering capacity over a wide range of pH values. The range of the pH gradient achieved depends on the pH at which the ion exchanger is pre-equilibrated and on the pH of the eluent buffer. The matrix most often used in chromatofocusing is a weak anion exchanger, which exhibits a high buffering capacity. This anion exchanger is pre-equilibrated at a high pH value. The sample is then applied, usually in the running buffer, whose pH is lower than that of the pre-equilibrated column. After sample application, the column is constantly percolated with the specially formulated eluation buffer. Due to its lower pH value relative to the initial column pH, this percolation results in the establishment of an increasing pH gradient down the length of the column.

Upon sample application, negatively-charged proteins immediately adsorb to the anion exchanger, while positively-charged proteins flow down the column. Due to the increasing pH gradient formed such positively-charged proteins will eventually reach a point within the column where the column pH equals their own pI values (their isoelectric points). Immediately upon further migration down the column such proteins become negatively-charged, as the surrounding pH values increase above their pI values—hence they bind to the column. Overall, therefore, upon initial application of the elution buffer, all protein molecules will migrate down the column until they reach a point where the column pH is marginally above their isoelectric point. At this stage they bind to the anion exchanger. Proteins of different isoelectric points are thus fractionated according to this molecular distinction. The pH gradient formed is not a static one. As more elution buffer is applied, the pH value at any given point along the column is continually increasing. Thus any protein that binds to the column will be almost immediately desorbed as once again it experiences a surrounding pH value above its pI, and becomes positively charged. Any such desorbed protein flows down the column until it reaches a further point where the pH value is marginally above its pI value, and it again rebinds. This process is repeated until the protein emerges from the column at its isoelectric pH. To achieve best results, the isoelectric point (pI value) of the required protein should ideally be in the middle of the pH gradient generated. Chromatofocusing can result in a high degree of protein resolution, with protein bands being eluted as tight peaks. This technique is particularly effective when used in conjunction with other chromatographic methods during protein purification. Most documented applications of this method still pertain to laboratory-scale procedures. Scale-up of this technique to industrial level is discouraged by economic factors, most notably the cost of the eluent required.

Protein chromatography based on aqueous two-phase separations

As previously outlined, protein partitioning in aqueous two-phase systems represents a convenient, underutilized method of achieving a certain degree of protein fractionation. The technique is based upon the fact that if a protein solution is mixed with two incompatible aqueous polymers, most proteins migrate to one or other of the polymer phases. One of the most popular two-phase polymer systems is the polyethylene glycol/dextran (PEG/Dx) system. This technique may also be adapted such that it may be used in the form of column chromatography. Construction of a liquid–liquid partition column requires the selective binding of one or other of the two incompatible aqueous phases to a suitable chromatographic support, forming the stationary phase which is then packed into a chromatographic column. The second aqueous phase, the mobile phase, is then percolated through the column. Fractionation of proteins applied to such a column depends upon their relative affinity for the two phases. Proteins exhibiting a higher affinity for the mobile phase than for the stationary phase elute more rapidly than proteins that exhibit greater interactions with the stationary phase.

Matrix supports generally used are incompatible with one or other of the phase-forming polymers. The support material will then repel the incompatible polymer and is coated by the second phase, effectively shielding the support from the incompatible phase. Cellulose is incompatible with PEG and is thus coated by dextran in PEG/Dx phases; however, due to its dense fibrous structure, its capacity is relatively low. A dextran-rich phase will also coat cross-linked polyacrylamide supports. Polyacrylamide beads, however, tend to swell in the presence of dextran. The resultant swollen particles exhibit poor mechanical stability and collapse under low pressures. This limits the use of polyacrylamide-based systems for such purposes.

An alternative approach to designing suitable stable support matrices for use in aqueous two-phase partition chromatography, involves the fixing of linear polyacrylamide chains on the surface of a suitable support matrix bead by the process of graft polymerization. The resultant support beads exhibit high mechanical stability and are readily coated by the dextran-rich phase. Support materials onto which polyacrylamide may be successfully grafted include silica and polyacrylated vinyl polymer.

Liquid–liquid partition chromatography in two-phase systems may be used to separate nucleic acids as well as proteins. To date most applications of this method involve the fractionation of serum proteins; it has not found wide application at a preparative level.

HPLC of proteins

Most of the chromatographic techniques described thus far are performed under relatively low pressures, generated by gravity flow or by low pressure pumps; this is known as low-pressure liquid chromatography (LPLC). Fractionation of a single sample on such chromatographic columns typically requires a minimum of several hours to complete as relatively slow flow rates must be maintained. Low flow rates are required because as the protein sample flows through the column, the proteins are brought into contact with the surface of the chromatographic beads by direct (convective) flow. The protein molecules then rely entirely upon molecular diffusion to enter the porous gel beads, which is generally a slow process, especially when compared to the direct transfer of proteins past the outside surface of the gel beads by direct liquid flow. If a flow rate significantly higher than the diffusional rate is used, then band spreading and loss of resolution will result. This is due to the fact that any protein molecules that have not entered the bead will flow through the column at a faster rate than the (identical) molecules that have entered into the bead particles. Such high flow rates will also result in a lowering of adsorption capacity, as many molecules will not have the opportunity to diffuse into the beads as they pass through the column.

One approach which allows increased chromatographic flow rates without loss of resolution entails the use of microparticulate stationary phase media of very narrow diameter, which effectively reduces the time required for molecules to diffuse in and out of the porous particles. Any reduction in particle diameter dramatically increases the pressure required to maintain a given flow rate. Such high flow rates may be achieved by utilizing high-pressure liquid chromatographic systems. By employing such methods, sample fractionation times may be reduced from hours to minutes. Preparative high-pressure liquid chromatography (HPLC), now frequently referred to as high performance liquid chromatography, has attracted considerable industrial interest owing to this reduction in processing times. Although significant advances have been made in this regard, the use of preparative HPLC systems for the purification of industrially important proteins is still in its infancy.

The successful application of HPLC was made possible largely by (a) the development of pump systems which can provide constant flow rates at high pressure, and (b) the identification of suitable pressure-resistant chromatographic media. Traditional soft gel media utilized in low-pressure applications are totally unsuited to high-pressure systems due to their compressibility.

An ideal support material should:

- Be mechanically and chemically stable.
- Exhibit a low degree of non-specific adsorption.
- Be readily reusable.
- Be inexpensive.
- Be available in small particle sizes with a narrow range of particle and size distribution.
- Exhibit a high degree of porosity.

Obviously no one material satisfies all of the above criteria. Silica gel fulfils some of the more important criteria, most notably its high mechanical stability. As such, it has found widespread application in high-pressure chromatographic systems. Silica gel is, however, unstable under alkaline conditions, and does display some non-specific adsorption characteristics. Its relative instability at alkaline pH values limits its usefulness in the purification of therapeutically important proteins as the current preferred method of depyro-genation (see Chapter 4) of chromatographic columns involves prolonged standing in solutions of 1 M NaOH. This disadvantage may be overcome by coating the silica particles with a suitable hydrophilic layer.

Various organic polymers such as cross-linked polystyrene have been developed as alternatives to silica. Although many such substances are chemically more stable, some are difficult to produce with diameters as low as the silica particles normally used. Typically, stationary phase particle diameters for HPLC range between $5\,\mu m$ and $55\,\mu m$, a tenth of the diameters of particles utilized in low-pressure liquid chromatographic applications.

Most analytical HPLC columns available have diameters 4–4.6 mm and lengths of 10–30 cm. Preparative HPLC columns currently available have much wider diameters, typically up to 80 cm, and can be longer than a metre (Figure 3.20). Various chemical groups may be incorporated into the matrix beads, thus techniques such as ion-exchange, gel filtration, affinity, hydrophobic interaction and reverse-phase chromatography are all applicable to HPLC.

Many small proteins, in particular those that function extracellularly such as insulin, growth hormone and various cytokines, are quite stable and may be fractionated on a variety of HPLC columns without significant denaturation or decrease in bioactivity. Preparative HPLC may be used in industrial-scale purification of insulin and of interleukin 2. In contrast, many larger proteins are relatively labile and in some cases, loss of activity due to limited protein denaturation may be observed upon fractionation by HPLC.

At both preparative and analytical levels HPLC exhibits several important advantages compared with low-pressure chromatographic techniques. HPLC offers superior resolution due to the reduction in bead particle size. The diffusional distance inside the matrix particles is minimized, resulting in sharper peaks than those obtained when low-pressure systems are employed. With increased flow rates, HPLC systems also offer much improved fractionation speeds, typically of the order of minutes rather than hours, and HPLC is amenable to a high degree of automation. The major disadvantages associated with HPLC include cost and, to a lesser extent, capacity. Even the most basic HPLC systems are expensive compared with low-pressure systems. Solvent systems, plus other disposables required for the day-to-day operation of such systems, are also expensive. At a preparative level, the installation and maintenance costs of a HPLC system must be balanced against the financial savings associated with the overall reduction in processing times achieved by utilization of such a system. In some instances HPLC column capacity may also be an important deciding factor, although some modern preparative systems may fractionate 100 g or more of protein per production cycle.

For both technical and economic reasons, preparative HPLC is employed almost exclusively in downstream processing of low-volume, extremely high-value proteins, mostly intended for therapeutic use (Figure 3.21). For example, HPLC is extensively used in the purification of interleukin 2 (IL-2), a protein secreted by activated T-helper cells promoting the proliferation of various lymphocytes and playing a central role in the immune response. Due to its powerful immunostimulatory properties, IL-2 is of potential therapeutic relevance to a range of medical conditions. IL-2 may be isolated in trace amounts from the supernatant fluids

Figure 3.20. Preparative HPLC column (15 cm in diameter) used in processing of proteins required for therapeutic or diagnostic purposes. Column manufactured by Prochrom, Nancy, France (photograph courtesy of Affinity Chromatography Ltd) →

of activated lymphocytes by reverse phase HPLC (RP-HPLC). IL-2 is now being produced in large quantities by various microbial hosts including *E. coli* and yeast by the application of recombinant DNA technology. IL-2 produced by microorganisms such as *E. coli*

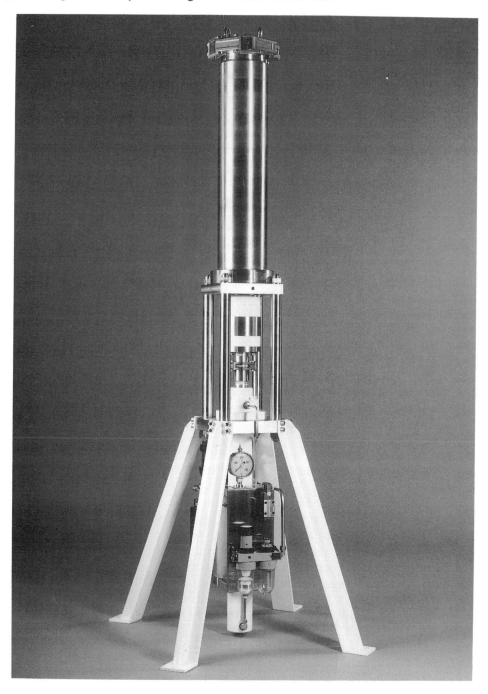

Figure 3.21. A preparative high performance liquid chromatography system, manufactured by Prochrom, Nancy, France (photograph courtesy of Affinity Chromatography Ltd)

generally accumulates intracellularly in the form of insoluble, partially denatured aggregates in inclusion bodies (Chapter 2). Upon homogenization inclusion bodies may be collected and washed using low-speed centrifugation. Such inclusion bodies, which predominantly consist of IL-2, may then be extracted by suitable solvents, such as alcohols and acetic acid, and purified directly by RP-HPLC. Due to its strongly hydrophobic character, IL-2 is eluted from the reverse-phase column considerably later than any impurities present. Preparative scale systems can yield up to 100 mg of IL-2 per production run. Human epidermal growth factor (urogastrone), due to its wound and ulcer healing properties, has considerable therapeutic potential and is now produced as a heterologous protein in recombinant *E. coli* and purified by preparative HPLC.

HPLC systems will most probably find wider biotechnological applications at an analytical level. The high resolution, sensitivity and in particular speed afforded by this technique renders the method suitable for routine in-process purity assessments, and in finished product quality control checks. HPLC systems are already widely used in such applications, most notably in the purification of human insulin produced semisynthetically from porcine insulin, in addition to human insulin produced in microbial systems (Chapter 6).

An alternative chromatographic system to HPLC has now been introduced. Termed fast protein liquid chromatography (FPLC), this technique employs operating pressures significantly lower than those used in conventional HPLC systems. Lower pressures allow use of matrix beads based on polymers such as agarose. FPLC chromatographic columns are constructed of glass or inert plastic materials. Conventional HPLC columns are manufactured from high-grade stainless steel. In many cases, FPLC systems are economically more attractive than HPLC. Despite their operation at lower pressures they still combine high resolution with enhanced speed of operation when compared with conventional liquid chromatography systems. FPLC systems are particularly suited to research and development and in analytical applications. For larger-scale applications, for example pilot-scale studies or small-scale production, FPLC systems, such as Pharmacia's BioPilot system, are most appropriate. The BioPilot system is a fully automated liquid chromatography system based on FPLC technology (Figure 3.22). The system consists of a control unit and a fractionation or separation unit. The computerized control unit allows control of parameters such as flow rates and eluate composition and controls the ultraviolet monitor and chart recorder, which detect and record protein elution profiles from the column. The separation unit consists of all the buffer reservoir systems in addition to the chromatographic columns. Several different chromatographic columns may be housed in the BioPilot system and are interconnected by a series of connection tubes and valves. The system may be operated with column volumes ranging from 1 ml to 30 litres, and a wide variety of column types, such as ion exchange, gel filtration, hydrophobic interaction and affinity columns, may be used in the system. Integrated high-precision pumps allow maximum flow rates of up to 100 ml per minute.

Larger automated chromatographic systems have been designed to cater for large-scale production of high-value proteins.

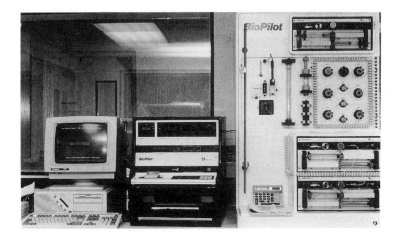

Figure 3.22. The BioPilot system manufactured by Pharmacia Ltd (photograph courtesy of British Biotechnology Ltd)

Pharmacia's BioProcess system consists of an integrated control panel, process pump and separation unit, which may consist of several different types of chromatographic columns. Flow rates of up to 400 litres per hour are achievable and this system can purify proteins from gram to kilogram quantities per production cycle. As in the case of the BioPilot system, the BioProcess system is designed to allow rigorous sanitizing, which is required to produce proteins successfully for *in vivo* diagnostic or therapeutic applications. The high degree of process automation to which such systems lend themselves contributes enormously to overall process economy. This helps offset the initial large capital outlay required to purchase such systems.

ENGINEERING PROTEINS FOR SUBSEQUENT PURIFICATION

Proteins may be separated from each other by various techniques which exploit physicochemical differences between the different molecules. Genetic engineering techniques now facilitate the incorporation of specific peptide or protein tags in the protein of interest. A tag is chosen which confers on the resultant hybrid protein some pronounced physicochemical characteristic facilitating its subsequent purification. Such a molecule is normally produced by fusing a DNA sequence which codes for the peptide or protein tag to one end of the genetic information encoding the protein of interest. Protein tags that allow rapid and straightforward purification of the hybrid protein by techniques such as ion-exchange, hydrophobic interaction or affinity chromatography have been designed and successfully used. Incorporation of an enzymatic or chemical cleavage site between the protein of interest and the purification tag allows subsequent selective release of the tag, yielding the protein molecules of interest.

A variety of protein purification tags may be used. Addition of a polyarginine (or polylysine) tag to the C-terminus of a protein confers a strong positive charge on the resultant hybrid protein. This resultant protein may then be purified by cation exchange chromatography. This approach has been used in the purification of proteins such as interferon and urogastrone. Addition of a tag containing a number of hydrophobic amino acids confers on the resultant molecule a strongly hydrophobic character, which allows its effective purification by hydrophobic interaction chromatography. Some amino acids such as histidine are known to bind various metals, thus a purification tag consisting of polyhistidine may be employed to purify proteins by metal chelate chromatography, which is carried out on a suitable chromatographic matrix to which metal ions such as copper or nickel have been immobilized.

Peptide and polypeptide tags which facilitate protein purification by affinity chromatography have also been developed. The gene coding for protein A may be fused to the gene or cDNA encoding the protein of interest. The resultant hybrid may be purified using a column containing immobilized IgG. Immunoaffinity purification may be employed if antibodies have been raised against the tag utilized. One synthetic peptide known as Flag has been developed specifically for use as an immunoaffinity tag. The Flag peptide includes a cleavage site for the protease enterokinase. Monoclonal antibodies have been generated which specifically bind the Flag peptide. One such monoclonal antibody has been identified which, when immobilized on a chromatographic matrix, allows desorption of the Flag–protein hybrid under mild conditions. The Flag system may be used to purify a wide range of proteins. Various enzymes may also be used as purification tags. The use of β-galactosidase is particularly popular as it facilitates purification by readily prepared affinity chromatographic systems, and providing its enzymatic capabilities are not destroyed in the fusion protein, it may be conveniently used to assay the hybrid protein.

Upon purification of the hybrid protein, it is necessary in most cases to remove the tag used. This is especially critical if the purified protein is destined for therapeutic use, as the tag may be immunogenic. Selective cleavage of the tag must be followed by subsequent separation of the tag from the protein of interest. This may require a further chromatographic step. In some cases, removal of the tag may be necessary to facilitate the efficient functioning of the protein of interest.

Removal of the peptide or polypeptide tag is generally carried out by chemical or enzymatic means. This is achieved by designing the tag sequence so that it contains a cleavage point for a specific protease or chemical cleavage at the protein–tag fusion junction. Sequence-specific proteases that are often employed to achieve tag removal include the endopeptidases trypsin, factor Xa and enterokinase. Exopeptidases such as carboxypeptidase A are also sometimes utilized. Generally speaking, endopeptidases, which cleave internal protein peptide bonds, are used to remove poly-peptide tags, whereas exopeptidases are used most often to remove short-chain peptide tags. The exopeptidase carboxypeptidase A, for example, sequentially removes amino acids from the C-terminus of a protein until it encounters a lysine, arginine or proline residue. Chemical cleavage of specific peptide bonds relies on the use of chemicals such as cyanogen bromide or hydroxylamine.

Although several methods exist that can achieve tag removal, most such methods suffer from inherent drawbacks. One pre-requisite for any method of tag removal is that the protein of interest must remain intact after the cleavage treatment. The required protein therefore should not contain any peptide bonds susceptible to cleavage by the specific method chosen. Chemical methods, for example, must generally be carried out under harsh conditions, often requiring high temperatures or extremes of pH. Such conditions can have a detrimental effect on normal protein functioning. Proteolytic removal of tags is often less than 100% efficient, and any traces of additional contaminating proteases in the protease preparation used often results in proteolytic degradation of the protein of interest; a situation clearly to be avoided.

Most heterologous proteins that are produced at high levels and accumulate intracellularly in *E. coli* inclusion bodies are generally solubilized by denaturants such as various detergents or urea. Protein tags may not function effectively in the presence of such denaturants. Many affinity tags will facilitate affinity purification of the hybrid protein only if the tag is present in its native conformation. Use of ionic denaturants such as guanidinium hydrochloride will adversely affect purification if the tag employed facilitates this purification step by ion exchange.

Most methods of tag removal employed to date suffer from at least one of the above drawbacks, which has thus far limited the application of this technique at an industrial level.

BULK ENZYME PREPARATION

Bulk enzyme preparations, often referred to as "industrial-scale" enzymes, are generally subjected to little downstream processing compared with the purification procedures employed in the case of therapeutic and diagnostic protein products. The vast majority of enzymes produced in bulk are biopolymer-degrading enzymes, such as amylases, proteases and pectinases. These are produced extracellularly by submerged fermentation of various microbial species—particularly species of *Bacillus* or *Aspergillus*. Many such bulk enzyme preparations simply consist of concentrated, often dried, cell-free fermentation products.

A generalized downstream processing scheme typically utilized in the production of such bulk enzyme preparations from microbial sources is outlined in Figure 3.23. Once maximum yield of the protein has been achieved in the fermentation process, the fermentation medium is cooled to under 5 °C. This stabilizes the protein product and discourages further microbial growth. The microbial cells must then be removed from the culture medium, typically by centrifugation or by filtration. If the enzyme is extracellular, as is usual, then filtration tends to be the method of choice. If the enzyme is intracellular, cell harvesting by continuous centrifugation is usually undertaken. In such cases, after harvesting the cells are resuspended and disrupted in order to release their intracellular

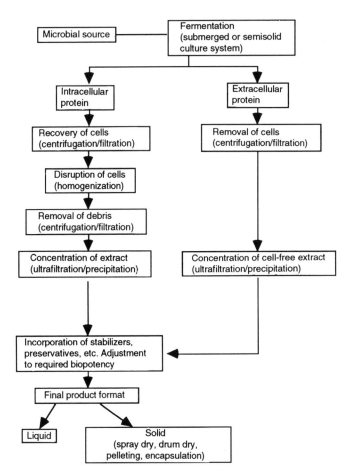

Figure 3.23. Bulk protein production from microbial sources—a generalized outline

contents. The cell homogenate must then be centrifuged in order to remove any intact cells, in addition to cellular debris. The protein of interest should then be present in the supernatant.

In both cases the next step in downstream processing is usually concentration, which is necessary to reduce the process volume to more manageable volumes, especially in the case of extracellular enzymes present in large volumes of fermentation medium. Concentration is often achieved by low-temperature evaporation or, more recently, ultrafiltration. The concentrate may then be subjected to an additional filtration step, in order to remove any intact microorganisms still remaining in the extract. The most common filter employed at this stage is a cellulose fibre depth filter. Any microorganisms present are trapped in the cellulose fibres as the concentrate passes through the filter.

If the crude enzyme preparation is presented in liquid form, the final steps of the downstream processing procedure normally involve the addition of various stabilizers and preservatives in order to increase the product shelf-life. The final product volume is then adjusted as appropriate in order to ensure that the enzyme activity falls within its specified limits.

Many bulk enzymes are sold in solid form. Perhaps the simplest method of producing a powdered enzyme preparation is by spray-drying the enzyme concentrate after the final filtration step. Spray-drying essentially involves the generation of an aerosol of tiny droplets from the enzyme-containing liquid which are directed into a stream of hot gas. This results in evaporation of the water content of the droplets, leaving behind solid, enzyme-containing particles. Spray-dried enzyme preparations may subsequently be dried in vacuum ovens if required. The dried preparation may then be ground into a fine powder and the activity level adjusted to the required bioactivity by the addition of suitable powder diluents.

Powdered enzyme preparations suffer from the disadvantage of a high level of dust formation during handling. Exposure of downstream processing personnel or end product users to such enzyme-containing dust may cause severe allergic reactions. Widespread allergic responses to enzyme dust were experienced in the 1960s when proteases were first added in large quantities to household detergents. Initially this resulted in the withdrawal of the protease preparations from the detergents, with a resultant slump in world protease sales. The problem was largely overcome by encapsulating the enzymes.

Enzyme encapsulation technology allowed the successful reintroduction of proteases into detergent preparations and now a wide range of bulk enzymes are handled and utilized safely on an industrial scale. Solid bulk enzyme preparations are normally prepared by a method somewhat analogous to tablet manufacture in the pharmaceutical industry. This involves mixing the fine enzyme powder with a binding material such as carboxymethyl cellulose, an inert filler, usually a salt, and water—thus forming a paste-like material. Specialized equipment converts the bulk paste into small, spherical structures up to 3 mm in diameter. The spheres are then dried and encapsulated using a suitable water-soluble wax.

Bulk enzyme preparations marketed in dried form include a variety of microbial amylases used in the starch processing industry, pectinases used in the clarification of fruit juice and proteases incorporated into detergent preparations (Chapter 8). These preparations may be quite impure and the desired enzymatic activity may represent no more than 2–5% of the total protein content of the final product. If a higher degree of purity is required the desired protein may be precipitated by agents such as inorganic salts, alcohols or acetone as part of the downstream processing procedure. All enzyme products are now normally tested for toxicity and allergenicity.

PURIFICATION OF PROTEINS FOR THERAPEUTIC AND DIAGNOSTIC USES

Proteins destined for diagnostic or therapeutic applications must be purified to a very high degree. This is particularly true in the case of any protein preparation that is to be administered parenterally, i.e. by direct injection or infusion into the body. The high level of purity demanded in such instances is necessary in order to minimize or eliminate the occurrence of adverse clinical reactions against trace contaminants in the protein product. This is discussed in detail in the next chapter. Many proteins such as antibodies or enzyme preparations, used for *in vitro* diagnostic or analytical purposes,

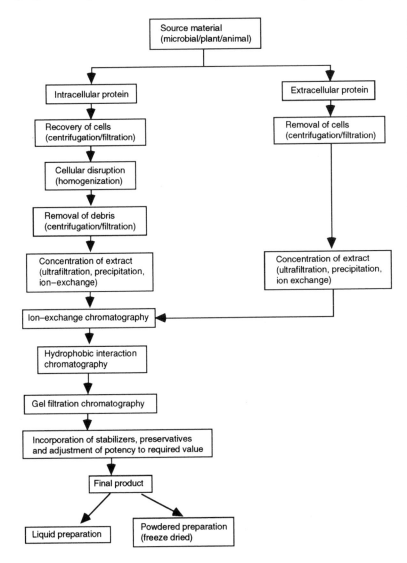

Figure 3.24. Production of protein products destined for therapeutic or diagnostic application (generalized outline). Additional steps may be required to assure complete removal of specific (non-proteinaceous) contaminants which may be of medical significance (see Chapter 4). Other chromatographic columns employed include affinity chromatography or chromatofocusing. Final protein products must generally be more than 95% pure if the product is destined for parenteral administration (an injectable). Numerous quality control steps are carried out on both in process samples and finished product. Environmental conditions within the manufacturing area are closely controlled

must also be at least partially purified in order to remove any contaminating biomolecules which would otherwise interfere with their diagnostic applications.

The downstream processing of such proteins invariably entails not only preliminary treatments but also several purification steps performed on various chromatographic columns. A typical overall purification process is outlined in Figure 3.24. If the protein of interest is secreted into the extracellular medium, which may be a fermentation or culture medium, or if it is a liquid such as blood, the initial purification step involves the removal of whole cells by centrifugation or ultrafiltration. If the protein of interest is intracellular, cell harvesting is followed by cell disruption by homogenization. If the intracellular protein has accumulated in soluble form, the next step involves the removal of any remaining whole cells and cellular debris, generally by centrifugation. If, however, the protein has accumulated in the form of an inclusion body, homogenization is followed by a low-speed centrifugation step which collects the inclusion bodies. After a washing step, in which the inclusion bodies are suspended in buffer or water and subsequently centrifuged, the inclusion bodies are solubilized by a suitable method (Chapter 2).

In the case of extracellular proteins, often enzymes or antibodies, a concentration step is often employed first. Concentration is achieved by precipitation, ultrafiltration or batch ion-exchange procedures. In the case of most intracellular proteins this concentration step is unnecessary, although techniques such as precipitation and, in particular, batch ion-exchange procedures are often used in such cases to "clean up" the extract. This involves removing contaminants such as particulate matter or lipids which might subsequently clog chromatographic columns. It may also function as an initial purification step. In order to achieve high-resolution purification column chromatography is usually the next step. The number of chromatographic steps will depend primarily on the resolution achieved by each individual step and on the overall level of purity required. It is important to optimize the conditions employed in each step in order to achieve maximum resolution and maximum product recovery. For economic reasons the number of chromatographic steps utilized to achieve the desired level of purity should be kept to a minimum.

A minimum of three chromatographic steps are usually needed for the purification of a protein intended for parenteral administration. Although there are many different chromatographic techniques available, by far the most common techniques used industrially are those of ion-exchange (most common), affinity chromatography, hydrophobic interaction chromatography and gel filtration chromatography. Gel filtration chromatography is often the last "polishing" step, for a number of reasons. It effectively fractionates the required protein molecules from dimeric or higher molecular weight aggregates, which may form during the earlier purification procedures. It may also effectively remove other molecules such as ligands which may have leached from affinity columns. Gel filtration columns have, however, a low loading capacity. They also must be run at relatively low flow rates, and hence are often expensive in terms of their operation; the old maxim that "time is money" is particularly true when applied to downstream processing. By employing the relevant irrigation buffer gel filtration may also be used to change the final purified protein into the buffer system most suitable for its final processing or storage.

As yet, most chromatographic systems used are run under low pressure conditions, although preparative systems which operate under intermediate pressure, such as FPLC, and preparative high-pressure systems are assuming increasing industrial significance. The degree of purification achieved after each chromatographic step may be monitored by assessing the quantity of protein of interest present relative to the total protein content by activity assay, or analysis by electrophoresis or by HPLC.

A number of guidelines to be followed when designing a downstream protein purification scheme are often quoted. Such guidelines, which are little more than statements of common sense, are general in nature and as such may not be applicable or appropriate to some purification schemes. However, these guidelines include the following:

- Initial characterization of the starting material should be undertaken in detail, so that a rational purification scheme may be developed. The final product purity required should be clearly defined.
- The purification scheme should be kept as simple and as straightforward as possible, ensuring that the highest end-product yield is obtained within the shortest time possible.
- Use fractionation techniques that separate the protein molecules from each other on the basis of the most dramatic differences in the parameters of molecular distinction.
- Select techniques and conditions under which the techniques are used such that they exploit the greatest differences in physicochemical properties between the protein of interest and other impurities.
- Use a highly selective step as early as possible in the purification scheme. This may well reduce the number of subsequent steps that must be used to achieve the desired level of purity.
- If possible, leave the most expensive processing step until last.

Through advances in modern biotechnological techniques, an ever increasing number of protein products are now purified on a preparative scale. Detailed procedures used in such purifications remain confidential, for understandable commercial reasons. Examples of highly purified proteins that are commercially available are listed in Table 3.4. Many of these protein products are discussed in detail in the following four chapters, and the specific purification procedures, where possible, are discussed.

Table 3.4. Some commercially available proteins destined for therapeutic administration, which thus must meet stringent purification specifications

Type of protein	Examples
Hormonal preparations	Insulin, glucagon, growth hormones
Blood products	Blood factors, serum albumin
Cytokines	Interleukins, interferons
Enzymes	Urokinase, asparaginase, tPA
Vaccines	Hepatitis B surface antigen, HIV antigens
Monoclonal antibodies	Raised against specific antigens

SCALE-UP OF PROTEIN PURIFICATION SYSTEMS

Scale-up of protein purification systems poses many difficulties and pitfalls, not only for scientific personnel but also for the process engineer. The process engineer is charged with designing and installing all of the equipment required to process the protein product successfully on a preparative scale. As can be seen from Figure 3.25, process design is far from simple.

Most large-scale process equipment items such as holding vessels and transfer pumps are constructed from stainless steel or plastics such as polypropylene. Glass vessels, so commonly used on a laboratory scale, are seldom used for large-scale preparative work, mainly due to their lack of structural strength. The grade of stainless steel or the plastic type to be used in the manufacture of process vessels must be selected with care. Materials used must be inert, and resistant to the corrosive action of any chemical used during the process. They should not allow any leaching of potentially toxic metals or chemicals into the product stream.

Liquid transfer, so easily achieved on a laboratory scale by physically pouring the liquid from one vessel into another, is achieved on a preparative scale by pumping the liquid via a series of stainless steel or plastic pipes. Only certain pump designs are

Figure 3.25. Bioreactor at a Boehringer Mannheim manufacturing facility (photograph courtesy of Boehringer Mannheim UK)

suitable for transfer of liquid protein solutions. Some pump systems tend to entrap air during their pumping action, which can lead to protein denaturation.

Preparative-scale chromatographic equipment also often differs in design and appearance from laboratory-scale systems. Most small-scale chromatographic columns are manufactured from glass or from transparent plastics and range in capacity from 1 ml to 300 ml in volume. Preparative-scale chromatographic columns range in capacity from hundreds of millilitres to several litres and, in some cases, up to several hundred litres. While some such columns are manufactured from glass, the majority are manufactured from reinforced plastic or stainless steel. Scale-up is normally achieved by increasing the column diameter as opposed to the column height. Columns several metres in diameter are available commercially. In the case of some chromatographic systems such as gel filtration, the degree of separation achieved is proportional to the column length. Scale-up of such systems is made more difficult by the compressible nature of many gel beads. Such difficulties may be overcome by employing more recently developed gel types which exhibit enhanced structural characteristics, or by altering column design. Stack columns are often used for preparative-scale gel filtration of many industrially important proteins such as insulin.

Radial flow chromatography represents an alternative column design. In such systems the sample flows across the column rather than down its length. The applied sample is distributed in a narrow cavity around the column's outer periphery. It then flows radially through the packing material, with the eluate flowing into a central channel through which it exits the column. This column design generally facilitates higher flow rates than to those attainable using columns of traditional design.

Complete chromatographic systems are designed, manufactured and marketed by several companies. Most such purification systems consist not only of the process equipment required but also of a computerized control unit, which facilitates a high degree of system automation.

In general, innovations in downstream processing techniques are introduced infrequently, largely for two reasons: firstly, regulatory authorities such as the Food and Drug Administration (FDA) in the USA review licence applications to produce and market proteins designed for use in the food or health-care industries. Such applications are more likely to be looked upon favourably if proposed downstream processing procedures are based upon established methodologies. Secondly, many protein products produced in a highly purified form command a high sale price. Many manufacturers have therefore tended simply to add on the cost of downstream processing to the final product price. As the level of competition between rival product manufacturers increases, downstream processing costs, which can account for up to 80% of the overall production costs, will become more and more critical in determining product competitiveness.

PROTEIN STABILIZATION AND FINISHED PRODUCT FORMATS

Loss of the biological activity of proteins may occur during downstream processing, or subsequently during storage of the finished product. Loss of biological activity may be caused by a myriad different factors, as outlined in Table 3.5, and is obviously

Table 3.5. Some factors that may adversely affect the biological activity of proteins

Chemical	Detergents Urea Guanidinium chloride Organic solvents Oxidizing agents Heavy metals
Physical	Extremes of pH Elevated temperatures Freezing and thawing Vigorous agitation Irradiation
Biological	Proteolytic activity Carbohydrase activity (on glycoproteins)

undesirable on economic grounds. Various strategies must be adopted in order to reduce or eliminate such loss of activity. The most effective strategies can be developed if the cause of the reduction in activity is identified.

Decreasing activity may be caused by physical, chemical or biological factors. Numerous chemicals can inactivate proteins, for example detergents may lead to rapid protein denaturation by the partial or complete unfolding of the natural conformation of a protein molecule. Denaturation results in loss of biological activity as this invariably depends upon the integrity of the native protein conformation. Other denaturing agents include high concentrations of urea or guanidinium chloride. Many organic solvents may also promote protein denaturation as such solvents alter the dielectric constant of aqueous solutions. This adversely affects the non-covalent forces which serve to maintain the protein conformation in its native biologically active state. Many organic solvents also interact directly within hydrophobic amino acid residues, thus further destabilizing the protein's natural conformation.

In theory, protein denaturation may be fully reversed simply by removing the denaturant from the solution. This, however, is not always observed in practice, for a number of reasons, as follows:

- Denatured proteins often aggregate. Denaturation exposes many hydrophobic amino acid residues which are normally buried inside the protein when folded in its native conformation, and hence are shielded from the outside aqueous environment. Exposed hydrophobic residues tend to associate in order to minimize their exposure to the surrounding aqueous phase, resulting in aggregation.
- Denaturation renders proteins more susceptible to proteolytic cleavage.

- Some denatured molecules, particularly those of a high molecular weight, may refold incorrectly, thus not renaturing completely.

In many instances the extent of denaturation is dependent upon the concentration of the denaturing chemical present. At lower concentrations denaturation may be minimized or eliminated. In fact, in some cases such as with some organic solvents, chemical denaturants may actually stabilize protein structure if present in appropriate concentrations. Different proteins also differ in their susceptibility to denaturants. Chemically-induced denaturation may be totally eliminated by removing or replacing the offending chemical. If loss of activity due to chemical denaturation is suspected, it would be prudent to assess the effect of the various chemicals to be used in downstream processing on the bioactivity of the protein before scale-up commences.

The presence of oxidizing agents may also result in protein inactivation, due to the oxidation of susceptible amino acid side chains such as cysteine or methionine. Such oxidative reactions may be minimized by the addition of suitable reducing agents such as dithiothreitol (DTT). The presence of heavy metals, especially metals such as lead or mercury, may also result in protein inactivation. Heavy metals react with several amino acid side chains, in particular with sulphydryl groups ($-SH$ groups). Heavy metals may be introduced into a protein solution by leaching from unsuitable holding tanks or other process equipment, or via the addition of chemicals which are of an inadequate grade of purity.

Extremes of pH may also result in protein inactivation. Different proteins may have widely differing pH stability profiles. The majority of proteins are susceptible to inactivation at extremes of pH. Incubation of proteins in strongly acidic solutions may result in chemical alteration of certain protein groups, such as deamination of glutamine or asparagine residues, in addition to hydrolysis of peptide bonds. Protein inactivation by exposure to extremes of pH may be minimized by using a suitable buffering system during the processing steps.

Loss of protein activity may also be induced by a number of physical parameters. Elevated temperatures, for example, quickly denature most proteins. Most mammalian proteins exhibit decreasing biological activity when exposed to temperatures significantly above normal body temperature, and are generally inactivated if exposed to such elevated temperatures for any significant periods of time. On the other hand, some proteins are extremely heat-stable. Many microbial proteins are particularly resistant to heat denaturation, in particular proteins obtained from microorganisms which normally grow in environments where elevated temperatures are encountered, as in the region of hot springs.

The deleterious effects of heat on the bioactivity of proteins may be eliminated by maintaining strict temperature control during protein handling and purification. Although such a statement seems self-evident, localized increases in temperature during processing may sometimes be overlooked. Cellular disruption by homogenization invariably leads to increased solution temperatures which may be detrimental to proteins unless stringently controlled. Granulation also involves exposure of the protein in question to transient elevated temperatures.

Most proteins are maximally stable at low temperatures, and for this reason most processing procedures take place at 4 °C, while prolonged storage of many protein solutions may best be undertaken at temperatures below 0 °C. It is worth remembering, however, that some proteins may actually be destabilized by decreasing the solution temperature below a given value.

Freezing and thawing may also result in loss of bioactivity or potency. During the freezing process dramatic changes may occur in the microenvironment of the protein. Crystallization of water molecules leads to increases in the concentration of the various solutes present. As the temperature decreases, some buffer components may also crystallize more rapidly than others, which may cause dramatic alteration in local pH values. Such effects may lead to significant levels of protein inactivation and may be minimized by ensuring that freezing is accomplished as rapidly as possible, for example by immersion in liquid nitrogen or an ice–ethanol bath.

Various mechanical forces to which protein solutions may be subjected during downstream processing can also result in protein inactivation. Vigorous agitation or pumping of such solutions often introduces significant levels of air or gas into the aqueous product, resulting in foaming. This substantially increases the gas–liquid interface area available to the protein. Proteins tend to align themselves along such interfaces and this may promote protein unfolding such that hydrophobic amino acids are exposed to the gaseous phase. Bulk transfer of protein solutions via pumps and transfer tubing may also result in protein inactivation due to shear forces generated within such systems. An additional physical parameter which may result in protein damage is that of exposure to radiation. For this reason sterilization of protein solutions is generally achieved by methods other than irradiation.

Loss of protein activity can be caused in many cases by a variety of biological agents, most notably by proteases. Proteolytic enzymes may be classified as endopeptidases or exopeptidases. Endopeptidases catalyse the hydrolysis of internal peptide bonds in a polypeptide chain and are usually sequence-specific, in that they recognize a specific amino acid sequence and cleave a specific peptide bond within that sequence. Exopeptidases catalyse the sequential removal of amino acid residues from either the N-terminus or the C-terminus of proteins. These may also be quite specific, catalysing the removal of only certain amino acid residues.

All cell types synthesize a variety of proteolytic enzymes. Under normal functional conditions intracellular proteases play a critical role in many metabolic processes, such as maintaining protein turnover. Many microorganisms secrete a significant repertoire of proteolytic enzymes into their external environment. Such proteases serve to degrade proteinaceous material in their immediate environment, resulting in the liberation of amino acids and short-chain peptides. Unlike larger polypeptides, such molecules can readily be assimilated by the microorganisms. Such extracellular proteolytic activity thus serves a critical nutritive function.

The presence of proteolytic enzyme activity in protein solutions often greatly complicates downstream processing. In some cases it may result in substantial or total degradation of the protein of interest. High levels of proteolytic activity are invariably present in extracellular protein extracts. Appropriate steps must therefore be taken in order to minimize their effect on the extracellular protein of interest. In some instances the desired protein is itself a protease. Some proteins are resistant to proteolytic cleavage because presumably they have no amino acid sequences exposed on their surfaces that are prone to proteolytic attack. Virtually all proteins become more susceptible to proteolysis upon partial or complete denaturation. Denaturation can unmask any amino acid sequences prone to such proteolytic attack.

A wide variety of proteolytic enzymes are associated with the intracellular contents of unicellular and multicellular organisms. Homogenization results in the release not only of the protein of interest but also of these proteolytic enzymes. Loss of biological activity due to proteolytic degradation during downstream processing may be minimized by adoption of a number of strategies. Enzymes are considerably less active at low temperatures, thus purification steps (especially the initial steps) are best carried out at 4 °C. Fractionation should be performed as rapidly as possible in order to minimize contact between the protein of interest and proteolytic enzymes. Various protease inhibitors may be incorporated into the protein solution. This strategy must be adopted with caution, in particular if the desired product is destined for administration to animals or humans. Several inhibitors of proteolytic enzymes are highly toxic: the inhibitor diisopropylfluorophosphate (DFP) also inhibits acetylcholinesterase. Obviously the presence of such inhibitors in the final product would be unacceptable. This somewhat limits the choice of protease inhibitors which may be used. If any inhibitor, or other chemical, added to the product during processing is not recognized as being safe, it must be proved that such undesirable molecules are removed during subsequent processing steps.

Loss of biological activity of proteins due to proteolytic enzyme action is most likely to occur during the initial stages of a protein purification procedure. As the purification procedure progresses, most if not all the proteases should be separated from the protein of interest. The possibility always exists, however, that some proteolytic enzyme activity will copurify with the desired protein. It is therefore considered prudent to extend the precautions adopted against unwanted proteolytic attack until it has been established with certainty that all proteolytic enzyme activity has been removed.

Proteolytic enzymes represent by far the most significant biological agents that may cause loss of protein activity. Other enzymatic activities may also adversely effect protein molecules. Various carbohydrases, for example, may remove or alter the carbohydrate content of glycoproteins. The presence of micro-organisms during downstream processing must also be minimized as they may secrete proteases or other enzymes which could have a detrimental effect on the biological activity of the desired protein.

STABILIZATION OF THE FINISHED PRODUCT

Many proteins lose their biological activity during storage. Maximization of stability is of crucial importance as most proteins are sold on the basis of their biological activity rather than on a weight basis. In some instances, the addition of specific stabilization agents may enhance the stability of the finished product. Protein products are generally available in a number of different formats and may be sold as solutions, suspensions or dried powders.

A variety of different stabilizing agents may be incorporated into products in order to prolong their shelf-life. Such additives may be of a general or specific nature. Specific additives include substances such as substrates, cofactors, or some other ligand species which may interact with the protein in question, thereby stabilizing its native conformation. Appropriate additives and the most suitable concentrations in which they should be used must be determined empirically. In certain cases, some ligands may actually destabilize the protein product.

A variety of non-specific chemical additives are often incorporated into a finished product formulation containing proteins. Glycols such as polyethylene glycol, glycerol or various alcohols exert a considerable stabilizing influence upon some proteins. Various sugars such as sucrose in addition to neutral salts such as ammonium sulphate or sodium chloride, may also be added to the final product. Bulk enzymes, such as acid proteases and α-amylases, marketed as liquid preparations often contain 20% sodium chloride as preservative. Various other proteins are sold as suspensions in ammonium sulphate, or in solutions containing 30–50% glycerol. Increased protein purity often signals a decrease in protein stability. Highly purified proteins such as those destined for use in the health-care industry often present additional problems relating to stability. In many cases protein stability may be enhanced by addition of a high concentration of some other protein such as bovine serum albumin to the final product. Many forms of insulin, for example, are marketed as suspensions containing protamines which are a group of basic proteins normally found in association with nucleic acid in the sperm of certain fish. Protamines employed in the preparation of protamine–zinc–insulin suspensions are purified from the sperm or testes of certain fish.

Protein stability is also dependent upon additional factors, such as the pH of the final solution. This should be adjusted to a value at which the protein exhibits maximum stability. Storage temperatures will also greatly influence the stability and hence the shelf-life of the final product. In general, protein solutions are more stable at lower temperatures and thus are often stored at $4\,°C$, or in the frozen state at temperatures as low as $-70\,°C$. The presence of proteases in the product, even in trace quantities, may result in significant or total loss of protein activity during storage. Proteolytic activity is reduced at lower temperatures, although certain proteases may still exhibit a low level of activity even at temperatures some degrees below $0\,°C$. Many proteins exhibit increased stability if immobilized on a suitable solid matrix. Protein immobilization is discussed more fully in subsequent chapters.

Some proteins may be stabilized by the inclusion in the final product of antibody preparations raised against the protein in question. Antibody preparations utilized may be polyclonal but preferably should be monoclonal. Presumably the bound antibody stabilizes the overall protein conformation and/or masks protein sequences which may be susceptible to proteolysis. Incubation with antibodies, however, may be counterproductive, if for example the antibody binds to critical functional regions of the protein, such as an active site or a ligand binding site. It has been demonstrated that this strategy may afford some proteins substantial protection against a wide variety of inactivating factors such as extremes of pH or temperature, oxidation or proteolysis. The cost of producing appropriate antibodies precludes their use as stabilizing agents for most protein preparations. The antigenic nature of such antibodies also precludes their use in preparations destined for parenteral administration.

Many processes leading to inactivation or denaturation of proteins in solution may be minimized or eliminated by removal of the solvent water by drying. Drying may be achieved by evaporative techniques. Evaporative techniques employed on an industrial scale include spray-drying, drum-drying, vacuum-drying and fluid-bed drying. These methods are relatively harsh and are thus unsuitable for drying labile proteins. During spray-drying, protein-containing droplets are exposed to high temperatures for a very short period, typically 1 second or less. Exposure to such high temperatures can damage or destroy many proteins. Labile proteins are best dried by a technique known as freeze-drying or lyophilization. This technique was perfected during the Second World War, at which time it was used with great success to preserve and store large quantities of biological products such as vaccines and various blood products.

Lyophilization.

Lyophilization involves the drying of materials directly from the frozen state. This is achieved by firstly freezing the protein solution in a suitable, unstoppered container, normally the final product container. A vacuum is then applied and the temperature is increased in order to promote sublimation of the ice which occurs under conditions of reduced pressure. The ice is drawn off directly as water vapour. The containers are sealed following completion of the freeze-drying process. Many freeze-dried proteins may be stored at room temperature for prolonged periods with little or no loss of biological activity. Some freeze-dried products, however, exhibit significant loss of activity if stored under such conditions, and thus must be stored at lower temperatures. Some advantages and disadvantages associated with the freeze-drying technique are listed in Table 3.6.

Many proteins, in particular high-value, low-volume products such as vaccines, therapeutic enzymes, hormones, antibodies and diagnostic reagents, are often marketed in freeze-dried form. The average moisture content of a freeze-dried protein preparation is of the order of 3%. However, domains often exist within the product which contain a much higher moisture content. This can contribute to product inconsistency.

The first step in the freeze-drying process involves freezing the protein solution in the finished product containers, generally glass vials. As the temperature decreases, ice crystals begin to form. Such crystals contain only water molecules. As the ice crystals grow, the protein concentration, and the concentration of all other solutes present in the remaining liquid phase, steadily increases. Any solute species, such as salts, buffer components, other chemical additives and proteases present in the product are concentrated many times. The greater the solute concentration, the greater the reaction rate between such solutes. Proteins that are damaged by high concentrations of such solutes may be inactivated at this stage of the

Table 3.6. Some advantages and disadvantages associated with the freeze-drying of protein products

Advantages	Disadvantages
Freeze-drying represents one of the least harsh methods of protein drying	Equipment required at an industrial scale is extremely expensive
It yields a lightweight product which reduces shipping and distribution costs	Running costs are high
Freeze-dried proteins can be rapidly reconstituted (rehydrated) prior to use	Freeze-drying entails long processing times (typically 3–5 days)
Freeze-drying is accepted by regulatory authorities as a technique suitable for the preservation of finished products destined for parenteral administration	Some proteins exhibit an irreversible decrease in biological activity upon freeze-drying
	Freeze-dried products sometimes exhibit uneven moisture distribution in the final product

freeze-drying process. Such inactivation may be due to chemical or biological modifications of the protein, or may be caused by protein aggregation.

As cooling continues, some of the solutes present in the concentrated solution may also crystallize and hence are removed from the solution. As the temperature decreases still further, the viscosity of the unfrozen solution increases dramatically, and it becomes more and more rubber-like. Eventually, the unfrozen solution will change from a rubbery consistency to that of glass. The temperature at which this occurs is known as the glass transition temperature ($T_{g'}$). The glass consists of all the uncrystallized solute molecules, including the protein, as well as all the uncrystallized water molecules present in association with these solutes. Structurally, this glass is not a solid but actually a liquid which exhibits a very slow flow rate of the order of micrometres per year. The same is true of glass formed from any material—this includes window glass or spectacle glass. Mechanically, however, it may be regarded as a solid. Molecular mobility of all solutes within the glass, which contains up to 50% water, is to all intents and purposes non-existent. Chemical mobility, and hence reactivity, within the glass phase essentially ceases. Upon initial formation of ice crystals it is therefore desirable to reduce the solution temperature below the glass transition temperature as quickly as possible, in order to minimize protein inactivation.

The glass transition temperature of any given solution depends upon its composition and may be determined experimentally by a technique known as differential scanning calorimetry. Differential scanning calorimetry involves heating the glassy product and subsequently plotting temperature versus its specific heat value. A sharp increase in heat flow is observed at the glass transition temperature. Determination of the $T_{g'}$ value of a protein solution facilitates the development of a more rational freeze-drying protocol for that particular protein preparation.

During the freeze-drying process a vacuum is applied to the system once a temperature below the glass transition temperature has been attained. For most protein solutions this involves decreasing the temperature to between $-40\,°C$ and $-60\,°C$. The temperature may then be allowed to increase, in order to promote sublimation of the crystalline water. This requires an input of energy. In most industrial freeze-drying systems, this is achieved by simply increasing the shelf temperature. Adequate temperature control is required to ensure maximum efficiency during the primary drying process. The shelf temperature must normally be maintained at values above $0\,°C$ in order to promote efficient sublimation. The internal vial temperature will still be well below $0\,°C$. If the temperature is maintained below the $T_{g'}$ value during primary drying, removal of crystallized water will leave behind a very well-defined protein cake. If, however, the product temperature increases above the glass transition temperature, which could occur if the shelf temperature were too high, the viscous solution left behind as the crystallized water sublimes can "flow". This results in the formation not of a light, fluffy cake but of a very dense cake in which the protein may be inactivated and which will rehydrate slowly, if at all. The glass transition temperature of any solution depends upon its composition. During the drying process the $T_{g'}$ value is continually changing, increasing as water vapour is removed from the product.

Upon completion of primary drying, the protein cake still retains appreciable quantities of moisture, the uncrystallized water molecules associated with the glass. This water is removed during secondary drying by allowing the internal vial temperature to increase still further, so that this water is removed by sublimation. Upon completion of secondary drying, a vacuum is released and the vials are then sealed.

Purified protein molecules rarely represent the only types of molecules present in a freeze-dried preparation. Various additives, often termed excipients, may be included in the solution prior to freeze-drying. Such excipients are added in order to fulfil specific functions such as enhancing the stability of the active component in the finished product, or they may protect the protein during the actual lyophilization process by acting as lyoprotectants (Table 3.7).

Table 3.7. Some lyoprotectants frequently used during freeze-drying operations. The actual lyoprotectant chosen depends upon not only the protein of interest but also upon its intended use

Proteins	Bovine serum albumin
	Human serum albumin
Sugars	Glucose
	Sucrose
	Trehalose
Amino acids	Lysine
	Arginine

The carbohydrate mannitol is often added in order to prevent product blow-out, which is the loss of product during lyophilization due to its removal in the water vapour stream. Excipients may be added to enhance protein solubility. Tissue plasminogen activator, for example, is poorly soluble at neutral pH though the addition of the amino acid arginine dramatically increases its solubility at such pH values. The incorporation of a broad range of such excipients in products destined for *in vitro* use is usually acceptable. Only certain excipients, which are regarded as safe, may be incorporated into protein products destined for therapeutic or *in vivo* diagnostic use. The method by which most excipients, in particular lyoprotectants, interact with and stabilize the protein species remains poorly understood.

Although several different models of freeze-dryers are available, all consist of a number of essential features. All possess a drying chamber into which the product is loaded and subsequently frozen. In the case of basic laboratory-scale systems, this often consists of a round-bottomed flask which may be connected to the freeze-dryer via a suitable port. A vacuum pump is required in order to evacuate the drying chamber and maintain a vacuum during drying. A suitable condenser is also required. During drying, the water vapour emanating from the product is collected in the condensing chamber. Due to the low internal temperature (typically $-50\,°C$ to $-60\,°C$), the water accumulates in the form of ice. Upon completion of the drying cycle, the chamber temperature is increased so that the ice melts. The water is then drained from the chamber before the next freeze-drying cycle begins.

Industrial-scale freeze-dryers usually cater for thousands of product vials during each drying cycle. Upon completion of the product fill, rubber stoppers are partially inserted into the mouth of each vial in such a way as to allow water vapour to flow freely from within the vial during the drying process. The vials are placed on a series of trays which are loaded onto shelves in the drying chamber. Each shelf may be electrically heated or cooled. After loading is completed, the drying chamber door is closed. The shelf

temperature is then decreased to temperatures in the order of
−40 °C to −60 °C in order to freeze the product. Various probes are
inserted into test vials so that the actual product temperature may
be accurately monitored.

When the predetermined temperature (which should be below the
product's glass transition temperature) has been reached, chamber
evacuation is initiated. Appropriate increases in shelf temperature
promote efficient primary and secondary drying under vacuum.
Upon completion of the freeze-drying cycle, the vacuum is released
and the vials are sealed *in situ*. Shelf design in modern freeze-drying
systems allows hydraulic upward or downward movement of the
individual shelves. As any one particular shelf moves upwards, the
partially inserted rubber seals will be inserted fully into the vials
when they come in contact with the shelf immediately above. In
modern freeze-drying systems, all variable parameters such as
temperature and reduced pressure levels may be preprogrammed on
a central control console. The freeze-drying cycle may thus be fully
automated. Parameters such as individual shelf temperature,
product temperature and vacuum level are continually monitored
and recorded during the process. Most lyophilization systems also
allow *in situ* heat sterilization of the empty drying chamber. This
greatly facilitates its subsequent use when freeze-drying proteins
which must be processed under sterile conditions.

LABELLING AND PACKING OF FINISHED PRODUCTS

Upon filling and sealing in their final containers, all products are
subsequently labelled and packed. Such operations are generally
highly automated and do not require significant technical input.
Labelling, however, is a critical operation in its own right.
Mislabelling remains one of the most frequent causes of product
recall. Information present on a product label should include the
name and strength or potency of the product, batch number, date
of manufacture and expiry date, in addition to the storage
conditions to be employed. Information detailing the presence of
any preservatives or other excipients may also be included, in
addition to a brief summary of the correct mode of product
usage.

FURTHER READING

Books

Creighton, T.E. (1989) *Protein Function, A Practical Approach*. IRL Press,
Oxford.

Dean, P. *et al.* (1985). *Affinity Chromatography, A Practical Approach.* IRL Press, Oxford.

Goding, J.W. (1986). *Monoclonal Antibodies, Principles and Practice*, 2nd edn. Academic Press, London.

Harris, E. & Angal, S. (1989). *Protein Purification Methods, A Practical Approach.* IRL Press, Oxford.

Harris, E. & Angal, S. (1990). *Protein Purification Applications, A Practical Approach.* IRL Press, Oxford.

Jakoby, W.B. (1984). Enzyme purification and related techniques. In *Methods in Enzymology*, p. 104, part C. Academic Press, New York.

Wiseman, A. (1987). *Handbook of Enzyme Biotechnology*, 2nd edn. John Wiley, Chichester.

Articles

Afeyan, S.P. *et al.* (1990). Perfusion chromatography, an approach to purifying biomolecules. *Bio/Technology*, **8**, 203–206.

Bonnerjea, S. (1986). Protein purification: the right step at the right time. *Bio/Technology*, **4**, 954–958.

Bruton, C.J. (1983). Large scale purification of enzymes. *Phil. Trans. R. Soc.* (London), **B300**, 249–261.

Cazes, J. (1988). Centrifuged partition chromatography. *Bio/Technology*, **6**, 1398–1401.

Clonis, Y.D. (1987). Large scale affinity chromatography. *Bio/Technology*, **5**, 1290–1293.

Franks, F. *et al.* (1991). Materials science and the production of shelf stable biologicals. *Pharm. Tech. Int.*, **3**, 9, 24–28.

Geisow, M.J. (1991). Stabilizing protein products: coming in from the cold. *Trends in Biotechnol.*, **9**, 149–150.

Huddleston, J. *et al.* (1991). The molecular basis of partitioning in aqueous two-phase systems. *Trends in Biotechnol.*, **9**, 381–388.

Jones, K. (1991). A review of biotechnology and large scale affinity chromatography. *Chromatographia*, **32**, 469–480.

Knight, P. (1990). Bioseparations: media and modes. *Bio/Technology*, **8**, 200–201.

Knight, P. (1989). Chromatography: 1989 report. *Bio/Technology*, **7**, 243–255.

Knight, P. (1989). Downstream processing. *Bio/Technology*, **7**, 777–782.

Lambert, P.W. & Meers, J.L. (1983). The production of industrial enzymes. *Phil. Trans. R. Soc.* (London), **B300**, 263–299.

Muller, W. (1990). Liquid–liquid partition chromatography of biopolymers in aqueous two-phase systems. *Bioseparation*, **1**, 265–282.

Muller, W. (1990). New ion exchangers for the chromatography of biopolymers. *J. Chromatog.* **510**, 133–140.

O'Carra, P. & Barry, S. (1972). Affinity chromatography of lactate dehydrogenase. *FEBS Lett.* **21**, 281–285.

Ohlson, S. *et al.* (1989). High performance liquid affinity chromatography: a new tool in biotechnology. *Trends in Biotechnol.*, **7**, 179–185.

Pilak, M.J. (1990). Freeze drying of proteins, part I: process design. *BioPharm.*, **3**, 8, 18–27.

Pilak, M.J. (1990). Freeze drying of proteins, part II: formulation selection. *BioPharm.*, **3**, 9, 26–30.

Safer, G. (1986). Current applications of chromatography in biotechnology. *Bio/Technology*, **4**, 712–715.

Sassenfeld, H.M. (1990). Engineering proteins for purification. *Trends in Biotechnol.*, **8**, 88–93.

Seipke, G. *et al.* (1986). High-pressure liquid chromatography (HPLC) of proteins. *Angew. Chem. Int. Ed. Engl.*, **25**, 535–552.

Shamey, E.Y. *et al.* (1989). Stabilization of biologically active proteins. *Trends in Biotechnol.*, **7**, 186–190.

Simpson, J.M. (1988). Preliminary design of ion-exchange columns for protein purification. *Bio/Technology*, **6**, 1158–1162.

Sjödahi, J. (1989). Protein engineering in Japan: towards atomic biology. *Trends in Biotechnol.*, **7**, 144–147.

Suralia, A. *et al.* (1982). Protein A: nature's universal anti-antibody. *Trends Biochem. Sci.*, **7**, 74–76.

Van Brunt, J. (1988). Scale-up: how big is big enough? *Bio/Technology*, **6**, 480–482.

Wheelwright, S.M. (1987). Designing downstream processes for large scale protein purification. *Bio/Technology*, **5**, 789–793.

Chapter 4

Therapeutic proteins: special considerations

- Parenteral administration of biopharmaceuticals.
- Guide to good manufacturing practice.
    - International pharmacopoeias.
    - Hygienic aspects: cleaning, decontamination and sanitation.
    - Clean areas.
    - Terminal sterilization and aseptic filling.
    - Water for biopharmaceutical processing.
    - Documentation.
    - Process validation.
- The range and medical significance of product impurities.
    - Microbial contaminants.
    - Viral contaminants.
    - Pyrogenic contaminants.
    - DNA contaminants.
    - Protein contaminants.
    - Chemical and miscellaneous contaminants.
- Further reading.

Proteins and other pharmaceutical preparations derived from living cells are often termed "biologics". Biologics include a broad range of toxoids and vaccines, various hormone preparations such as insulin, a variety of blood products such as blood factors, albumin and antisera, and certain enzymatic preparations of therapeutic value.

The advent of recombinant DNA technology renders the production and use of biologically-based therapeutics among the most exciting and rapidly growing areas of pharmaceutical science. In principle, this technology facilitates the production of unlimited quantities of any protein of clinical significance. While this area of scientific endeavour still remains in its infancy, several notable recombinant proteins have become available over the last number of years. Examples include recombinant insulin and interferon, both of which have been approved for sale by relevant regulatory agencies. A more extensive listing of such products is presented in Table 4.1.

Preapproval investigational studies are required for pharmaceutical preparations. These are comprehensive, time-consuming and represent a significant financial commitment by the company developing the product. Such investigational studies involve both extensive preclinical and clinical trials which, between them, may last up to 9 years. The drug approval process, as currently implemented in the USA, is briefly summarized below.

The drug under investigation, biological or non-biological, is initially characterized as comprehensively as possible in *in vitro* studies. The consequences of its administration to laboratory animals is also assessed. Such preclinical tests yield valuable information regarding the efficacy and safety of the product and may take up to 3 years to complete.

If the drug appears to be safe and effective in attaining its intended therapeutic purpose, an investigational new drug (IND) application is lodged with the relevant regulatory authority. In the USA, this is the Food and Drug Administration (FDA). Such an IND application details all the information generated during preclinical trials. If, upon review of the application, the FDA are satisfied that further studies are warranted, IND status is granted and clinical testing upon humans begins. Clinical trials may be segregated into three consecutive phases.

During phase I trials, the drug is administered to a small group of healthy volunteers in order to determine its pharmacological properties, its appropriate dosage level and route of administration, how it is metabolized and excreted and any side-effects. Such investigational studies are usually conducted within 1 year.

If satisfactory results are recorded, phase II trials may proceed. These trials, which may take up to 2 years to complete, assess the drug's safety and effectiveness when administered to volunteer patients, i.e. people suffering from the condition which the drug is claimed to alleviate or treat. More extensive animal and human trials continue in order to assess the drug's overall safety and efficacy.

During phase III trials the drug is administered to several thousand patients in order to assess further therapeutic potential and safety. These extensive clinical trials may last for up to 3 years.

Upon satisfactory completion of phase III trials, a new drug application (NDA) file is submitted to the FDA. The NDA file contains all the information generated by the preclinical and clinical trials in addition to other pertinent information such as the method of production. The FDA review of the NDA may take up to 3 years. During the review procedure, clarification or further information may be sought regarding various points of information submitted. If the information contained in the NDA file satisfies the FDA with

Table 4.1. Medicines from biotechnology: recently approved drugs and vaccines

Product name	Producing company	Medical applications	US development status
Actimmune® Interferon gamma-1b	Genentech (San Francisco, CA)	Management of chronic granulomatous disease	Approved December 1990
Activase® Alteplase, recombinant	Genentech (San Francisco, CA)	Acute myocardial infarction Acute pulmonary embolism	Approved Novermber 1987
Alferon® Interferon alfa-n3 (injection)	Interferon Sciences (New Brunswick, NJ)	Genital warts	Approved October 1989
Engerix-B® Hepatitis B vaccine (recombinant)	SmithKline Beecham (Philadelphia, PA)	Hepatitis B immunization	Approved September 1989
EPOGEN® Epoetin alfa	Amgen (Thousand Oaks, CA)	Treatment of anaemia associated with chronic renal failure, including patients on dialysis and not on dialysis, and anaemia in Retrovir®-treated HIV infected patients	Approved June 1989
PROCRIT® Epoetin alfa	Ortho Biotech (Raritan, NJ)	Treatment of anaemia associated with chronic renal failure, including patients on dialysis and not on dialysis, and anaemia in Retrovir®-treated HIV infected patients	Approved December 1989
Humatrope® Somatropin (rDNA origin for injection)	Eli Lilly (Indianapolis, IN)	Human growth hormone deficiency in children	Approved March 1987
Humulin® Human insulin (recombinant)	Eli Lilly (Indianapolis, IN)	Insulin-dependent diabetes	Approved October 1982
Intron A® Interferon alfa-2b (recombinant)	Schering-Plough (Madison, NJ)	Hairy cell leukaemia Genital warts AIDS-related Kaposi's sarcoma Non-A, non-B hepatitis	Approved June 1986 Approved June 1988 Approved November 1988 Approved February 1991
Leukine™ Sargramostim (GM-CSF)	Immunex (Seattle, WA)	Autologous bone marrow transplantation	Approved March 1991
Proline™ Sargramostim (GM-CSF)	Hoechst-Roussel (Somerville, NJ)	Autologous bone marrow transplantation	Approved March 1991

Neupogen® Fibrastim (rG-CSF)	Amgen (Thousand Oaks, CA)	Chemotherapy-induced neutropenia	Approved February 1991
ORTHOCLONE OKT® 3 Muromonab-CD3	Ortho Biotech (Raritan, NJ)	Reversal of acute kidney transplant rejection	Approved June 1986
Protropin® Somatrem for injection	Genentech (San Francisco, CA)	Human growth hormone deficiency in children	Approved October 1985
RECOMBIVAX HB® Hepatitis B vaccine (recombinant) MSD	Merck (Rahway, NJ)	Hepatitis B immunization	Approved July 1986
Roferon®-A Interferon alfa-2a recombinant/Roche	Hoffman-La Roche (Nutley, NJ)	Hairy cell leukaemia AIDS-related Kaposi's sarcoma	Approved June 1986 Approved November 1988

Reproduced by kind permission of the American Pharmaceutical Manufacturers Association.

regard to the product's safety and effectiveness, marketing approval is normally granted and the product goes on sale (Table 4.2).

Thus, from the initial application, it may require an investment of tens of millions of dollars and up to 10 years to obtain marketing approval. This explains in part why most new drugs originate in large, well-funded pharmaceutical companies and why profit margins applied often appear excessively high.

Table 4.2. Drug development process. Only those drugs which satisfactorily pass testing in any one phase may advance to the next phase of testing. Upon successful completion of phase III, a new drug application (NDA) is filed with the relevant regulatory authority. Approval, from this stage, if granted, may take a further 3 years. The total cost can be up to US $200 million

Development status	Evaluation undertaken	Average time required
Preclinical testing	*In vitro* studies and animal studies	3 years
Phase I trials	Safety testing in healthy human volunteers	1 year
Phase II trials	Efficacy and safety testing in patients	2 years
Phase III trials	Large scale trials in substantial numbers of patients Verification of efficacy and long term safety	3 years

After the introduction of the product to the marketplace, the producer is normally required to submit periodic reports on drug performance to the FDA. Such reports detail any instances of adverse reactions to the drug in question, in addition to providing information on production, quality control and distribution.

Under certain mitigating circumstances, some drugs may be approved for clinical use without being subjected to such exhaustive investigational studies. Such instances are rare, however, and occur only if a serious illness is identified against which no existing approved therapeutic agent is effective.

Well over a hundred new protein products are currently under clinical investigation or are awaiting FDA approval. Monoclonal antibodies constitute the single largest category of such biological products. Other categories of products are listed in Table 4.3. Over half of these investigational products aim to treat cancer or cancer-related conditions. Others aim to treat conditions such as haemophilia, AIDS, herpes, growth deficiency, arthritis, cystic fibrosis and septic shock. Many such products are discussed individually in the following two chapters. If approved by the regulatory authorities, most such products will be introduced into the marketplace within the subsequent 3–5 years. The widespread availability of such a spectrum of such therapeutic proteins may revolutionize the future management of a wide range of human diseases and conditions.

Table 4.3. Some biopharmaceutical products undergoing clinical trials (see Chapters 5 and 6)

Clotting factors	Colony stimulating factors
Erythropoietins	Growth factors/hormones
Interferons	Interleukins
Tumour necrosis factors	Vaccines

PARENTAL ADMINISTRATION OF BIOPHARMACEUTICALS

Certain precautions must be observed when producing therapeutic proteins destined for parenteral administration. Parenteral preparations may be defined as sterile products that are intended for administration by injection, infusion or implantation. Such modes of administration can obviously introduce the protein or drug directly into the blood stream. Bodily defence mechanisms associated with natural assimilation of molecular species through, for example, the skin or gastrointestinal tract are thus bypassed. For this reason, every aspect of the manufacturing protocol of such therapeutic proteins must attain high standards of safety. Virtually all therapeutic peptides and proteins are delivered parenterally. The physicochemical properties of these molecules preclude delivery by alternative modes such as oral administration. The majority of proteins are denatured and degraded by the conditions and enzymatic activities associated with the gastrointestinal tract. Very small quantities of some peptides or proteins are known to cross the epithelial barrier; however, such quantities would not constitute a therapeutically significant dose.

Several alternative peptide or protein delivery systems remain the subject of intense research. Nasal delivery of products may constitute a viable alternative. Some low molecular weight, highly potent peptides, such as luteinizing hormone releasing hormone or oxytocin, can successfully be administered by nasal delivery. Larger peptides and polypeptides cross the nasal epithelial membrane with greater difficulty. Additional factors which militate against effective nasal delivery systems include an active nasal mucociliary clearance system, which quickly clears drugs from the absorption area, and the presence in the nasal cavity of proteolytic enzymes. The ease of administration and large surface area available for drug absorption encourages further research.

Yet another potential delivery system currently under investigation involves the use of microspheres. Encapsulation of proteins in microspheres which remain stable in the upper digestive tract, particularly in the stomach, protects the therapeutic product from enzymatic degradation. Release of the peptide or protein in the small intestine maximizes the chances of absorption. This approach is regarded with sceptism by many for a variety of reasons, including the fact that the level of absorption attainable of large peptides and polypeptides from the small intestine is questionable. Nonetheless several companies are investigating such an approach.

Transdermal delivery of biopharmaceuticals is another approach under active research. Transdermal patches have proven to be effective in the delivery of steroids such as oestrogen and of nicotine. Delivery of peptides and small polypeptides by such means has proved to be more difficult. Disappointing results are due mainly to the large molecular dimensions of peptides and the fact that most are lipid-insoluble. Application of a low-level electrical current can effectively push peptides through the skin via hair follicles and sweat glands by simple charge repulsion. This technique, known as iontophoresis, is the subject of intensive research by several commercial concerns.

GUIDE TO GOOD MANUFACTURING PRACTICE

The standards applied when producing biopharmaceutical products are based on those already applied to the production of traditional pharmaceutical substances. Additional standards relevant to specific aspects of biopharmaceutical production must also be applied when purifying biologics. Owing to the paramount importance of product safety, all regulatory authorities exhaustively review a proposed production process step-by-step. Authorities insist on stringent compliance with appropriate standards of product safety, quality and efficacy.

Many of the principles that ensure that an overall manufacturing process results in the production of a pharmaceutical product of appropriate quality are summarized in publications which detail "good manufacturing practice". Good manufacturing practice (GMP) is concerned with all aspects of production and quality control of pharmaceutical products. National regulatory agencies must satisfy themselves that the method of manufacture used in the production of any pharmaceutical product within their jurisdiction satisfies good manufacturing principles. This is essential to the granting or renewal of a manufacturing licence. The regulatory authorities inspect the manufacturing facilities at regular intervals to ensure continued compliance with GMP.

Publications entitled *Guide to Good Manufacturing Practice* explain in detail the principles of GMP. The Commission of the European Communities, for example, publishes the *EC Guide to GMP for Medicinal Products*. This publication is used by all national regulatory authorities of EC member states in assessing applications for manufacturing authorization and as the basis for inspection of facilities manufacturing medicinal products. Each chapter of this guide deals with a particular concept or principle (Table 4.4). A variety of supplementary guidelines dealing with the manufacture of sterile medicinal products are also included in the publication.

Table 4.4. Topics discussed and principles outlined in the *EC Guide to GMP for Medicinal Products*

Quality management	Complaints and product recall
Documentation	Premises and equipment
Contract manufacture and analysis	Quality control
Personnel	Safety inspection
Production	

Guides to GMP detail principles of a general nature which, when enforced, help the attainment of total quality assurance during the manufacturing process. The principles outlined in such guides are in most instances equally applicable to production of classical pharmaceutical and biopharmaceutical products. Owing to the growing importance of biopharmaceuticals within the pharmaceutical industry, several documents have been published which relate specifically to the production of biotechnology-based products. Examples include the *Points to Consider* series, published by the FDA. Most such publications discuss specific points or details unique to the production of certain biopharmaceutical products—such as the production of monoclonal antibodies or the production of heterologous proteins in microbial species.

Clearly the exact method employed to produce any biopharmaceutical varies from product to product. Because of the proprietary nature of most biopharmaceutical products, detailed industrial manufacturing protocols are considered highly confidential. Hence, in most cases such details remain unpublished. The efficacy and safety of such biologically based substances are examined particularly closely before product authorizations are granted. In general it is the responsibility of the manufacturer to satisfy the regulatory authorities that the proposed method of production for any product complies with the principles of GMP, and will consistently yield a product of the intended standard in terms of safety and efficacy. Furthermore, it must be shown that the standards applied are themselves appropriate and adequate. It is not intended to present an exhaustive discussion detailing the entire content of the *Guide to Good Manufacturing Practice;* however, certain specific aspects of GMP which have a profound effect on the manufacture of biopharmaceutical products are detailed below. Topics include hygienic aspects of production, clean-room technology, generation of high-purity water, aseptic processing, documentation and validation. First, however, publications termed "international pharmacopoeias" are discussed, as these are inter-linked with GMP in relation to the production of medicinal substances.

International pharmacopoeias

Many pharmaceutical substances are manufactured to the exacting standards laid down in publications termed "pharmacopoeias". There are more than two dozen pharmacopoeias published by relevant bodies worldwide. The most notable of such publications include the *British Pharmacopoeia* (BP), the *European Pharmacopoeia* (EurPh) and the *United States Pharmacopoeia* (USP). All such pharmacopoeias are regularly updated. New editions are usually published every 5–8 years. The first edition of the *British Pharmacopoeia* was published in 1864, upon the merging of the London, Edinburgh and Dublin pharmacopoeias. The fifteenth edition was published in 1993. The BP is compiled by the British Pharmacopoeia Commission, a body which is appointed by the Secretaries of State for Health and Agriculture in Great Britain.

A pharmacopoeia is a book containing monographs or formal descriptive passages, detailing pharmaceutical and other medicinal substances. Special emphasis is placed on quality and safety. It provides an authoritative set of standards relating to the quality to which medicinal substances shall be manufactured. Pharmacopoeias are usually legally enforceable documents. Most of the monographs detail actual pharmaceutical substances, although monographs on subjects such as surgical dressings also appear. The majority of pharmaceutical substances currently available are chemical rather than biological in nature. This fact is reflected in pharmacopoeial content. The bulk of monographs detail individual pharmaceutical drugs or finished product formulations. Information relayed in a typical monograph includes the name, chemical formula and physicochemical properties of the drug in question, the active ingredients present, the level of purity and potency of such active ingredients acceptable, specific methods of assaying the ingredients, appropriate method of sterilization, where applicable, and the appropriate storage conditions required.

Monographs on established biologically based products such as insulin, various blood products and vaccines are also included in the various pharmacopoeias. Additional monographs detailing biologically based drugs are appearing in response to the growing number of recombinant proteins of therapeutic value which are produced by the biotechnology industry. A monograph on human insulin produced by recombinant methods or by enzymatic modification of porcine insulin was included in the 1988 edition of the BP. Additional monographs such as these will be published in future editions as more and more biopharmaceutical products become available.

Any product, chemical or biological, that meets all of the quality and other requirements laid down in a pharmacopoeia is entitled to include the initials of that pharmacopoeia in its product name. Insulin injection BP therefore represents an injectable insulin preparation which meets the requirements of the insulin injection monograph in the *British Pharmacopoeia*. Manufacture of products to pharmacopoeial standard is usually required as it assures the end user that the product attains a level of quality which is internationally recognized as being appropriate for its intended application.

Another publication of relevance to pharmaceutical and biopharmaceutical manufacturing is *Martindale: The Extra Pharmacopoeia*. This publication provides concise, unbiased reports detailing the characteristics, uses and actions of the vast majority of therapeutic drugs currently in use. Each drug is detailed in monograph format and the information includes chemical characteristics, adsorption and fate, uses and administration, adverse effects and suitable dosage levels of the drug or substance in question. This text thus provides an invaluable encyclopaedic reference source of drug information to those involved in the pharmaceutical and health-care industries.

The first edition of *The Extra Pharmacopoeia* was published in 1883, compiled by William Martindale. The twenty-ninth edition, containing monographs detailing almost 5000 substances, was published in 1989. Although the majority of such monographs detail chemicals, recent editions carry increasing numbers of monographs on biopharmaceutical products. This merely reflects the growing importance of such substances within the health-care industry.

Hygienic aspects: cleaning, decontamination and sanitation

Attainment of a high degree of hygienic standards is particularly important when manufacturing any product destined for parenteral use. Cleaning may be defined as the removal of "dirt". Dirt, as defined here, includes all organic and inorganic material which may accumulate in process equipment and processing areas during routine downstream processing. In this regard any protein or other molecules retained in a chromatographic column subsequent to elution of the protein of interest may be regarded as dirt. Decontamination generally refers to the inactivation and removal of undesirable substances such as endotoxins and other pyrogenic compounds, in addition to other harmful substances such as viruses. (Pyrogenic substances are any materials that, when injected into the body, result in a fever). Sanitation refers to the removal and inactivation of viable organisms which may be present. In many but not all instances, cleaning procedures are effective decontamination and sanitation methods. Such procedures are applied to both process equipment and to the area in which the manufacturing process is undertaken. Cleaning, decontamination and sanitation of process equipment form an integral part of any manufacturing procedure (Table 4.5).

Table 4.5. The range of substances that cleaning, decontamination or sanitation of production equipment aims to remove (production equipment utilized in most biopharmaceutical manufacturing facilities includes chromatographic columns, holding and other vessels, in addition to centrifuges, filtration systems and driers)

Cellular debris	Viral particles
Precipitated/aggregated materials	Microorganisms
General particulate material	Traces of protein or other molecules from previous production runs
Pyrogenic substances	
Lipids and related substances	

Process equipment which may be detached or dismantled can be cleaned and sanitized relatively easily. Processing vessels may be cleaned simply by rinsing or scrubbing thoroughly with high-purity water. Detergents may be required but if used, all traces must be washed away with purified water. The grade of water used, in particular for the final rinsing, should ideally be sterile and pyrogen-free. Generation of water of such purity is discussed later. Detachable tubing may be cleaned in a similar fashion. Pumps and other equipment such as continuous flow centrifuges should be dismantled where possible to ensure a thorough cleaning of the constituent parts. Such cleaning procedures will remove virtually all dirt and will significantly reduce the level of viable organisms present—the bioburden. It will not, however, guarantee sterility. Components may be sterilized by a number of methods including heat (moist or dry), irradiation or chemical means. Sterilization of any process equipment with chemical agents such as formaldehyde, hypochlorite or peroxides must be undertaken with caution, owing to the possibility that traces of such chemicals would remain in the sanitized equipment; this would contaminate products during subsequent production runs. Most processing equipment may be effectively sterilized by autoclaving, as may ancillary equipment such as flexible tubing and other plastic materials. Prior to their use, most filters are assembled in their final housings, usually stainless steel modules, and are also sterilized by autoclaving.

The internal surfaces of process fixtures such as metal pipework or some vessels may be sterilized by passing "live" steam through for an appropriate period of time. Cleaning and sanitation of fixtures without their disassembly or removal from their normal functioning positions is usually termed "cleaning in place". While all process equipment should be cleaned at regular intervals, only sterilization of those items of equipment that come into direct contact with products is generally necessary.

Chromatographic systems require regular cleaning, decontamination and sanitation and are generally cleaned in place. Prior to pouring or repouring a column, the column and gel matrix may be cleaned or sterilized separately. Prior to sterilization, the chromatographic column should be disassembled and scrupulously cleaned, firstly with an appropriate detergent, followed by rinsing thoroughly with high-purity water. Depending upon the material from which it is manufactured, the column may subsequently be sterilized by autoclaving or by treatment with chemicals such as sodium hydroxide (NaOH). Most chromatographic media may be sterilized by autoclaving, though chemicals such as hypochlorite or peroxides may also be used under certain circumstances.

Upon packing columns with relevant gel materials it becomes necessary to devise a suitable protocol to achieve effective cleaning in place (CIP) of the chromatographic system. Several cleaning and sanitizing agents may be employed. The specific agents chosen will depend largely upon factors such as the chemical compatibility of the substance and the gel, in addition to the level of cleaning required and the composition of the sample to be applied to the column. The nature of the sample determines the range of impurities which might be retained in the chromatographic matrix. Chromatography of plant extracts may result in column fouling by chlorophyll and other pigments, or with negatively-charged polyphenols or various enzymes, some of which may be capable of digesting dextran and other carbohydrate-based chromatographic media. Molecules retained upon chromatography of microbial extracts may include lipids and endotoxins. Chromatography of animal tissue extract may introduce contaminants such as lipids and viruses into the gel media. The loading of dirt and other contaminants retained on a column after each cycle is thus largely dependent upon the source of the protein. This contaminant load will determine how often CIP procedures must be carried out. Normally column cleaning and sanitizing is required after one to ten column runs. Effective CIP systems significantly reduce the risk of carry-over of contamination from one product run to the next, and also extend the useful life of the column.

Upon completion of a production run, a thorough flushing of the column with several column volumes of water or suitable process buffer may flush out many column contaminants. Rinsing of the gel with a concentrated solution of neutral salts such as KCl or NaCl is often effective in removing precipitated or aggregated proteins, or other molecules which may be weakly bound to the gel. Inclusion of a chelating agent such as EDTA in the buffer may help remove any metal ions which remain associated with the gel. Subsequent washing with a solution of NaOH, or in certain cases with detergents, may be required in order to remove tightly bound material. In some instances where column fouling by lipids is particularly problematic, increasing the column temperature to 40–60 °C may be effective. Most lipids are liquids at such temperatures. This approach may be adopted only if it has been shown that subjecting the chromatographic system to such changes in temperature will not adversely affect its fractionation characteristics.

Sodium hydroxide is extensively used as a CIP agent. It is effective in removing many contaminants which bind tightly to the gel matrix, in addition to removing or destroying pyrogenic material such as bacterial endotoxins, viral particles and bacteria. Sodium hydroxide is popular not only because of its efficiency as a cleaning agent, but also because of its ready availability and the fact that it is inexpensive.

The real possibility of the presence of pathogenic viral particles in biological materials such as animal tissue, blood or in mammalian cell culture systems is a matter of grave concern. Many viruses are known to carry oncogenes, cancer-causing genes, while others represent the causative agents of serious illnesses such as hepatitis or AIDS. NaOH will inactivate HIV within minutes.

The effectiveness of NaOH as a cleaning and sanitizing agent is dependent upon both time and concentration. Generally speaking, NaOH is applied at concentrations of 0.5–1 M, and should have a gel contact time of an hour or more. However, NaOH may be used only if it is compatible with both the chromatographic gel and all of the column components. Most chromatographic media may be exposed to NaOH for reasonable periods without adverse effect. Silica gel is an exception as it quickly dissolves at pH values greater than 8.0. Following exposure to NaOH, the chromatographic system should immediately be thoroughly rinsed with sterile, pyrogen-free water or an appropriate buffer prepared with such water. Efficient removal of NaOH subsequent to completing a CIP protocol is necessary in order to prevent column deterioration due to prolonged exposure to alkaline conditions, and to ensure that residual NaOH does not contaminate the product during the next production run.

Routine application of an effective CIP procedure is greatly simplified if the overall chromatographic system has been designed with process hygiene in mind. The choice of gel and column type will determine what CIP agents may be utilized. Design of the chromatographic column and ancillary equipment will determine the susceptibility to microbial contamination. Valves and pipe connections represent danger points when considering the risk of introducing microbial contaminants. Pipe connections in particular should be designed such that no "dead leg" zones are present.

Clean areas

The above section detailing hygienic aspects of manufacturing processes concentrates on the maintenance of a high degree of hygiene in respect to process equipment. Maintenance of high hygienic standards in the general environment in which the process is undertaken also forms an integral part of GMP when considering injectable products. Clean-room technology is of central importance in this regard.

Clean areas are specially designed rooms in which general environmental conditions are tightly controlled. High-efficiency particulate air filters (HEPA filters) are installed in the ceilings of such rooms. All air entering the clean room is filtered through the HEPA units. Filters may be classified according to their ability to remove particulate matter. Air normally exits a clean room via specialized outlets incorporated into the walls just above floor level. HEPA filters are evenly spaced over the ceiling area in order to generate a relatively uniform downward current of filtered air throughout the room (Figure 4.1).

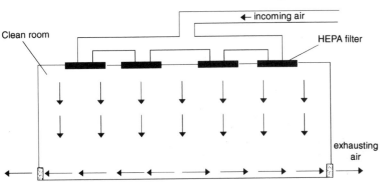

Figure 4.1. Generalized representation of HEPA-filtered air flow through a clean room

HEPA filtered air enters the clean room at a rate that constantly maintains a positive air pressure within the room. This ensures that no air from environmentally uncontrolled adjacent areas may enter the clean room area. In most cases, the equivalent of 20–25 room volumes of filtered air should be pumped through the room per hour. The overall effect of such clean-room technology is that particulate matter, which often harbours microorganisms, is continuously flushed from the room. Various levels of environmental cleanliness may be attained by using HEPA filters of differing filtration efficiency. In Europe, clean rooms are classified in order of increasing cleanliness as grade D, C, B or A. In the USA, clean rooms are classified as class 100 000 (equivalent to European grade D), class 10 000 (grade C) or class 100 (grades A and B). In both instances, the classification relates to the maximum permitted number of particles and viable microorganisms per cubic metre of air within the clean room area. Thus according to the parameters laid down in the *EC Guide to GMP for Medicinal Products*, grade A laminar air work-stations must contain (per m^3) a maximum of 3500 airborne particles of 0.5 μm, no particles of 5.0 μm size and statistically less than 1 viable microorganism per m^3. At the other end of the scale, grade D (class 100 000) clean rooms are permitted to contain up to 3 500 000 particles of 0.5 μm, up to 20 000 particles of 5.0 μm particles and up to 500 viable microbes per m^3 (Table 4.6).

Table 4.6. Air classification system for manufacture of sterile products as outlined in the *EC guide to GMP for medicinal products*

Grade	Maximum permitted number particles m^3		Maximum permitted number of viable microorganisms per m^3
	0.5 μm	5.0 μm	
A (laminar airflow work station)	3 500	0	Statistically less than 1
B	3 500	0	100
C	350 000	2 000	100
D	3 500 000	20 000	500

The air handling system constitutes only part of the design of a clean-room facility. Internal room layout and fixtures should facilitate effective cleaning and sanitation. Uncleanable recesses are to be avoided, thus corners and joinings between walls and floors are generally rounded. Pipework should be installed in such a way that the area around them may be easily cleaned. All exposed surfaces should be of a smooth, sealed, unbroken finish, so that accumulation of particulate matter or microorganisms is minimized. No unnecessary equipment or other fixtures should be housed in the room. Documented cleaning procedures should be carried out before and after each processing cycle. Cleaning procedures should effectively and safely remove dust, particulate matter and product stains from the floor and equipment within the clean area. After removal of excess dirt, the floor, walls and equipment—including items such as workbenches and chairs—may be decontaminated and sanitized by rubbing down or rinsing with an appropriate disinfectant solution. Fumigation of the clean room offers an effective final sanitation step, especially in relation to bioburden reduction or elimination in poorly accessible areas. Fumigation is generally achieved by filling an aerosol generating device with an appropriate bactericidal agent, with subsequent fogging of the area.

It is important to ensure that the cleaning and sanitizing protocols employed in no way result in contamination of the product, or of equipment which will come in direct contact with the product, by the cleaning chemicals. Efficient HEPA-filtered air delivery in conjunction with thoughtful room design and effective cleaning and sanitizing procedures ensure that high-quality environmental conditions may be attained within clean areas. Microorganisms and particulate matter may, however, be introduced into the clean area via several routes during processing operations. The most significant sources of environmental contamination are (a) the operating personnel, (b) inadequately cleaned process containers and instruments, and (c) the process itself.

Operating personnel represent a major potential source of process contaminants such as particulate matter and microorganisms. Product contamination by personnel is minimized by ensuring that all operators entering clean room areas wear protective one- or two-piece outfits made from suitable non-shedding material (Figure 4.2). The operators should change into presterilized protective clothing in a suitably designed changing room which leads directly into the clean room. Gloves and face masks are also worn, particularly when working in class A, B or C environments. A high standard of personal hygiene is of critical importance and all operators entering the clean area should receive regular training in all disciplines relevant to GMP when operating in a clean room environment. Only the minimum number of operators required should be present in the clean room.

During any processing procedure, it becomes necessary to transfer raw materials or other substances into the clean room. If not carefully controlled, such transfers may result in carryover of microorganisms or particulate matter into the clean area. Chemical and other raw materials are generally weighed or transferred into

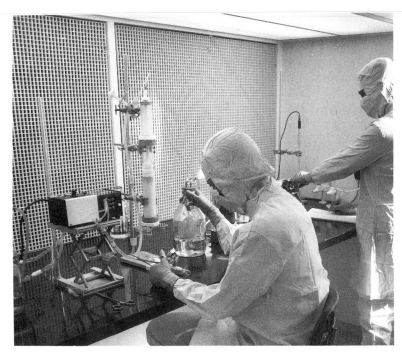

Figure 4.2. Protein purification (pilot scale), as may be undertaken under clean room conditions. In this case product manipulation is occurring within a vertical laminar flow hood. The minimum amount of process materials and personnel are in place. Personnel wear suitable protective clothing. Clothing is designed to minimize potential contamination of product with bioburden or particulate matter derived from operators (photograph courtesy of Wellcome Research Laboratories)

suitable clean and, if possible, sterile weighing bins. In order to minimize introduction of unwanted contaminants, all such materials are introduced into the clean room via suitably designed transfer locks. The transfer lock acts as a buffer zone between the clean room itself and the outside, uncontrolled environment.

In most pharmaceutical facilities, some or all of the relevant product processing steps are performed in a class C clean room. The filling of many such products into their final containers is subsequently carried out under grade A laminar flow conditions—using a portable laminar flow hood which is placed within a grade B clean room. Material movements between two clean rooms of different classes (e.g. D to C or C to B) usually also occurs via transfer locks. The higher the grade of the clean room, the higher the positive internal air pressure needed to ensure that any air flow occurring between adjacent clean rooms always flows from a higher graded clean room to one of a lower grade.

Many processing steps themselves tend to generate particulate matter within the clean area. Large-scale manufacture of buffers entails handling large quantities of powder. Care must be taken to minimize atmospheric dust formation under such circumstances. Other processing equipment such as continuous flow centrifuges and some pump types tend to generate aerosols. Thus the process and equipment used should be designed to keep such aerosol or dust formation to a minimum. This is particularly important if the aerosol or dust-forming material might endanger the product or the health of the operator.

Terminal sterilization and aseptic filling

Many, if not most, pharmaceutical products are relatively heat-stable. Terminal sterilization refers to the sterilization of a product after it has been filled into its final product container and the container has been sealed. Wherever possible heat sterilization remains the method of choice when producing terminally sterilized pharmaceuticals. Other methods used include dry heat and various forms of irradiation.

Terminal sterilization assures product safety from a microbiological standpoint, as the sealed container prevents recontamination of the product while the integrity of the seal is maintained. It is important to ensure that all practicable steps are taken to minimize the microbial load of the product during each stage of processing. The presence of a high microbial load during processing will have an adverse effect on the product and its safety. Many microorganisms could metabolize or otherwise alter the product constituents and could secrete various enzymes and other substances into the product. Autoclaving a product with a high microbial content will result in the generation of bacterially derived pyrogens and other substances. Control of microbial and other potential contaminants is best achieved by adherence to the principles of good manufacturing practice as applicable to the particular substance under manufacture. Processing within clean rooms of an appropriate grade is particularly important in this regard. Many solutions destined for terminal sterilization are filtered through 0.45 μm filters immediately prior to filling into their final containers. This step removes particulate matter and the vast majority of any microbial species that may be present.

All products manufactured as sterile preparations are filled into containers of suitable quality. These must also be resistant to the effects of the sterilization method employed. Filling of product into containers is normally achieved on an automated filling line. The various parts of such filling systems which come into direct contact with the product are normally autoclavable and, hence, may be sterilized before commencement of each fill. Filling of most parenteral products should be carried out in a laminar flow workstation within a grade C clean room.

Preservatives such as methylhydroxybenzoate, propylhydroxybenzoate, cresol, chlorocresol, phenol and benzethonium chloride are included in many pharmaceutical products. The particular preservatives used must be compatible with the product in question, and must be added at appropriate concentrations. Preservatives are of particular importance in products which are filled into multidose containers, as loss of product sterility is risked during removal of each dose.

Unlike traditional pharmaceutical preparations, most biopharmaceutical substances are heat-labile. Many proteins are inactivated at temperatures significantly above 40–50 °C; terminal sterilization of product by heat is therefore not an option. In many cases, sterilization by irradiation is also impractical due to lack of appropriate irradiation facilities, in addition to the fact that many biopharmaceutical substances can be adversely affected by irradiation.

Most protein and other biopharmaceutical solutions are sterilized by filtration and subsequently introduced into their final containers using strict aseptic techniques. Sterilization by filtration is achieved using a filter of 0.22 μm pore size or smaller. Such filters will remove bacteria and fungi, but are not guaranteed to remove all viruses or mycoplasmas. In-line 1.0 μm or 0.45 μm filters are usually utilized immediately prior to the sterilizing (0.22 μm) filter. Filters used should in no way affect the product; thus fibre-shedding filters should be avoided, and it must be verified that no product ingredient is altered or removed during the filtration step. The integrity of all filters should be tested upon completion of a fill.

Successful aseptic manipulation depends largely upon the skills and training of the operators involved. It is necessary to presterilize all process equipment that will come into contact with the product during and after the filtration step. The filter is often fitted into the filter housing and subsequently autoclaved, along with all of the tubing and filling heads that will come into contact with the product during the product fill. Upon completion of the autoclave cycle these system components may be aseptically assembled in the filling area, which should be of grade A status. The containers into which the product is to be filled must also be sterile. Many plastic containers such as infusion bags are effectively sterilized by the moulding conditions used in their manufacture. Alternatively, such containers may be sterilized by irradiation. Glass vials are often sterilized by heat prior to the filling operation.

The level of risk of contamination in relation to aseptically prepared products depends upon three parameters: (a) the concentration of airborne microorganisms; (b) the neck diameter of the container to be filled with product; and (c) the time period that the open container is exposed to the environment. Reduction of any one of these parameters decreases the likelihood of accidental microbial contamination during an aseptic fill. The concentration of airborne microorganisms is minimized by carrying out all manipulations within a grade A clean room, and minimizing the number of personnel present during the aseptic operation. In this regard, design and installation of highly automated aseptic filling systems should greatly reduce the dangers of product contamination. A number of automated filling systems are available commercially and some may be cleaned in place and subsequently sterilized by steaming in place. The neck diameter of the container to be filled should be minimized and the filling operation should be operated at maximum speed in order to reduce further the likelihood of accidental microbial contamination. Sterile aqueous preparations which are filled aseptically may contain antimicrobial preservatives added to an appropriate final concentration. Insulin injections, for example, generally contain 0.1–0.25% phenol or cresol as preservative.

Water for biopharmaceutical processing

Generation of water of suitable purity for pharmaceutical and biopharmaceutical processing is central to any such operation. All pharmaceutical plants must either purchase or produce in-house water of appropriate grade or purity. Because of the large quantities of water required and economic considerations, virtually all pharmaceutical facilities have their own water purification facilities. The complexity of such facilities depends upon the intended application of the water required and upon the quality of the incoming raw water.

Potable or drinking water is obtained mainly from surface sources such as lakes and rivers, or from underground sources such as wells or springs. The level and type of natural impurities found in such water varies greatly and is largely dependent upon the source from which the water was drawn. Most raw water supplied to domestic dwellings and industrial sites is first treated by the local

Table 4.7. Range of impurities found in potable water. Even extremely low levels of any such impurities renders this grade of water unacceptable for pharmaceutical processing purposes

Particulate matter	Soil particles
	Dirt particles
	Particles derived from decaying organic matter, such as leaves
	Particles derived from leaching of internal surfaces of water pipes
Various dissolved substances	Substances derived from decaying organic matter
	Traces of agricultural run-off
	Minerals, leached into the source water
	Various polluting substances
Viable organisms	Various microorganisms

authority or other relevant body. Such treatment may introduce additional contaminants into the water. Furthermore, other contaminants may leach from distribution pipes into the water supply. Potable water is therefore unsuitable for use in most pharmaceutical applications (Table 4.7). No one method of purification will efficiently remove all impurities.

Incoming water, which is generally potable water, is subjected to a number of purification techniques before it is used in pharmaceutical processing (Figure 4.3). The initial purification steps often include depth filtration, membrane filtration and activated carbon treatment.

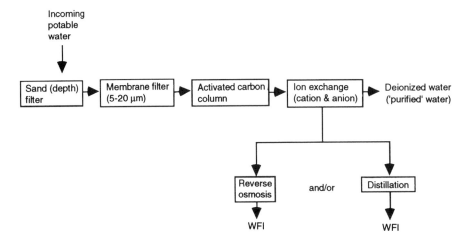

Depth filtration remains one of the most popular methods of removing particulate matter from the incoming water supply. Depending upon the water quality, membrane filtration may be employed in addition to, or in some cases instead of, a depth filter. One of the most commonly utilized substances in depth filtration is sand. Sand beds or sand filters are inexpensive to install and maintain. Periodic backwashing will free the filter of trapped particulate matter. Upon completion of the filtration step the process water is often passed through a bed of activated carbon. Many organic and other contaminants are effectively adsorbed to the activated carbon and hence are removed from the bulk water. Dissolved organic substances often impart a characteristic colour, taste or odour to water. Some contaminants, such as chlorine, are removed by the activated carbon beds through a catalytic mechanism. Activated carbon itself is relatively inexpensive and may be replaced as required.

Many water purification systems use a combination of anion and cation exchange resins in order to deionize incoming water. Modern ion exchangers are extremely efficient and remove virtually all ionic

Figure 4.3. Generalized system by which potable water is purified in order to facilitate its use in pharmaceutical and biopharmaceutical processing. Water passed through cation and anion exchange columns is termed "purified" or "deionized" water. Deionized water may be utilized in the manufacture of some oral and topical pharmaceutical preparations, but is unsuitable for the manufacture of injectable preparations. Water employed in the generation of such parenteral preparations must undergo further purification. Most commonly this involves a subsequent distillation step or reverse osmosis step, thus generating water for injections (WFI). It is water of this quality that is always utilized in the manufacture of injectable pharmaceutical and biopharmaceutical products

components present in the feed water. Ion-exchange resins require regular regeneration. This is achieved by flushing with concentrated solutions of HCl and NaOH, respectively. Regeneration also sanitizes the exchangers, although flushing with formaldehyde may be carried out as a periodic CIP procedure. Automatic regeneration is often triggered when the electrical resistivity of the treated water falls below a certain predetermined value. Deionized water finds a variety of applications in research and development, quality control laboratories and in the production of some pharmaceutical substances destined for oral administration.

Reverse osmosis is widely used as a water purification technique. Reverse osmosis is recognized as an efficient and relatively economical method of removing not only viruses, microorganisms and microbial products such as pyrogens, but also virtually all inorganic impurities from water.

Osmosis occurs if two solutions of different concentrations are separated by a semipermeable membrane which is permeable to solvent molecules but not to solute molecules. Under such conditions, solvent molecules travel across the membrane from the solution of lower solute concentration to that of the higher concentration. The force causing this movement is termed "osmotic pressure". The magnitude of such pressure is dependent upon the difference in concentration between the two solutions.

During the osmotic process, there is a natural flow of solvent molecules from the high solvent concentration side of the membrane to the lower solvent concentration side. However, if a pressure greater than the natural osmotic pressure is applied to the system from the higher concentration side, solvent molecules will flow in the opposite direction (e.g. from the high concentration side to the low concentration side). This process is termed reverse osmosis. Reverse osmosis may be used to concentrate or purify solutions.

Reverse osmosis removes in excess of 90% of all inorganic salts and virtually all microorganisms, viruses and dissolved organic contaminants from the feed water. Membranes employed have traditionally been manufactured from materials such as cellulose acetate. Many modern reverse osmosis units employ polyamide or other such membranes; these membranes are more resistant to chemical and microbiological degradation. Most such membranes are utilized in a hollow fibre or spiral-wound configuration.

Distillation remains one of the most popular techniques of water purification. This process removes most impurities present in water; the impurities remain behind in the heating chamber. The process, however, makes heavy demands upon energy input, as the liquid water must be converted into vapour. Cooling water is also required to promote subsequent condensation of the vapour to purified liquid water. Volatile impurities including various organic substances and ammonia will carry over into the distillate.

Water of various grades of purity is required for a variety of purposes during the manufacture of pharmaceutical substances. The two most important grades of water are *purified water* and *water for injections* (WFI). Standards of purity defining these grades of water are listed in monographs in the various international pharmacopoeias.

Purified water may be prepared from potable water by a variety of techniques, in particular by percolating through ion-exchange resins. Purified water is generally used as the solvent during the manufacture of a wide variety of aqueous oral dosage formulations. It is not intended for use in the preparation of parenterally administered products. Purified water must meet various specifications relating to pH, maximum permitted concentrations of chloride, sulphate, ammonia, calcium, carbon dioxide, heavy metals, oxidizable substances, total solids and bacteriological purity. Such specifications are detailed in the various pharmacopoeias.

Water for injections is defined in the *British Pharmacopoeia* as sterilized, distilled water, free from pyrogens. The *United States Pharmacopoeia* includes reverse osmosis as a suitable means of producing water for injection. WFI may be produced directly from potable water but is normally produced from purified water. WFI is thus very highly purified water and must meet exacting pharmacopoeial standards. This grade of water, which contains no added substances, is required during the manufacture of aqueous-based parenteral preparations. WFI is used extensively during biopharmaceutical operations. It is required when suspending or redissolving biological substances such as proteins during downstream processing. Buffers and other aqueous solutions utilized during the processing of such products should be manufactured using WFI. Equipment used during downstream processing is often flushed with WFI during pre- and postoperational cleaning procedures.

Water for injections may be collected in suitable containers, autoclaved and stored for future use. In many pharmaceutical facilities, however, freshly prepared WFI is fed directly into a sealed storage tank; from there it is circulated continuously throughout the processing facility via a closed network of pipes, termed a ring main system. Various points from which WFI may be drawn are incorporated into the ring main system at suitable locations within the facility. Such outlets are found mostly in clean room areas in which downstream processing of parenteral products takes place. The ring main system is designed such that no "dead legs" exist in which water could stagnate. The circulating water is normally maintained at temperatures greater than 70 °C in order to prevent microbial growth. The ring main must be drained at regular intervals and the empty piping sterilized in place, usually by passing steam through the system.

The design and range of water purification equipment installed in the water treatment area of pharmaceutical plants will vary from facility to facility. A generalized purification system would often consist of the following procedures. Potable water is first passed through a sand filter and subsequently through an activated carbon column. The resultant water is then deionized using twin cation and anion exchange resins. This deionized water, which should meet pharmacopoeial requirements relating to purified water, is often fed directly into a suitable storage tank from where it may be circulated through relevant areas of the pharmaceutical facility via a pipework system. An ultraviolet irradiation system may be incorporated in order to maintain the water in a sterile state. Purified water may also be fed directly to stills. Such distilled water should meet the pharmacopoeial requirements in relation to water for injections. WFI may be circulated throughout the manufacturing facility via the ring main system at temperatures in excess of $70\,^\circ$C.

Documentation

The pharmaceutical industry is characterized by a very extensive requirement for documentation. The principles of GMP require that a comprehensive range of written documentation be available such that errors associated with verbal communication are minimized, and that traceable histories relating to each batch of product manufactured are available and easily accessed. All documents must be written in a concise, unambiguous fashion and should be clearly legible. In some facilities, a variety of documents are stored on computer. The term "documentation" embraces a wide range of written statements, the most important of which are (a) specifications; (b) standard operating procedures (SOPs); and (c) methods and records of manufacture.

Specifications detail the standard of quality and purity to which any material utilized in the production process, and the finished product itself, should conform. Specification sheets relating to all raw materials as well as products are therefore required. The aim of such specification sheets is to help ensure that the product is manufactured to the highest degree of quality intended. Many products are manufactured to the specifications set out in one of the international pharmacopoeias.

Standard operating procedures are written instructions detailing various procedures of a general nature which form an integral part of pharmaceutical manufacturing, but which are often independent of any one product or process. Examples include SOPs detailing cleaning procedures for clean rooms or for specific items of equipment. Other SOPs detail the correct mode of operation of a specific item of equipment such as an homogenizer, freeze-drier or product-filling machine. Yet other SOPs detail essential maintenance procedures relating to process equipment or indeed to the facility itself. Some SOPs may relate directly to personnel, such as detailing the correct step-by-step procedures which must be followed within a changing room when preparing to enter a clean room.

All pharmaceutical products must be manufactured according to an approved, documented manufacturing procedure. Such written manufacturing procedures must impart step-by-step manufacturing instructions such that, in theory, a qualified person unfamiliar with the production process could successfully manufacture a product on the basis of the written manufacturing instructions alone. Each product must be manufactured according to a master formulation and the manufacturing instructions. When a batch of the product is to be manufactured, this document is usually copied and is signed by the responsible operator after each written step, upon the completion of that step. This copy then forms the batch manufacturing record which is filed and may be consulted later if necessary.

Each time a batch of any particular pharmaceutical or biopharmaceutical product is manufactured, that batch is assigned a specific batch number. A batch number is a combination of letters and/or numbers which are unique to that particular batch. Batch numbers thus serve as identification tags which may be used to pinpoint any particular batch, and trace its history of production. This is particularly important in cases where a specific batch of the product is found to be unsafe or to deviate from any of its product specifications. In such instances, the batch must be rejected, and destroyed or reworked such that the reworked product falls within specification.

In cases where a substandard batch has already been dispatched and may be on the market, an immediate and effective recall procedure must be initiated. The presence of a batch number on the label of each final dosage container, such as vials, ampoules and tubes, renders such recall possible.

Process validation

Validation refers to the act of proving that any given process or procedure employed during the manufacture of a pharmaceutical substance actually achieves its intended effect. Process equipment and procedural methodologies must be satisfactorily validated before any new production procedure is approved. Periodic validation studies are subsequently carried out to ensure the process or equipment continues to meet the level of performance expected. Clean-room HEPA filters, for example, should be validated on a regular basis to ensure that they continue to perform to their specification. Other process equipment such as autoclaves also require periodic assessment for validation.

A variety of process parameters are also subject to validation procedures. Downstream processing procedures producing a protein for parenteral administration should also be capable of removing any undesirable impurities such as viruses or DNA from the protein of interest. This capability may be tested by spiking the initial raw material with the offending substance and subsequently monitoring the effectiveness with which the procedures remove such contaminants.

THE RANGE AND MEDICAL SIGNIFICANCE OF PRODUCT IMPURITIES

Protein products destined for parenteral administration must be free from all impurities that might have an adverse effect on the wellbeing of the recipients. Any purification process producing such proteins must be designed with this in mind. Impurities most commonly encountered are outlined in Table 4.8. Additional impurities may be present depending upon the source of the raw material, and on the various additives and stabilizers which may be incorporated into the product during downstream processing. In most instances, these may also be regarded as process impurities and often must be removed during subsequent purification steps.

Table 4.8. The range and medical significance of potential impurities present in biopharmaceutical products destined for parenteral administration

Impurity	Medical consequence
Microorganisms	Potential establishment of a severe microbial infection—septicaemia
Viral particles	Potential establishment of a severe viral infection
Pyrogenic substances	Fever response which in serious cases culminates in death
DNA	Significance is unclear—could bring about an immunological response
Contaminating proteins	Immunological reactions. Potential adverse effects if the contaminant exhibits an unwanted biological activity

The majority of techniques utilized during downstream processing are designed to separate protein molecules from each other, i.e. to purify the protein of interest from the hundreds or, in many cases, thousands of other proteins present in the starting material. In many instances non-proteinaceous impurities may be efficiently removed by one or more of the protein fractionation steps of downstream processing. In other instances, it may be necessary to include specific steps in order to remove some impurities whose presence in the final product would be considered unacceptable.

Details of the various potential product impurities, their medical significance and methods to minimize or eliminate these from the final protein product form the subject matter of the remainder of this chapter.

Microbial contaminants

Pharmaceutical products intended for parenteral administration must be sterile, with the exception of live bacterial vaccines. The presence of microbiological contaminants in the final product could result not only in microbial degradation of the product, but also in the establishment of a severe microbial infection in the recipient patient. Direct introduction of viable microorganisms into the blood stream could induce a range of serious illnesses such as septicaemia. Septicaemia is characterized by widespread damage or destruction of various tissues, due to adsorption from the blood stream of pathogens or toxins derived from such pathogens.

Low levels of microbial contaminants are frequently found in association with many protein products during downstream processing. In some cases such microorganisms may have produced the protein of interest. In others, the microbial contaminants are introduced into the product from the general environment, from non-sterile processing equipment or from downstream processing personnel. The level of microbial contamination must be kept to a minimum by strict adherence to the principles of GMP.

In some circumstances the product may be subjected to one or several in process filtration steps. Samples of product may be withdrawn at various junctures during downstream processing and subjected to microbiological analysis in order to assess the microbial load or bioburden at that stage of the process. A decision to include a filtration step may be taken on the basis of this analysis. Agar plates are also usually placed at strategic working locations during downstream processing operations and are exposed under normal working conditions. Detection of a high microbial count on any such plate suggests that the product may be subject to microbial contamination at that point. Appropriate action must then be taken to identify, reduce or eliminate the source of such microbial contamination.

Viral contaminants

Traditional pharmaceuticals manufactured from chemical substances are usually considered free of viral contaminants. Biopharmaceutical products are more likely to contain such contaminants. It is conceivable that viruses may be introduced into the product during downstream processing from infected personnel and contaminated equipment. This, however, is unlikely if proper GMP procedures are followed. Viral particles most probably originate from infected source materials.

Products obtained from human tissue, blood or urine may contain a number of human viral contaminants (Table 4.9). The human immunodeficiency virus, HIV, and the hepatitis B virus are perhaps the two most likely contaminants, although a variety of other pathogenic viruses may be present, including human T-cell leukaemia viruses HTLV-I and HTLV-II, herpes simplex virus, human papillomavirus, cytomegalovirus or Epstein–Barr virus. In many cases proteins derived from human sources are purified from pooled raw material obtained from hundreds or thousands of individual donors. Inclusion of even one infected sample in the pooled material will render the entire batch infected. Under such

Table 4.9. Viral contaminants that may be present in source biological materials obtained from human volunteers

Virus	Medical significance
Human immunodeficiency virus (HIV)	Causative agent of AIDS
Hepatitis B virus	Causative agent of hepatitis B
Human T-cell leukaemia viruses (HTLV-I, HTLV-II)	Can cause leukaemia
Herpes simplex virus (HSV)	HSV-1 is the major causative agent of: herpetic stomatitis herpes labialis (cold sore) keratoconjunctivitis HSV-2 represents the major causative agent of genital herpes
Human papillomavirus	Causative agent of common wart Also implicated in anal canal carcinoma
Cytomegalovirus (CMV)	Generally symptomless: however, represents a serious opportunistic infection in immunocompromised individuals
Epstein–Barr virus (EBV)	Causes infectious mononucleosis Linked to Burkitt's lymphoma and nasopharyngeal carcinoma

circumstances, it is considered prudent to screen all donations individually prior to pooling.

Many proteins destined for parenteral administration are obtained from bovine sources. Examples include insulin and follicle stimulating hormone. In addition, bovine serum or fetal calf serum is often included in cell culture media. Many cell lines grown in such media produce proteins destined for parenteral administration. Hybridomas produce monoclonal antibodies of therapeutic value. Use of virally contaminated serum effectively contaminates the cell culture supernatant from which such therapeutic proteins must be purified.

As in the case of proteins derived from human sources, proteins obtained from bovine tissue are usually purified from a large number of pooled samples. The presence of even one infected sample thus contaminates all the starting material. Contaminant viruses most likely to be present in bovine-derived biological material include infectious bovine rhinotracheitis virus, bovine immunodeficiency virus, and the causative agents of diseases such as bovine spongiform encephalopathy (BSE).

Animal cell culture also represents an important source of some therapeutically beneficial proteins. Mammalian cell lines often used include murine hybridoma cell lines as well as Chinese hamster ovary cell lines. A wide range of murine viruses such as lymphocytic choriomeningitis virus may be potential contaminants of mono-clonal antibody preparations derived from murine hybridomas.

Elimination of viral contamination and the associated potential health risks to recipients may be most effectively achieved by utilizing raw material sources that are free from viral contamina-tion. As previously mentioned individual donor tissues should be tested in order to eliminate any diseased samples from subsequent purification procedures. Donor herds should have disease-free status. The development of animal cell culture that facilitates cellular growth in serum-free media represents an important breakthrough. This effectively eliminates the possibility of such cellular products being contaminated with bovine-derived viruses. The production of proteins of therapeutic interest in recombinant microbial systems, however, offers the greatest degree of assurance that the final therapeutic product will be free of clinically significant viral contaminants.

Two strategies are generally employed to reduce the risk of final product contamination by viruses. The first strategy involves the incorporation of specific steps capable of viral destruction in the downstream processing procedure. The second involves ensuring that one or more of the purification steps employed is effective in separating viral contaminants from the protein of interest.

It is not necessary to include specific viral inactivation steps in many purification procedures. Certain buffer components or other chemical additives used may inactivate a range of fragile virus particles. Specific inactivation steps which may be used include heat and irradiation. Maintenance of elevated temperatures (35–60 °C) for several hours has been shown to inactivate most viruses. Heat steps are used extensively to inactivate bloodborne viruses. Some more recent studies have investigated the effectiveness of treatment at high temperature for a short time. Ultraviolet irradiation has also been used as a method of inactivation. Several other methods continue to be evaluated. These include exposure to extremes of pH and multiple repeat filtration steps through 0.1 μm or 0.2 μm filters. Some of these methods may also have a detrimental effect on the protein of interest. Such a possibility should be investigated prior to adoption of a suitable protocol.

Many steps used in protein purification systems effectively separate viral particles from proteins. This is particularly true with regard to chromatographic fractionation steps. Viral particles and protein molecules exhibit widely differing physicochemical properties. In some instances viral contaminants may adsorb to the chromatographic matrix and are removed from the product stream. Such a possibility renders regular column CIP operations essential. In other instances, viral particles percolate through the column at different rates to those of the proteins and hence are separated from such proteins. Gel filtration offers a particularly effective viral clearance step.

Central to successful viral detection studies is the availability of sensitive assay systems. Most assays currently available detect a specific virus or a group of related viruses. In many cases, therefore, it becomes necessary to identify the viral particles whose potential presence is most likely in the source material and to assay specifically for those viral species.

A variety of assay types may be utilized. Some such assays use a virus-specific DNA probe. DNA probes are often used to identify certain human retroviruses such as HTLV-I or HTLV-II. Other viral assays rely on immunologically-based diagnostic tests such as enzyme-linked immunosorbent assay (ELISA) (Chapter 7). The human immunodeficiency virus is usually detected by ELISA, which detects the presence of antigens associated with HIV.

Some *in vitro* viral assays involve the incubation of potentially infected samples with a detector cell line sensitive to a range of human or murine viruses. Upon incubation, the sensitive cell line is examined for any associated cytopathic effects for 14 days or longer.

A range of mouse, rat or hamster antibody production tests (MAPs, RAPs or HAPs) may also be used to detect the presence of various viruses. In such systems the mouse, rat or hamster is injected with the test sample. Approximately 4 weeks later serum samples are collected and screened for the presence of antibodies recognizing a range of viral antigens.

Extensive validation studies must be carried out to ensure that the protein purification protocols used during downstream processing effectively inactivate any contaminant viruses, or separate such viruses from the final protein product. Validation studies normally involve spiking a sample of raw material from which the protein of interest is purified with a known amount of viral particles. The sample is then subjected to the complete purification procedure. The purification process is normally scaled down to laboratory size during such studies. The level of viral inactivation or clearance may be monitored after each purification step, and an associated viral reduction factor may thus be calculated.

Typically, a mixture of at least three different viral types are used during such validation studies. The viruses chosen depends upon a number of factors, including knowledge of the viral types most likely to be present in the source material. The chosen viruses should exhibit a range of physicochemical properties and should all be easily detected.

The levels of viruses used in spiking vary, however, quantities of up to 1×10^{10} viral particles are commonly used. The cumulative inactivation or removal effects of the scaled-down purification procedure should lead to a reduction in the viral population such that the likelihood of a single dose of final product containing a single viral particle should not be greater than one in a million.

Pyrogenic contaminants

Pyrogenic substances represent another group of potentially hazardous contaminants, whose presence in parenteral products is highly undesirable. Pyrogens may be described as substances which, when administered parenterally, influence the hypothalamic regulation of body temperature. Most pyrogenic substances when administered to humans cause a fever. Various chemicals and particulate matter can be pyrogenic. Contamination of finished products by such pyrogenic substances is avoided by implementation of comprehensive raw material and in-process quality control testing, and by strict adherence to the principles of good manufacturing practice. Most solutions destined for parenteral administration are passed through a $0.45\,\mu$m or $0.22\,\mu$m filter directly prior to filling into final product containers. Such filtration steps will remove particulate matter which otherwise could elicit a pyrogenic response in recipient patients. By far the most common pyrogenic substances likely to contaminate products are bacterially derived endotoxins. The presence of endotoxins is difficult to control owing to their ubiquitous nature.

Many microorganisms produce toxins which may be classified into one of two main categories: exotoxins and endotoxins. Exotoxins are proteins secreted extracellularly by microorganisms. Because of their proteinaceous nature, exotoxins are heat-labile, they are generally very potent and have defined modes of action. Diphtheria toxin for example, halts protein synthesis in susceptible cells by inactivating elongation factor 2, a factor required to sustain translation of mRNA. Botulinum toxin exerts its toxic effect by preventing the secretion of acetylcholine, which is essential for normal functioning of neuronal cells. Cholera toxin is an enterotoxin exerting its effects on the small intestine where it promotes permanent activation of adenylate cyclase, resulting in sustained elevated levels of cAMP. This results in massive fluid and electrolyte loss from affected intestinal cells into the intestinal lumen, causing diarrhoea. Most enterotoxins are exotoxins.

Endotoxins are lipopolysaccharide (LPS) substances originating from the cell wall of Gram-negative bacteria. Unlike the cell wall of Gram-positive organisms, which consists of a thick protective layer of peptidoglycan, the cell envelope of Gram-negative bacteria consists of an underlaying layer of peptidoglycan covered by an outer membranous structure. This outer membrane is composed mainly of phospholipids, lipopolysaccharides and proteins. The lipopolysaccharide, which is the endotoxin, consists of a core polysaccharide attached to a lipid A moiety which serves to anchor the structure in the membrane, and polysaccharide O antigens, which project outward from the membrane surface. The core polysaccharide usually contains a variety of seven-carbon sugars, heptoses, in addition to hexoses such as galactose, glucose and N-acetylglucosamine. The O polysaccharides also contain a variety of hexoses such as glucose, galactose and mannose, in addition to more unusual dideoxy sugars such as abequose and paratose. The lipid structure does not contain glycerol and the fatty acids are linked by ester linkages to N-acetylglucosamine. It is believed that the lipid structure component is mainly responsible for the toxic properties of the lipopolysaccharide molecules while the polysaccharides render the molecule water-soluble. It must also be stressed that not all lipopolysaccharide molecules found in association with Gram-negative bacteria exhibit toxic effects.

Lysis of Gram-negative bacteria results in the liberation of free endotoxin molecules. The presence, and in particular lysis of Gram-negative bacteria during downstream processing may result in contamination of the product with large quantities of endotoxin. Endotoxins are less potent than most exotoxins and relatively large quantities of endotoxin-contaminated product must be administered before adverse clinical symptoms arise. It is therefore particularly important that large-volume parenteral products such as infusion fluids be free from endotoxin contamination, as endotoxins elicit a number of undesirable clinical responses.

One of the most notable responses is that of fever, i.e. a pyrogenic response. This effect is indirect. Endotoxin serves as a potent stimulant to interleukin 1 (IL-1) production by macrophages. IL-1 is also known as endogenous pyrogen and initiates the fever response. Endotoxin is also a potent activator of the complement system, and is considered to be instrumental in the development of septic shock. This may be triggered at least in part by injected endotoxin, though it is often initiated in response to the presence of high circulatory levels of endotoxin due to Gram-negative bacteraemia (the presence of Gram-negative bacteria in the blood). Septic shock is a particularly serious medical condition, which in severe cases results in the patient's death (Figure 4.4).

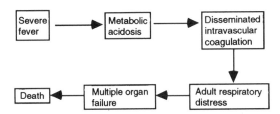

Figure 4.4. Clinical stages associated with septic shock

The presence of pyrogenic substances in parenteral preparations may be detected by a number of methods. Historically, the rabbit assay constituted the most widely used method. This test involves parenteral administration of a sample of the product in question, with subsequent monitoring for changes in the rabbit's temperature. While this test is capable of detecting a wide range of pyrogenic substances, it suffers from a number of limitations. The process is laborious, space- and time-consuming and requires experienced animal technicians. Excitation of the rabbits used can affect the experimental results obtained. Large-scale rabbit testing using different rabbit colonies is also subject to variations, which can lead to the possibility of variable standards. The rabbit test has in most cases been replaced by an *in vitro* test based on endotoxin-stimulated coagulation of amoebocyte lysate obtained from horseshoe crabs.

It has been known since the mid-1950s that Gram-negative bacterial infections cause intravascular coagulation in the American horseshoe crab, *Limulus polyphemus*. The factor responsible for coagulation exists within the crab's circulating blood cells (termed amoebocytes). The presence of bacterial endotoxin results in activation of this clotting agent. Scientific investigations have revealed that endotoxin activates a coagulation cascade, not unlike the coagulation cascade system found in higher animals. Activation of the cascade is dependent not only upon the presence of endotoxin, but also on the presence of divalent cations such as magnesium or calcium. The terminal stage of the cascade system (Figure 4.5) involves the proteolytic conversion of an inactive proenzyme to an activated clotting enzyme. This enzyme catalyses the proteolytic cleavage of the clotting protein, coagulogen, forming coagulin and free peptide C. Free coagulin molecules interact non-covalently resulting in clot formation.

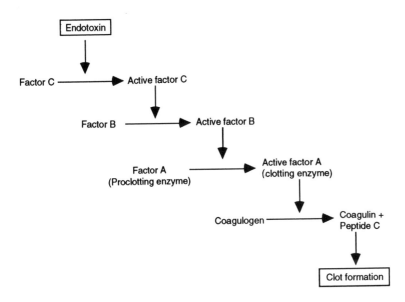

Figure 4.5. Endotoxin-mediated intravascular coagulation as occurs in the American horseshoe crab. The lysate of the crab's amoebocytes contains all the constituents of the coagulation cascade. Incubation of such a lysate with a test sample will result in coagulation (gel formation) if such a test sample contains endotoxin. If the sample is devoid of endotoxin coagulation is not observed

The *Limulus* amoebocyte lysate (LAL) assay is based upon this coagulation cascade. LAL reagent is prepared by extracting and washing *Limulus* amoebocytes, followed by induction of cellular lysis. The *Limulus* lysate, which is commercially available, is incubated with the test preparation in pyrogen-free test tubes, usually for a period of 1 hour. Any endotoxin present in such test samples will result in activation of the coagulation cascade with resultant gel or clot formation in the test tube.

The LAL assay is widely used to detect the presence of endotoxin in parenteral products, in bulk reagents such as WFI used in the manufacture of parenterals and in biological fluids such as serum or cerebrospinal fluids. The popularity of the test is a reflection of its sensitivity, specificity and reproducibility. The LAL test has largely replaced the rabbit test as the method of choice for detection of endotoxin in water for injections and finished product parenteral solutions, and is specified by many international pharmacopoeias.

The sensitivity of the LAL test remains unsurpassed as it can detect a few picograms of endotoxin per millilitre of test sample. Its extreme specificity, however, is cited by some as a disadvantage.

The LAL assay will not detect non-endotoxin pyrogenic substances, but additional miscellaneous substances that generate a pyrogenic response should be adequately controlled by implementation of GMP.

Validation of the LAL assay system, as applied to any biopharmaceutical process, must also be undertaken. It is particularly important to show that gel formation is not inhibited by any constituents present in the sample being tested. This may be achieved by spiking the test sample with known quantities of endotoxin. The presence of inhibitory substances would prevent clotting.

The LAL assay system has been adapted to a chromogenic test format. This involves incubation of the test sample with amoebocyte lysate and an additional substrate, normally a short peptide to which a chromogen has been attached. Activation of the coagulation cascade by endotoxin results in the proteolytic degradation of the substrate, with release of free chromogen. Unlike the bound chromogen, the free chromogen absorbs specific wavelengths of visible light. The level of absorption reflects the quantity of endotoxin present in the test sample. This method is therefore quantitative. Colour formation is rapid and the test may be completed in 15 minutes.

Endotoxin may be destroyed or removed from process equipment, or product by a number of means. However, lipopolysaccharides are heat-stable and are not easily destroyed by thermal treatment. Depyrogenation of heat-resistant process equipment and test tubes used in LAL assays may be achieved using dry heat (180 °C for 3 hours, or 240 °C for 1 hour) or moist heat (three consecutive autoclave cycles).

Chromatographic columns are normally depyrogenated by exposure to NaOH during CIP procedures, or by exhaustive rinsing with pyrogen-free water or buffer until the eluate is shown to be pyrogen-free.

Endotoxin levels in the final parenteral product may be minimized by adherence to the principles of GMP. Product exposure to bacterial contaminants during downstream processing should be minimized. This is most effectively achieved by

- Ensuring that all relevant purification procedures are undertaken in areas of controlled environmental conditions, such as clean rooms.
- Ensuring that equipment coming into contact with the product is cleaned, sanitized and, if possible, sterilized before commencement of the operations.
- Preventing bacterial build-up in the process during various stages of downstream processing. Particular care must be taken with regard to endotoxin control if the protein of interest is sourced from a Gram-negative bacteria.

Endotoxin molecules exhibit physicochemical properties that differ greatly from the physicochemical characteristics of most proteins. For this reason, many protein fractionation techniques result in effective separation of endotoxin from the protein of interest. Inclusion of a specific endotoxin removal step as part of a downstream processing protocol is unnecessary in many cases. Specific removal steps, if required, normally take advantage of the charge or molecular weight characteristics of the endotoxin molecule. While individual molecules of lipopolysaccharide exhibit a molecular weight of less than 20 kDa, such molecules aggregate in aqueous conditions giving rise to supramolecular structures of 100–1000 kDa. Gel filtration chromatography may thus effectively separate such endotoxin molecules from the protein of interest if the molecular weight of the protein in question is below 100 kDa. Conversely, if the protein of interest has a molecular weight greater than 100 kDa, chelating agents such as EDTA may be included in the buffers used, as such agents promote depolymerization of endotoxin molecules to the monomeric form. Under such conditions, gel filtration may be used to separate endotoxin from protein. Separation on this basis may also be achieved using an ultrafiltration membrane of appropriate pore size.

Lipopolysaccharides possess a high negative charge and thus most ion-exchange steps will separate endotoxin from protein molecules. Endotoxin may also be removed from bulk solutions such as water by passing the solution through a membrane filter exhibiting a positive surface charge. This depyrogenation method is unsuitable if the solution contains other negatively-charged important molecules in addition to endotoxin.

DNA contaminants

The significance of DNA as a contaminant in a final product preparation destined for parenteral administration remains unclear. Theoretically, entry of contaminant DNA into the genome of recipient cells, if such a process could occur, could have serious clinical implications. Such implications could include alteration of the level of expression of cellular genes, or expression of a foreign gene product. Many cell lines used to produce certain proteins, such as hybridomas producing monoclonal antibodies, are known to contain active oncogenes. Although health risks associated with the presence of naked DNA in parenteral preparations is considered at worst to be minimal, the presence of oncogenes in an injectable product would be considered inappropriate.

In many cases there is little need to incorporate specific DNA removal steps in downstream processing procedures. Nucleic acids present in crude cell extracts are often degraded by endogenous nucleases. Exogenous nucleases can be added if considered appropriate. Nuclease treatment will result in degradation of nucleic acids yielding nucleotides, which are much less of a potential hazard than DNA segments corresponding to intact genes. Most protein purification steps subsequently remove nucleotides from product-containing fractions. Nucleic acid molecules may also be precipitated from solution by addition of positively-charged polymers such as polyethyleneimine. Such an approach is possible as an alternative, or in addition, to nuclease treatment.

Most chromatographic steps will result in effective separation of DNA from proteins, owing to major differences in their physico-chemical properties. As in the case of endotoxin, the strong negative charges ensure that an ion-exchange step is particularly effective in the removal of DNA contaminants from the product stream. Levels of contaminating DNA are normally measured using a species-specific DNA hybridization assay. If the protein of interest is purified from *E. coli*, total *E. coli* DNA is radiolabelled and used as a probe to detect any *E. coli* DNA contaminating the product.

DNA validation studies are normally carried out in order to illustrate that the downstream purification process used is capable of reducing the level of DNA in the final product to within acceptable limits. The definition of what constitutes a suitable upper limit is somewhat arbitrary, but quantities of up to 10 pg of residual DNA per therapeutic dose are generally considered acceptable.

DNA validation studies are carried out by methods very similar to those of viral validation studies. This involves spiking a raw material sample with a known amount of DNA and applying this to a scaled-down version of the proposed purification system. The DNA distribution profile may be monitored after each purification step, together with the fractionation behaviour of the protein of interest. In this way, a DNA reduction factor may be calculated.

DNA obtained from the same source as the protein of interest is normally used in validation studies. DNA from other sources may sometimes be used in model studies, as DNA from different sources should behave in an essentially identical manner. The DNA employed to spike samples may itself be radiolabelled, allowing direct detection. Alternatively, the DNA applied may be unlabelled, and in such circumstances a specific labelled DNA probe must be used to detect the spiking material. The actual quantity of DNA used must be carefully considered, as excess DNA may itself adversely affect the purification procedure. Large quantities of exogenous DNA added to the sample could, for example, bind anion exchange resins and therefore significantly reduce their capacity to bind other molecules. The physical characteristics of the DNA employed should also receive some thought, as DNA molecules present in any given starting sample may vary widely in terms of molecular weight. It may thus be considered prudent to carry out validation studies using DNA for spiking that exhibits a wide range of molecular weights.

Protein contaminants

The majority of purification steps included in downstream processing protocols are designed specifically to fractionate different protein molecules from one another. Precipitants such as neutral salts, ethanol and organic solvents fractionate proteins on the basis of differential solubilities. Chromatographic steps are utilized which separate proteins on the basis of size, charge, isoelectric points, hydrophobicity, immunological properties or some other parameter of molecular distinction. Such techniques frequently result in effective separation of the protein of interest from non-proteinaceous contaminants such as viral particles or endotoxins.

Despite the availability of a wide range of fractionation techniques, it remains a difficult task to purify a specific protein to homogeneity, while obtaining an economically viable yield. The range of potential protein contaminants is dependent upon the source of the protein, its method of production and the downstream processing procedures used.

Proteins (other than the protein of interest) present in the raw material may be termed endogenous proteinaceous contaminants. If the protein of interest is intracellular, cell lysis will liberate the entire intracellular contents. This results in the production of a crude cell homogenate, usually containing many thousands of different proteins, from which the single desirable protein type must be purified. If the protein is extracellular, the subsequent purification task is usually rendered less complex as there are fewer proteins in the extracellular medium.

In addition to protein contaminants emanating from the source material, several other proteins may be introduced during the production stages or purification procedures. For example, animal cell culture media is usually supplemented with either fetal calf serum, or with a defined cocktail of proteins, containing elements such as bovine serum albumin, transferrin, growth factor and various hormones such as insulin. Such exogenous proteins can add further to the overall level of protein contamination.

Several additional proteinaceous contaminants may be introduced during downstream processing. Nucleases are frequently added to cell homogenates in order to degrade contaminating nucleic acids. The presence or introduction of microbial cells during downstream processing may lead to contamination of the product stream with foreign proteins derived from such sources. Implementation of GMP should, however, minimize introduction of such microbially-derived contaminants. Other foreign proteins such as protein A or G may leach from affinity columns into the product stream. Immunoaffinity columns may result in contamination of product with immunoglobulin molecules. Affinity chromatography is often characterized by a low level of leakage of immobilized ligands from the affinity matrix; however, innovations in this area continue to decrease the magnitude of this problem.

Modified forms of the protein product itself may also be considered as impurities. All proteins are susceptible to a variety of modifications which may alter their biological activity or immunological characteristics (Table 4.10). In many instances, such altered products are generated during downstream processing and their effective separation from the intact parental molecules can prove quite difficult. Aggregated or extensively degraded molecules may be removed from the intact product by gel filtration chromatography, as can some molecules with extensively altered glycosylation patterns. Oxidized or deaminated proteins may be separated by techniques such as ion-exchange chromatography or isoelectric focusing.

Table 4.10. Some modifications to which most proteins are susceptible. Many such modified forms of proteins exhibit altered biological characteristics and hence must be removed from the final protein product

Aggregation

Oxidation of methionine residues

Deamination of asparagine/glutamine residues

Incorrect disulphide bond formation

Proteolysis by endo- or exo-acting proteases

Enzymatic alteration of post-translational modifications (glycosylation)

As previously outlined in Chapter 2, many heterologous proteins expressed intracellularly in *E. coli* at high concentrations form insoluble complexes termed inclusion bodies. Inclusion bodies consist of partially folded molecules of the recombinant protein. Recovery of active protein from such inclusion bodies firstly necessitates their solubilization by suitable denaturants. Subsequent removal of the denaturant generally allows refolding of the protein molecules, yielding their native, active form. Invariably, some of these protein molecules will undergo incorrect folding, in effect generating contaminant molecules.

The major clinical significance of protein impurities relates to their antigenicity. Some protein contaminants may also display a biological activity deleterious either to the protein product or to the recipient patient. The potential for adverse immunological reactions depends not only upon the immunogenicity of the product or contaminants but also upon the route of administration and in particular upon the frequency of administration. Contaminating proteins also act as adjuvant-like materials, thereby further increasing the immunogenicity of the protein. Immunological complications are less likely if a particular therapeutic product is administered as a "one-off", in contrast to a product such as insulin, which must be repeatedly administered to the patient.

Upon their administration all injected proteins run the risk of eliciting an immune response. Some protein vaccines or toxoids are administered specifically for this reason. Many therapeutically important proteins are derived from human sources and as such are non-immunogenic. Other proteins often share extensive structural homology with the natural human product. Porcine insulin, for example, differs from human insulin by only one amino acid residue. Bovine insulin differs from human insulin by three amino acids. As a result, administration of bovine and in particular porcine insulin normally does not elicit a strong immunological response.

Different protein molecules differ in their intrinsic ability to stimulate an immune response. It is not yet possible to predetermine if any particular protein will initiate a strong immunological reaction. In general, however, the less homology exhibited between the protein of interest and the analogous human protein, the greater the possibility that an immunological reaction will be observed. Larger polypeptides or proteins also tend to exhibit increased immunogenicity, as do proteins exhibiting extensive post-translational modifications which differ from modifications observed in their human counterparts.

Immunological reactions may be targeted against the product itself or against contaminants. As summarized above, the more closely related the product to the human protein, the less likely it is to elicit such a response. It follows that no immunological response should be initiated if the product contains no contaminants and is either purified from a human source or is identical to the human protein. Production of a product with such desirable properties is rarely, if ever, achieved. In practice it is extremely difficult to purify a protein to absolute homogeneity while obtaining a yield which renders the process economically viable. Furthermore slight modifications of the protein itself may occur during downstream processing or during subsequent storage. Such modified molecules are often immunogenic.

Purification of most therapeutically important proteins from human sources is an unrealistic option. In theory, however, a product identical to the native human protein may be produced in recombinant systems into which the relevant human gene or cDNA has been inserted and subsequently expressed. In practice this may not be the case. Many if not most human proteins of therapeutic value are subjected to post-translational modifications. Bacterial systems are unable to reproduce such post-translational modifications. While heterologous yeast systems are capable of carrying out certain modifications, such modifications may not be identical to those associated with the native human protein. In addition to this, all proteins produced by bacteria contain an *N*-formyl-methionine as their *N*-terminal amino acid residue. Thus, any human protein expressed in a bacterial system will contain such a modified methionine residue. Production of heterologous proteins in *Aspergillus* currently offers a variety of post-translational modifications most closely resembling those of mammalian proteins.

Immunological responses to administered proteins are clinically undesirable for a number of reasons, unless the product is to be used as an immunogen. In cases where repeat administrations of product are required, antibodies may be raised against the "foreign" protein. These antigen-specific antibodies may decrease or nullify the potency of the product. This effectively means that the patient develops clinical resistance to the protein drug, or that with time increasing quantities of the usually expensive protein must be administered in order to sustain a particular level of clinical responsiveness. Binding of antibody to the therapeutic protein administered may also distort the dose response curve. While initial binding of antigen by antibody may decrease the product's apparent potency, the reversibility of antibody–antigen binding usually results in a slow, sustained release of free protein in the blood stream. Binding of the therapeutic protein by antibodies may also affect normal degradation of the product. Many hormones such as insulin are removed from general circulation by a process of receptor-mediated endocytosis, and binding of antibody may severely restrict this uptake mechanism. Patients may also develop allergic responses to the particular product or to impurities consistently found in association with the products. Such allergic reactions or hypersensitivity generally result from the action of IgE antibodies or from T-cell mediated cellular toxicity.

Determination of final product purity levels is only as accurate and sensitive as the analytical detection technique used. Traditionally, the technique most often utilized to detect protein contaminants is that of polyacrylamide gel electrophoresis. Protein-binding stains used in the detection of the proteins include coomassie brilliant blue, or PAGE blue. Such stains rarely detect protein bands containing less than microgram quantities of protein, hence they are rarely sufficiently sensitive to detect protein contaminants which may be of clinical significance. Staining with highly sensitive silver-based stains may increase the sensitivity of detection several hundredfold. Silver staining is thus more appropriate when endeavouring to detect low levels of proteins which may be of clinical significance.

Chromatographic methods, often HPLC, may also be utilized in assessing the level of purity of a protein preparation. Again however, this method is not sufficiently sensitive to detect all traces of clinically important contaminants that may be present. It is considered that contaminant levels as low as 10 ppm of an average therapeutic dose may be sufficient to initiate an immunological response.

Immunological assay techniques offer the most sensitive methods of detecting and quantifying protein impurities. Specific radio-immunoassays or enzyme-linked immunosorbent assays may be used to screen for the presence of known or likely contaminants. The detection of unknown proteins by immunological methods requires a more generalized approach. One such approach often adopted when the protein of interest is produced by recombinant DNA methods is that of a "blank run". This involves constructing a host organism lacking the genetic information encoding the desired product, but which is identical to the producing organism in all other respects. Crude extracts derived from this organism are then subjected to at least part of the purification protocol for the protein of interest. As it is devoid of the actual product, the partially purified material should contain the range of contaminants which may copurify with the protein of interest. Antibodies are then raised against this partially purified material and these serve to screen for contaminants in the actual product preparation.

Another analytical method currently receiving much attention is that of mass spectrometry. Various technical developments have rendered this technique suitable for the analysis of high molecular weight molecules such as proteins. This highly sensitive technique may now be used to estimate the molecular weight of a 100 000 Da protein with an accuracy of 0.01%.

One point sometimes overlooked is that many protein contaminants, be they modified versions of the protein of interest or different protein molecules, may be completely innocuous. Many would suggest that the presence of trace amounts of protein contaminants does not by itself automatically justify introduction of additional purification steps in order to achieve their removal. Preclinical and clinical trials are perhaps the only real means by which the medical significance of any such trace contaminants may be properly assessed.

Chemical and miscellaneous contaminants

A variety of miscellaneous contaminants are often associated with the product stream at various stages of purification. Some such contaminants, such as minor levels of lipid or polysaccharide, may be derived from the producing cells. The majority of miscellaneous contaminants, however, are introduced from exogenous sources. Examples of such potential exogenous contaminants are listed in Table 4.11. The nature of the purification procedure largely dictates which (if any) such contaminants might be present in the product stream. In some instances contaminants introduced during the initial stages of purification may subsequently be removed by one or more of the later purification steps. In other cases contaminants may copurify with the protein of interest. Under such circumstances it may be necessary to incorporate an additional purification step in order to achieve effective removal of the contaminants concerned. It is particularly important to ensure that all traces of potentially toxic, carcinogenic or otherwise unsafe contaminants are removed from the finished product. Final product containers of suitable quality should be used in order to reduce or eliminate the risk of leaching of chemical or other substances into the product during storage.

Table 4.11. Miscellaneous contaminants which may be introduced to the product stream

Buffer components	Detergents
Chaotrophic agents	Ethanol
Proteolytic inhibitors	Stabilizers
Salts	Ligands and other chromatographic breakdown products
Glycerol	
	Chemicals and ions leached from process equipment and pipework
Antifoaming agents	

Many chemical contaminants such as buffer components and ligands are low molecular weight compounds. In such instances, gel filtration or ultrafiltration may effectively remove this material. Higher molecular weight contaminants are often more difficult to separate from the product. Methods of removing such contaminants may be logically chosen if the physicochemical properties of both contaminant and protein product are known.

In some instances, for example if a particular likely contaminant is known to be toxic, it may become necessary to validate the purification system with regard to its ability to remove a specific contaminant. This is best achieved by spiking a sample of the material at the stage of purification where the putative contaminant is introduced, and subjecting this spiked sample to the remainder of the purification steps. An appropriate assay capable of detecting the contaminant in question must be available.

As an additional safety measure finished products are often subjected to abnormal toxicity or general safety tests. Test protocols are outlined in various pharmacopoeias such as the *European Pharmacopoeia* or the *United States Pharmacopoeia*. Safety tests normally entail the intravenous administration of a dose of up to 0.5 ml of product to at least five healthy mice. The animals are then observed for a period of 48 hours, and should exhibit no adverse symptoms other than the symptoms expected. Death of one or more of the test animals is considered to signal further analysis and in such instances the test may be repeated using a larger number of animals. Safety tests are undertaken in order to detect any unexpected or unacceptable biological activities associated with the product concerned.

FURTHER READING

Books

EEC Guide to Good Manufacturing Practice for Medicinal Products (1989). Commission for the European Communities, Brussels.

Harris, E. & Angal, S. (1990). *Protein Purification, Applications, a Practical Approach*. IRL Press, Oxford.

Martindale, *The Extra Pharmacopoeia* 29th edn. (1989), Pharmaceutical Press, London.

The British Pharmacopoeia 15th ed. (1993). HMSO, London

United States Pharmacopoeia XXII (1990). United States Pharmacopoeial Convention, USA.

Articles

Anicetti, V. *et al.* (1989). Purity analysis of protein pharmaceuticals produced by recombinant DNA technology. *Trends in Biotechnol.*, **7**, 342–348.

Biotechnology Medicines in Development (1991). *Survey Report*. Pharmaceutical Manufacturers Association, Washington, DC, USA.

Fox, J.L. (1987). The US regulatory patchwork. *Bio/Technology*, **5**, 1273–1278.

Ganzi, G. (1984). Preparation of high purity laboratory water. *Methods Enzymol.*, **104** part c, 391–403.

Geisow, M.J. (1991). Characterizing recombinant proteins. *Bio/Technology*, **9**, 921–924.

Hanna, L. *et al.* (1992). Removing specific cell culture contaminants in MA6 purification process. *Pharm. Tech. Int.*, **4**, 34–39 (originally appeared in *BioPharm.* (1991), **4**, 9, 33–37).

Hodgson, J. (1993). Pharmaceutical screening: from off-the-wall to off-the-shelf. *Bio/Technology*, **11**, 683–688.

Illum, L. (1991). The nasal delivery of peptides and proteins. *Trends in Biotechnol.*, **9**, 284–288.

Kahn, D. (1990). Automatic flexible aseptic filling and freeze drying of parenteral drugs. *Pharm. Tech. Int.*, **2**, 41–46.

Konrad, M. (1989). Immunogeneity of proteins administered to humans for therapeutic purposes. *Trends in Biotechnol.*, **7**, 175–178.

Poggiolini, D. & Donawa, M.E. (1990). EEC pharmaceutical regulation: the multistate procedure and CPMP. *Pharm. Tech. Int.*, **2**, 21–25.

Rech, M. *et al.* (1991). Current trends in facilities and equipment for aseptic processing. *Pharm. Tech. Int.*, **3**, 48–52 (originally appeared in *Pharm. Technol.* (1991) **15**, 54–60).

Tai, J. & Liu, T.Y. (1977). Studies on *Limulus* amoebocyte lysate. *J. Biol. Chem.*, **252**, 2178–2181.

Thornton, R.M. (1990). Pharmaceutically sterile clean rooms. *Pharm. Tech. Int.*, **2**, 26–29 (originally appeared in *Pharm. Technol.* (1990), **14**, 2, 44–48).

Wallace, B. & Lasker, J. (1993). Stand and deliver: getting peptide drugs into the body. *Science*, **260**, 912–913.

Whyte, W. *et al.* (1992). Suggested modifications to clean room air standards of the E.C. Guide to GMP. *Pharm. Tech. Int.*, **4**, 17–20.

Chapter 5

Therapeutic proteins: blood products, vaccines, monoclonal antibodies and related substances

- Blood products.
  - Whole blood and blood plasma.
  - Blood coagulation factors.
  - Non-hereditary coagulation disorders.
  - Anticoagulants.
  - Thrombolytic agents.
  - Human serum albumin.
  - Alpha-1-antitrypsin.
  - Haemoglobin.
- Polyclonal antibody preparations.
- Vaccines.
  - Towards a vaccine for AIDS.
- Adjuvant technology.
- Monoclonal antibodies.
- Further reading.

The human vascular system generally contains 5–6 litres of blood. Whole blood consists of red blood cells, white blood cells and platelets, all of which are suspended in a fluid termed plasma. Plasma may be obtained by centrifugation of whole blood following the addition of a suitable anticoagulant to freshly drawn blood. The anticoagulant prevents clotting of the blood while the centrifugation step removes the cells suspended in the plasma. If blood is allowed to clot, the clot exudes a fluid termed serum. The clot consists of suspended cellular elements and platelets entrapped or enmeshed in an extensive cross-linked network of fibrin molecules. Fibrin is derived from fibrinogen—a plasma protein. Plasma is essentially serum in which fibrinogen is present.

The various components of blood serve a wide range of physiological functions within the body. Blood transports a wide range of substances within the body, such as nutrients, waste products, gases, antibodies, enzymes, parenterally administered substances, hormones and other regulatory factors. Blood also plays a vital role in several additional physiological processes, such as maintenance of tissue hydration levels and regulation of body temperature.

Many specific functions of blood are carried out by proteins found in plasma. Electrophoretic separation of plasma proteins reveals five major bands referred to as albumin, α_1-globulins, α_2-globulins, β-globulins and γ-globulins. Serum albumin represents the most abundant protein, accounting for more than 50% of total plasma protein. All globulin protein fractions contain a variety of different protein molecules.

BLOOD PRODUCTS

Specific blood proteins that are of therapeutic use include a range of factors involved in the blood clotting process, fibrinolytic agents which degrade clots, serum albumin and immunoglobulin preparations. Most such proteins in addition to whole blood and blood plasma have been available for many years. While these products have traditionally been obtained from blood donated by human volunteers, some are now produced by recombinant DNA technology. Recombinant DNA techniques are likely to have a major impact upon methods of production of existing blood products, in addition to making available other useful blood proteins which were unavailable in the past. Besides providing an alternative, inexhaustible and hopefully inexpensive source of blood proteins, recombinant DNA techniques will allow production of products free from infectious agents. This represents a major step forward, as all current blood donations must be screened for the presence of a range of potential infectious agents including the hepatitis B virus and the virus causing AIDS.

Whole blood and blood plasma

Whole blood is aseptically collected from human donors and is immediately mixed with an anticoagulant to prevent clotting. Suitable anticoagulants include heparin and sodium citrate. Whole blood may be used as a source of a variety of blood constituents (Table 5.1). The use of whole blood for such purposes would be considered wasteful if the specific purified component required is already available. Furthermore, fractionation procedures used to

Table 5.1. Therapeutic constituents of whole blood. Whole blood may be administered as a source of various specific blood constituents such as those listed

Red blood cells
Platelets
Various clotting factors
Immunoglobulins
Additional plasma constituents

produce specific purified blood products considerably reduce the risks of accidental transmission of disease from contaminated blood donations.

Blood obtained from donors must be screened for the presence of a variety of likely pathogenic contaminants, in particular HIV and hepatitis B viral species. Donated blood samples are therefore usually screened individually for the presence of hepatitis B surface antigen and the presence of anti-HIV antibodies. Whole blood is normally administered to patients following severe blood loss. Concentrated red blood cell fraction and plasma reduced blood is also available and may be administered for certain clinical conditions.

Blood plasma is prepared from whole blood by centrifugation and is then normally stored frozen until use. Plasma is normally employed clinically as a source of therapeutically important plasma proteins such as clotting factors, when the purified products are unavailable.

Yet another widely used fraction of whole blood is termed plasma protein solution or plasma protein fraction. This aqueous solution is normally prepared by limited fractionation of serum or plasma and consists predominantly of serum albumin with minor quantities of globulin proteins. Plasma protein solution is administered in cases of shock caused by a large decrease in the volume of blood. Such sudden blood loss may be as a result of internal or external bleeding, extensive burns or dehydration.

Blood coagulation factors

A variety of plasma proteins form an integral part of the blood clotting process. A genetic or induced deficiency in any one blood factor results in severely impaired coagulation ability. The vast majority of hereditary diseases characterized by poor coagulation responses result from a deficiency of blood factors VIII and IX. These blood factors have traditionally been purified from the blood of healthy human donors and subsequently administered to patients. Many such factors can be and are being produced by recombinant DNA methodologies. A variety of other non-hereditary clinical conditions, such as vitamin K deficiency, may also result in impairment of the blood coagulation process.

When a blood vessel is damaged or cut, specific elements in blood initiate the process of haemostasis—the curtailment and eventual cessation of blood loss. Haemostasis is characterized by the rapid attachment of platelets to the damaged area. Platelets also adhere to each other, and in this way often stem blood flow. The congregated platelets also secrete a variety of amines, such as adrenaline, which stimulate localized constriction of the blood vessels. The process of blood coagulation is also initiated, resulting in the formation of a blood clot or a thrombus at the site of damage.

The process of blood coagulation is dependent upon a number of associated blood clotting factors. Such factors are designated by Roman numerals, although each is also known by a common name. Blood factors are listed in Table 5.2. With the exception of factor IV (calcium ions), the factors are proteinaceous in nature.

Table 5.2. Blood clotting factors

Factor number	Common name
Factor I	Fibrinogen
Factor II	Prothrombin
Factor III	Thromboplastin (tissue factor)
Factor IV	Ca^{2+} (calcium ions)
Factor V	Labile factor (proaccelerin)
Factor VI*	
Factor VII	Proconvertin
Factor VIII	Antihaemophilic factor (von Willebrand factor)
Factor IX	Christmas factor (plasma thromboplastin component)
Factor X	Stuart factor
Factor XI	Plasma thromboplastin anticedent
Factor XII	Hageman factor
Factor XIII	Fibrin stabilizing factor

Activated forms of the above factors are designated by the addition of the letter "a" to the factor number, e.g. factor VIIa is activated factor VII.
*The protein originally termed factor VI was discovered to be factor Va, thus factor VI is now unassigned.

Many of the blood factors, for example factors II, VII, IX, X, XI and XII, exhibit proteolytic activity upon activation. The unactivated factors are protease zymogens. Activated (proteolytic) factors catalyse the proteolytic cleavage of the next factor in the clotting sequence, resulting in its activation. The clotting sequence may thus be regarded as a molecular cascade in which sequential activation of clotting factors is observed. Each single activated molecule will, in turn, catalyse the activation of numerous molecules of the next factor in the sequence. This results in considerable stepwise amplification of the initial signal.

Not all of the clotting factors listed in Table 5.2 exhibit proteolytic activity. The non-proteolytic accessory factors, factors III, IV, V and VIII, form essential components of the coagulation system. They are generally activated by one of the proteolytic factors, and upon activation, serve to enhance the rate of activation of other blood factor zymogens. The presence of certain phospholipid components released from damaged tissue or from platelets also serves to accelerate the rate of coagulation.

Two distinct blood coagulation pathways exist—the intrinsic and extrinsic pathways. The intrinsic pathway relies upon factors that are normally present in plasma. This pathway is initiated when factor VII is activated through contact with injured surfaces. Functioning of the extrinsic pathway requires, in addition to blood factors, the presence of tissue factor (factor III). Tissue factor, along with calcium (factor IV), factor VII and phospholipid, greatly stimulates activation of factor X. Tissue factor is an accessory protein present in a wide range of tissue types. It is particularly abundant in saliva in addition to lung and brain, and is an integral membrane glycoprotein. Tissue factor is released upon tissue damage along with phospholipid components of the membranes. Hence it can initiate the extrinsic coagulation cascade at the site of damage. The clotting process occurs most rapidly if initiated via the intrinsic pathway. While the initial activation sequences of these pathways differ, the terminal sequence of events is identical in both cases (Figure 5.1). These final (common) steps of the coagulation cascade involve the conversion of prothrombin (factor II) to thrombin (factor IIa). This proteolytic reaction is catalysed by activated Stuart factor (Xa).

Thrombin catalyses the proteolytic cleavage of soluble fibrinogen, yielding insoluble fibrin. Fibrin monomers subsequently interact forming a fibrin clot. Initially this interaction is of a non-covalent nature, yielding a large, aggregated, 'soft' clot. This soft clot is subsequently converted into a hard clot by factor XIII, also known as fibrin stabilizing factor (FSF). Activated factor XIII catalyses the formation of covalent cross-links between a lysine residue of one fibrin monomer and a glutamine residue

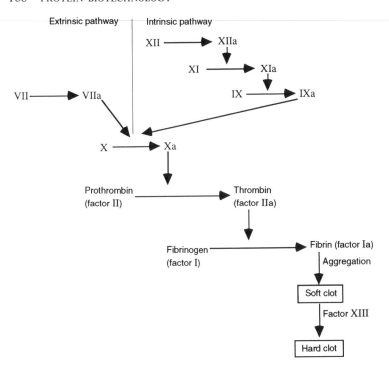

Figure 5.1. Simplified version of the intrinsic and extrinsic blood coagulation pathways

of an adjacent fibrin molecule. Factor XIII is itself activated by thrombin in the presence of calcium.

Genetic defects that significantly decrease the level of production or alter the amino acid sequence of any blood factor may result in serious illness, characterized by poor coagulational ability, with resulting prolonged haemorrhage. Defects in all clotting factors with the exception of tissue factor and calcium have been documented and characterized. Well in excess of 90% of all such defects relate to a deficiency of factor VIII. Many of the remaining cases are due to a deficiency of factor IX. The clinical disorders associated with deficiencies of factors VIII or IX include haemophilia A, von Willebrand's disease (vWD) and haemophilia B.

Factor VIII complex is considered to consist of two separate gene products. The smaller polypeptide exhibits coagulant activity and is often designated VIII:C. This polypeptide is coded for by the factor VIII gene. The larger polypeptide, designated von Willebrand factor (VIII:vWB), is predominantly associated with platelet adhesion. This factor is coded for by the vWB gene. Upon synthesis, individual von Willebrand factor polypeptides polymerize forming large multimeric structures. The product of the factor VIII gene (VIII:C polypeptide) then associates with the multimeric VIII:vWB protein, forming the overall complex structure (VIII:C and VIII:vWB which may be copurified from plasma).

Failure to synthesize VIII:C results in classical haemophilia (haemophilia A), while failure to synthesize VIII:vWB results in von Willebrand's disease. In the case of haemophilia A, the VIII:vWB gene product is synthesized as usual; however, von Willebrand's disease is characterized by the absence of both factors VIII:C and VIII:vWB. Patients suffering from von Willebrand's disease actually synthesize normal factor VIII:C; however, this polypeptide is rapidly degraded as stabilization of this factor requires its association with the vWB polypeptide (Figure 5.2).

Figure 5.2. Normal factor VIII:c and vWF synthesis and defective synthesis in haemophilia A and von Willebrand's disease (reproduced with permission, from Kumar, P., and Clark, M., (eds) (1990). *Clinical Medicine*, 2nd edn. Baillière Tindall, London)

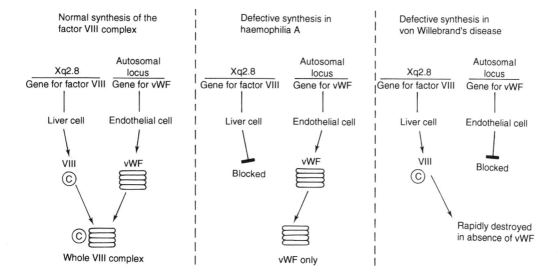

Haemophilia B, also known as Christmas disease, results from a deficiency of factor IX. Its clinical consequences are identical to those of classical haemophilia, but it does not occur as frequently as the latter disease.

The nature and severity of the clinical features of haemophilia depend upon the level of the factor in the plasma. Patients with very low levels (e.g. <1% of normal quantity) of factor VIII:C or factor IX are likely to experience frequent and spontaneous bouts of bleeding. Persons with higher levels of active factor (3–5% or above) experience less severe clinical symptoms.

Management of the above bleeding disorders is normally attained by administration of concentrates of the relevant factor. Factors VIII and IX are normally purified by suitable fractionation techniques from plasma obtained from healthy human donors. Factor IX, when purified by traditional fractionation procedures, usually contains appreciable quantities of factors II, VII, X and XI. This preparation therefore may also be used in clinical cases where deficiency of one of these additional factors is observed. The final product is normally sterilized by filtration, filled into final containers and freeze-dried. The containers are sealed under vacuum or under an oxygen-free nitrogen atmosphere in order to minimize the possibility of oxidative deterioration of the product. Anticoagulants such as heparin or sodium citrate are usually present in the final preparation, although such preparations do not contain antibacterial agents or other preservatives.

The importance of utilizing plasma free of detectable viral contaminants in the preparation of blood products cannot be over-emphasized. It is estimated that over 60% of haemophiliac patients have, at some stage, been infected by bloodborne pathogens from contaminated plasma or blood factor products. In addition to the screening of blood donations, several other approaches may be adopted in order to reduce further the likelihood of accidental transmission of infectious agents. Such approaches include addition of an antiviral substance to the final product and heat inactivation treatment. Purified preparations are also less likely to contain potential pathogens compared with whole plasma.

Elimination of the possibility of viral contamination is achieved by producing blood factors using recombinant DNA techniques. In the early 1990s, phase III clinical trials of at least two such recombinant factor VIII preparations were completed and marketing applications were submitted to the FDA. An application was also submitted with regard to production of recombinant factor IX. In both cases the recombinant glycoprotein was expressed in mammalian systems (Chinese hamster ovary cells and baby hamster kidney cells respectively). One of these systems contained the human factor VIII:C gene while the other contained the gene coding for the von Willebrand protein.

Expression in mammalian cells was undertaken in order to facilitate glycosylation of the recombinant protein, so that it resembled as closely as possible the natural human product. Factor VIII:C contains 25 potential glycosylation sites.

Variations in the glycosylation patterns could potentially render the protein immunogenic. This problem is not exclusively confined to recombinant products. Some patients, in particular those with severe haemophilia A, who synthesize little or no factor VIII:C, mount significant immune responses to injected factor VIII preparations. In such cases high levels of circulating neutralizing antibodies necessitate administration of increasingly high therapeutic doses of factor VIII to be effective. In such severe cases, administration of factor VIII may become ineffective. Several strategies to circumvent such complications may be adopted with varying degrees of success. Exchange transfusion of whole blood will transiently decrease circulating inhibitory antibody concentrations. Porcine factor VIII, which may not cross-react with the human factor antibodies, is sometimes administered. Immunosuppressive agents may also be administered concurrently. Another approach involves administration of activated factor X, which effectively bypasses the non-functional step of the coagulation cascade and hence promotes arrest of bleeding.

Development of circulatory antibodies to any administered blood factor represents a serious complication in terms of disease management, as repeated administration of such factors is required to manage the clinical condition.

Whole blood or blood plasma may be utilized as a source of blood factors, and hence administered to patients suffering from coagulation disorders such as haemophilia. Blood or plasma is, however, rarely administered in such cases if purified preparations of the factor required are available. The cryoprecipitate fraction of human plasma may be used directly as a source of factor VIII. The cryoprecipitate represents a protein fraction which precipitates upon cooling of the plasma. This fraction, however, may well contain viral agents present in contaminated blood donations.

Blood factors such as factors VIII and IX may be purified by a range of fractionation techniques, including precipitation or column chromatography. Many of the blood factor concentrates available are not purified to absolute homogeneity. Administration of such preparations would rarely elicit a strong immunological response, as any protein impurities present would also be of human origin, and hence would be recognized as "self". More stringent purification strategies must be adopted when producing the factor by recombinant methods. In such cases, any protein impurities present would be non-human in origin and hence are likely to be highly immunogenic. High-resolution purification methods include techniques such as immunoaffinity chromatography. One manufacturer of highly purified factor VIII employs an immunoaffinity column containing immobilized antifactor VIII monoclonal antibodies.

Non-hereditary coagulation disorders

Not all coagulation disorders are due to hereditary deficiencies in clotting factors. Most such acquired coagulation disorders stem from disordered liver function and/or vitamin K deficiency. Many blood factors, such as fibrinogen, prothrombin, and factors VII, IX and X, are synthesized in the liver. Vitamin K serves as an essential cofactor for a carboxylase enzyme. This enzyme catalyses the γ carboxylation of certain glutamate residues on factors II, VII, IX and X. This vitamin K-dependent post-translational modification must take place if these factors are to bind calcium ions.

Anticoagulants

Anticoagulants function to prevent blood from clotting. They are used clinically in cases where a high risk of blood clot formation is diagnosed and are also utilized to prevent the formation of further clots. Anticoagulants are thus often administered to patients who have suffered heart attacks or strokes in an effort to prevent recurrent episodes.

Thrombosis refers to the formation of blood clots. The blood clot itself is termed a thrombus. Thrombosis will most readily occur within diseased blood vessels. The formation of a thrombus in an artery will obstruct blood flow to the tissue it normally supplies. Formation of a thrombus in the coronary artery, coronary thrombosis, will obstruct blood flow to heart muscle. This results in a heart attack, which is usually characterized by death or "infarction" of part of the heart muscle—hence the term "myocardial infarction". The presence of thrombi which arrest blood flow to brain tissue usually results in a stroke. Furthermore, upon its formation, a thrombus or part of a thrombus may become detached and travel in the blood, only to become lodged in another blood vessel and obstruct blood flow at that point. This process is termed embolism, and may also induce heart attacks or strokes.

Some anticoagulants of therapeutic value are listed in Table 5.3. Heparin is a glycosaminoglycan-based anticoagulant and consists of sulphated polysaccharide chains of varying length, with molecular weights ranging between 6000 and 30 000 Da. It is synthesized and stored in many body tissues, especially in lung, liver, intestinal cells and cells lining blood vessels. Heparin preparations available commercially are normally prepared from porcine intestinal mucosa or from beef lung. This glycosaminoglycan exerts its anticoagulant activity by binding and thus activating antithrombin III, an α_2-globulin of molecular weight 60 000 Da which is in plasma. Activated antithrombin III inhibits a variety of activated

Table 5.3. Some anticoagulants used in therapeutics

Anticoagulant	Structure
Heparin	Glycosaminoglycan
Dicoumarol	Coumarin-based vitamin K antimetabolite
Warfarin	Coumarin-based vitamin K antimetabolite
Ancrod	Serine protease
Hirudin	Polypeptide capable of binding and inactivating thrombin

blood factors, including IIa, IXa, Xa, XIa and XIIa. Thus heparin, released naturally *in vivo* or administered therapeutically, inhibits the blood coagulation cascade. Administration of inappropriately large doses of heparin may result in severe haemorrhage. The anticoagulant activity of heparin is related to the molecular weight of the polysaccharide molecules. This fact has facilitated the development of low molecular weight heparins which retain anticoagulant ability while exhibiting decreased haemorrhagic effects.

Dicoumarol and warfarin are coumarin-based anticoagulants. Both may be administered orally. As previously outlined, these vitamin K antimetabolites exert their anticoagulant effect by inhibiting the vitamin K-dependent γ carboxylation of coagulation factors II, VII, IX and X. Gamma carboxylation of such factors is essential if they are to bind calcium ions, as is required during the normal coagulation process. The major adverse effect of administration of dicoumarol or warfarin is, predictably, the possibility of severe haemorrhage. Lethal doses of warfarin are employed as the active ingredient in many rat poisons.

Ancrod is a serine protease with anticoagulant activity. This enzyme has a molecular weight of approximately 35 000 Da and contains an appreciable quantity of carbohydrate. It is purified from the venom of the Malaysian pit viper. Ancrod catalyses the proteolytic cleavage of microparticulate fibrin molecules prior to clot formation. However, the enzyme has no effect on blood clots once they are formed.

The buccal secretion of leeches contain an anticoagulant termed hirudin. Components present in the saliva of leeches do not participate in the digestive process but function primarily to interact with, and inhibit, the host animal's haemostatic mechanism. Leech bites are thus characterized by subsequent prolonged bleeding, often lasting several hours.

Leeches have been used medically for centuries to promote blood-letting, and in instances where localized anticoagulant activity was required. Indeed the leech is enjoying somewhat of a comeback in medicine, where it is sometimes used to remove blood from inflamed areas and in procedures associated with plastic surgery.

The saliva of leeches contains a variety of peptides and low molecular weight polypeptides. Hirudin is recognized as the major anticoagulant associated with this saliva. Hirudin was first reported in the 1880s, though its characterization was not undertaken until the 1950s. The hirudin gene was cloned in the mid-1980s and has subsequently been expressed in a number of host systems. The polypeptide consists of 65 amino acids, having a molecular weight of 7000 Da. The molecule contains a sulphated tyrosine residue at position 63 and is also characterized by a high content of acidic amino acids towards the C-terminal end. The overall conformation of the molecule consists of a globular domain which is stabilized by three sites of intramolecular disulphide bridges and an elongated C-terminal region.

The anticoagulant activity of hirudin stems from its ability to bind thrombin, i.e. factor IIa, tightly. This results in inactivation of the thrombin molecule. Thrombin not only catalyses the proteolytic cleavage of fibrinogen thus forming fibrin and hence promoting clot formation, but also plays a role in activation of factors V, VIII and XIII. Binding of hirudin to thrombin masks both thrombin's fibrinogen binding site and its catalytic site.

Although the natural anticoagulant activity of hirudin has been recognized for some time, its lack of availability in adequate quantities rendered its widespread clinical use impractical. Over the past number of years recombinant hirudin has become available by production in *E. coli*, yeast and *Bacillus subtilis*. The amino acid sequence of the recombinant molecule is identical to that of the naturally occurring polypeptide. The recombinant molecules, however, do not possess the sulphated post-translational modification on tyrosine 63. The absence of the post-translational modification does not seem to be of any clinical significance.

Hirudin preparations manufactured by several different companies are currently undergoing extensive clinical trials. Hirudin appears to have a number of therapeutic advantages over some other anticoagulants (Table 5.4).

Table 5.4. Therapeutic advantages of hirudin as an anticoagulant

Hirudin acts directly on thrombin

It does not require a cofactor to exert its inhibitory effect

High doses are less likely to promote haemorrhage

It is a particularly weak immunogen

The saliva of some species of leech contains polypeptides other than hirudin that exhibit anticoagulant activity. One such anticoagulant is the protein antistasin. Antistasin has a molecular weight of approximately 15 000 Da and exerts its anticoagulant effect by inhibiting factor Xa. This protein is currently the subject of intensive study as it also displays anti-tumour activity. A number of other proteins obtained from various species of leech may also be of potential clinical significance. Specific examples include destabilase, an enzyme that catalyses the depolymerization of fibrin clots, and decorsin, which seems to inhibit platelet aggregation by interacting with a platelet surface glycoprotein.

Thrombolytic agents

Fibrinolysis forms an intrinsic part of the wound healing process
(Figure 5.3). This process refers to the enzymatic degradation and
removal of blood clots from the circulatory system. The process is
largely mediated by the serine protease plasmin. Plasmin catalyses
the enzymatic degradation of the fibrin strands present in the
clot. Plasmin is derived from plasminogen, its circulating zymogen. Human plasminogen is a glycoprotein of molecular weight
90 000 Da. It is synthesized in the kidney. Plasminogen consists
of a single polypeptide chain and contains numerous intrachain
disulphide linkages. Two natural forms exist which differ in carbohydrate content.

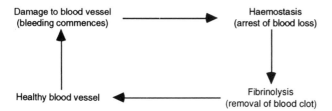

Figure 5.3. Simplified representation of wound healing

Plasminogen may be activated by a variety of specific serine
proteases, yielding active plasmin. Tissue plasminogen activator
(tPA) represents the most important physiological activator of
plasminogen (Figure 5.4). Tissue plasminogen activator, also
referred to as fibrinokinase, is a serine protease of molecular
weight 70 000 Da. It activates plasminogen by cleaving a single
Arg–Val bond. Tissue plasminogen activator found in human
plasma is predominantly formed in the vascular endothelium.
Two forms of tPA may be purified. Type I tPA is a single-chain
polypeptide, whereas type II consists of two polypeptide chains
connected by a disulphide linkage. Type II tPA is derived from
type I by proteolytic cleavage.

Fibrin contains binding sites for both plasminogen and tPA.
Activation of plasminogen by tPA forming plasmin occurs most
efficiently on the surface of blood clots. Active plasmin found free
in blood is quickly inactivated by another plasma protein termed

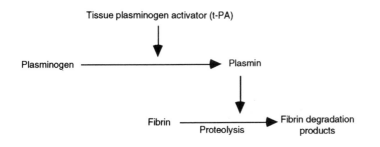

Figure 5.4. The fibrinolytic system, showing action of tissue plasminogen activator
(tPA). Other plasminogen activators
include urokinase, kallikrein and bacterial streptokinase

α_2-antiplasmin. This glycoprotein plays an important role in the regulation of plasmin activity.

Urokinase is a serine protease produced in the kidney. It also activates plasminogen, forming plasmin. Urokinase is produced as prourokinase and is found in both plasma and in urine. Two variants of the enzyme have been purified. High molecular weight urokinase has a molecular weight of 54 000 Da while the low molecular weight form is approximately 30 000 Da. Both forms exhibit similar biological activity. The low molecular weight form of the enzyme is considered to be produced from the high molecular weight form by proteolysis. Urokinase has been purified for therapeutic purposes from human urine by a variety of techniques. Chromatographic steps utilized include affinity chromatography and gel filtration. It is also produced commercially by tissue culture using human kidney cells.

Bacterial streptokinase may also be used to activate plasminogen. Streptokinase is produced by a variety of haemolytic streptococci and is generally purified from the filtrate obtained from cultures of several such organisms. The molecular weight of the purified protein is in the region of 45 000 Da. Due to its microbial source streptokinase is immunogenic and its administration sometimes results in adverse allergic reactions.

Unlike the plasminogen activators already discussed, streptokinase is devoid of proteolytic activity. This protein binds tightly to plasminogen which results in a conformational change in the plasminogen molecule rendering it proteolytically active. The activated plasminogen molecules are capable of catalysing the proteolytic activation of other plasminogen molecules, thereby yielding active plasmin.

Tissue plasminogen activator, urokinase and streptokinase are all utilized therapeutically as thrombolytic agents. As such they have been administered in a variety of situations including treatment of myocardial infarction, embolisms, strokes and deep vein thrombosis (which often develops subsequent to major surgery). Many such products would be administered over relatively short time periods subsequent to thrombus formation. Such treatment would then be followed by administration of a suitable anticoagulant for longer periods of time.

Human tPA preparations produced by recombinant DNA methods have been available commercially since the late 1980s. The recombinant product is produced by a specific murine cell line. Although the efficacy of this product is beyond doubt, it is quite expensive, especially when compared to streptokinase.

Several different tPA preparations are currently undergoing clinical trials. Recombinant tPA has most recently been produced in the milk of transgenic animals. Successful production in such systems may be achieved by utilizing an expression vector containing the promoter region of a highly expressed milk protein gene, such as the acid whey protein gene, linked to the cDNA or gene encoding the protein of interest.

Tissue plasminogen activator produced in transgenic animals is predominantly the two-chain isoform. The protein may be purified by a combination of techniques such as acid fractionation, hydrophobic interaction chromatography and immunoaffinity chromatography. The glycosylation pattern observed on the purified product differs somewhat from the patterns obtained when the protein is produced by the more conventional method of mouse cell culture.

In addition to proteins involved in the promotion or dissolution of blood clots, several other serum proteins are of considerable actual or potential therapeutic value. Examples of such proteins include human serum albumin, α_1-antitrypsin, haemoglobin and of course immunoglobulins, which are considered at length later in this chapter.

Human serum albumin

Serum albumin constitutes the most abundant protein present in serum, representing approximately 60% of total plasma protein. It is also one of the smallest known plasma proteins with a molecular weight of approximately 69 000 Da. It is one of a small number of plasma proteins that is devoid of a carbohydrate moiety. The protein is synthesized in the liver as preproalbumin. Removal of several amino acid residues from its amino terminus during passage through the endoplasmic reticulum yields mature albumin, which consists of 585 amino acids. The albumin molecule exhibits several intrachain disulphide linkages.

A major function of albumin is to provide most of the natural osmotic pressure of plasma. It also has a major transportational function and is especially important in transporting substances in aqueous media which are sparingly soluble. Most free fatty acids bind tightly to albumin and are transported in plasma in this manner.

Very little albumin is found in the urine of healthy individuals. Certain medical conditions, especially some forms of kidney disease, are characterized by secretion of large quantities of serum albumin into the urine. Some forms of liver disease may also result in significant decreases of hepatic protein synthesis, leading to a marked reduction in the concentration of several plasma proteins, most notably albumin.

Aqueous solutions of human serum albumin are available commercially. These range in concentration from 5% to 25%. Human serum albumin preparations are administered to patients suffering from some forms of kidney or liver disease, and are also used as a plasma volume expander for patients suffering from shock as a result of a decrease in the volume of blood. Such decreases are often associated with surgery or occur subsequent to serious injury.

Albumin is normally purified from serum, plasma or from placentas obtained from healthy donors. All raw materials should be screened for the presence of viral or other potentially pathogenic organisms prior to purification. The methods of purification utilized—precipitation and chromatography—must yield product which is 95–96% pure. Like most proteinaceous preparations the purified albumin is sterilized by filtration and subsequently aseptically filled into sterile containers. Although no preservatives are added, stabilizers such as sodium caprylate which protect the product, particularly against the effects of heat, are normally included in the final product. Subsequent to final filling, the product is subjected to heat which promotes inactivation of certain pathogens which may be present. This normally involves heating to 60 °C for a period of up to 10 hours.

While commercially available human serum albumin is currently obtained from blood or placentas, the protein has been produced as a heterologous product in a number of recombinant systems. Systems employed have included bacteria such as *Bacillus subtilis* and *E. coli*, yeast such as *Saccharomyces cerevisiae* and plants such as the potato.

Alpha-1-antitrypsin

Alpha-1-antitrypsin is a serum glycoprotein of molecular weight 52 000 Da. The protein consists of 394 amino acid residues and contains three glycosylation sites. Alpha-1-antitrypsin is synthesized in the liver and constitutes over 90% of the α_1-globulin fraction. It is normally present in serum at a concentration of 2 g per litre. A number of genetic variants of the protein have been described.

Alpha-1-antitrypsin constitutes the major serine protease inhibitor present in mammalian serum. It serves as a potent inhibitor of the protease elastase and as such prevents damage to lung tissue by neutrophil elastase. Neutrophils are a particular type of granulocyte, a type of white blood cell. Genetic deficiencies resulting in the absence of α_1-antitrypsin in human plasma have been described. Such deficiencies are particularly prevalent in persons of northern European descent. Persons suffering from such deficiencies often develop life threatening emphysema, resulting from unchecked damage to lung tissue by neutrophil elastase. Replacement therapy of α_1-antitrypsin delivered by intravenous infusion may arrest progression of this disease.

Alpha-1-antitrypsin preparations administered to patients suffering from congenital α_1-antitrypsin deficiency are normally obtained from pooled plasma fraction. The large quantities of inhibitor required (approximately 200 g per patient per year) and the possibility of accidental transmission of disease via infected source material has hastened the development of alternative recombinant sources. Alpha-1-antitrypsin has recently been successfully produced in the milk of both transgenic mice and sheep.

Haemoglobin

Much research has been carried out in an effort to develop a blood substitute or—perhaps more correctly—a red blood cell substitute. Development of such a substitute would decrease or eliminate concerns over viral contamination of blood cell preparations. It is hoped that any such substitute would also have a longer shelf-life than red cell preparations, which currently must be used within 42 days of collection. Furthermore, as antigens determining blood group type reside on the surface of red blood cells, the development of a cell-free substitute would allow its universal administration. Red blood cell concentrates are currently used clinically in the treatment of certain forms of anaemia and haemolytic disease.

A number of substances have been investigated as potential blood substitutes. Fluorinated hydrocarbons (fluorocarbons) have been assessed for this purpose, and have been used clinically in a limited number of cases. Such fluorocarbons can successfully transport both oxygen and carbon dioxide, and may be used to replace whole blood in some animal models. Inhalation of high oxygen concentrations is required if these compounds are to bind and transport clinically significant quantities of oxygen. In addition, some clinical investigations suggest that fluorocarbons may result in compromised immune function. Such disadvantages militate against their widespread use.

Purified haemoglobin has also been investigated as a potential red blood cell substitute. Haemoglobin is the principal protein found in circulating red blood cells or erythrocytes. It is a tetramer, consisting of two different polypeptide chains, α and β. The native molecule may be represented as $\alpha_2\beta_2$ with an overall molecular weight of approximately 64 000 Da. Each of the four polypeptide subunits also contain a haem prosthetic group which confers upon the molecules their oxygen binding properties (Figure 5.5). Binding of one oxygen molecule to one haem group greatly increases the affinity for oxygen of the remaining groups.

Such oxygen binding kinetics render haemoglobin ideally suited to its oxygen transporting role. Haemoglobin molecules alone, are not suitable as blood substitutes owing to (a) instability of the protein outside the environment of the red blood cell, and (b) its high affinity for oxygen. Additionally, haemoglobin molecules free in solution quickly dissociate into $\alpha\beta$ dimers. Such dimers are

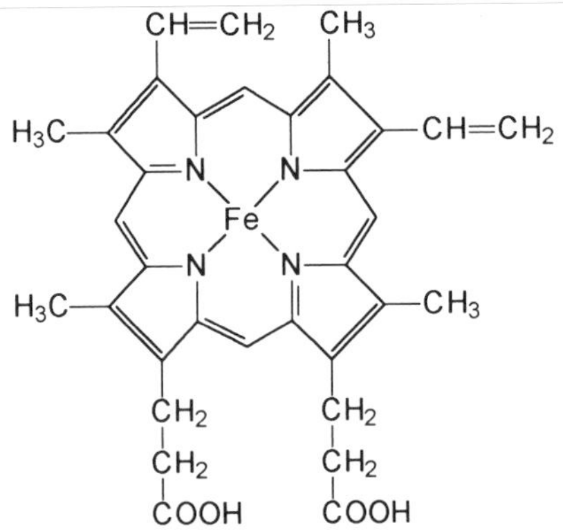

Figure 5.5. Chemical structure of the haem prosthetic group. The haem group is present not only in haemoglobin but also in myoglobin and a variety of other haem proteins

rapidly removed from the circulatory system. Normal human erythrocytes also contain large quantities of 2,3-diphosphoglycerate. Binding of 2,3-diphosphoglycerate to haemoglobin reduces the protein's affinity for oxygen. This promotes the release of bound oxygen, upon reaching oxygen-requiring tissue. The absence of 2,3-diphosphoglycerate renders the oxygen affinity of purified haemoglobin preparations too great to allow efficient oxygen release in such a manner.

Many of the problems relating to the stability of free haemoglobin and its high affinity for oxygen may be overcome, at least in part, by chemical modification. Treatment with glutaraldehyde or other agents that promote polymerization will prevent dissociation of the tetrameric molecule. Covalent attachment of pyridoxal 5'-phosphate groups reduces its affinity for oxygen. The introduction of specific cross-links between subunits will both stabilize individual molecules of haemoglobin and reduce its affinity for oxygen. All such modified haemoglobin molecules may be described as haemoglobin-based oxygen carriers (HBOC).

The vast majority of such HBOC preparations are derived from haemoglobin purified from human blood donations which have exceeded their designated useful shelf-life. Some studies have also been undertaken using preparations of bovine haemoglobin. Bovine haemoglobin exhibits a lower affinity for oxygen than does its human counterpart. Administration of such a "non-self" protein, however, may result in immunological or allergic complications. Porcine haemoglobin may be of interest in this regard as it more closely resembles human haemoglobin.

Like many other clinically interesting proteins, haemoglobin has been produced by recombinant methods. The haemoglobin α and β genes may be expressed in separate systems and intact haemoglobin may be reconstituted from the purified recombinant products. Alternatively, both genes may be expressed within a single host. This usually results in the automatic production of mature tetrameric haemoglobin molecules. Haemoglobin has been successfully produced in both recombinant *E. coli* and yeast systems and has been produced in the blood of transgenic animals.

Although production of recombinant human haemoglobin is technically feasible, outdated stocks of donated human blood still remain the major source from which it is extracted. This remains the most inexpensive method of production. Protein contaminants present in recombinant haemoglobin preparations would be non-human in nature, and thus immunogenic. This difficulty should not arise when haemoglobin is obtained from human red blood cells. Recombinant DNA technologies may, however, allow researchers to rationally develop and express human haemoglobin mutants with desirable characteristics such as improved stability or decreased oxygen affinity.

POLYCLONAL ANTIBODY PREPARATIONS

A wide variety of polyclonal antibody preparations are used clinically to afford immediate immunological protection against specific pathogens or other harmful antigenic substances. Administration of such specific antibody preparations is termed *passive immunization*. The purified antibody preparations administered are normally termed antisera. A distinction is often made between antisera and immunoglobulin preparations. The former term refers to antibodies isolated from animals, while the latter term refers to antibody preparations specifically obtained from human sources. In both cases the purified preparations consist predominantly of IgG molecules. Some of the more commonly used antisera and immuno-globulin preparations are listed in Table 5.5.

Passive immunization is generally used as a therapeutic measure if the patient is already suffering from a harmful condition caused or exacerbated by the presence of an antigenic substance. Such antigenic substances include viruses, microorganisms, and toxins or venom produced by certain spiders and snakes. The antibody preparations specifically recognize and bind the offending antigenic substances, thereby neutralizing or inactivating them. The exogenous antibody also helps to initiate a full immunological response against the foreign substance.

In contrast to this, the process of active immunization (the administration of a specific antigen, a vaccine, which stimulates the immune system to generate its own immunological response) is normally used as a prophylactic measure. The term "prophylaxis" refers to measures taken to prevent the future occurrence of specific diseases. Under certain circumstances, passive immunization may also be used as a prophylactic measure. For example, if a person is likely to come in contact with pathogens because of work or travel, prior administration of antibodies capable of

Table 5.5. Polyclonal antibody preparations most commonly used to induce passive immunity

Antibody preparation	Usual source	Antibody specificity
Normal immunoglobulin	Human	Exhibits a wide range of specificities against pathogens which are prevalent in the general population
Hepatitis B immunoglobulin	Human	Antibodies exhibiting a specificity for hepatitis B surface antigen
Measles immunoglobulin	Human	Antibodies exhibiting a specificity for measles virus
Rabies immunoglobulin	Human	Antibodies exhibiting a specificity for rabies virus
Cytomegalovirus immunoglobulin	Human	Antibodies exhibiting a specificity for cytommegalovirus
Varicella-zoster immunoglobulin	Human	Antibodies exhibiting a specificity for the causative agent of chicken pox
Tetanus immunoglobulin	Human	Antibodies exhibiting a specificity for the toxin of *Clostridium tetani*
Tetanus antitoxin	Horse	Antibodies raised against the toxin of *Clostridium tetani*
Botulism antitoxin	Horse	Antibodies raised against toxins formed by type A, B or E *Clostridium botulinum*
Diphtheria antitoxin	Horse	Antibodies raised against diphtheria toxoid
Gas gangrene antitoxins	Horse	Antibodies raised against the alpha toxin of *Clostridum novyi, C. perfringens* or *C. septicum*
Scorpion venom antisera	Horse	Antibodies raised against venom of one or more species of scorpion
Snake venom antisera	Horse	Antibodies raised against venom of various poisonous snakes
Spider antivenins	Horse	Antibodies raised against venom of various spiders, in particular the Black Widow spider

recognizing the pathogenic agent will afford transient immunological protection.

Antibody preparations available commercially have been raised
against a wide variety of harmful substances. As mentioned above,
such substances include a variety of pathogenic bacteria and
viruses, microbial toxins and poisonous antigens present in the
venom of some snakes and spiders. Antibody preparations raised
against venom are termed antivenins. Active principles present in
snake venoms are predominantly proteinaceous in nature, with
many exhibiting enzymatic activity.

Specific antiserum preparations are raised in large animals, of
which horses are the most popular. This is achieved by injecting

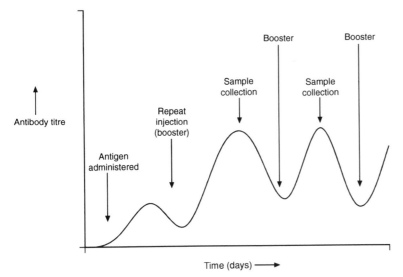

Figure 5.6. Schematic representation of antiserum production applicable to a variety of different animals. Administration of the antigen of interest results in a specific immune response. Serum IgG levels increase. Repeat injections of antigen (booster doses) maintain elevated levels of antibody. Careful monitoring of these levels facilitates bleeding (antibody collection) at the most appropriate junctures. Frequency of bleeding may be once every 7–14 days

the antigen of interest into the animal, thus initiating an immunological response (Figure 5.6). Booster doses of the antigen in question may be administered subsequently in order to heighten the antibody response. Samples of blood are withdrawn from the immunized animal at regular intervals and the antibody titre measured by an appropriate assay. When large quantities of antibodies which specifically recognize the antigen of interest are detected, the animal is normally bled. When large animals such as horses are used, several litres of blood may be withdrawn during each bleed. The blood may then be allowed to clot and the resulting antiserum subsequently recovered. Alternatively, the blood may be withdrawn directly into containers containing suitable anticoagulants and centrifuged to remove blood cells, thus producing plasma. The serum or plasma is then subjected to fractionation in order to recover a purified antibody preparation. Traditional purification protocols employ several sequential precipitation steps, usually using ethanol and ammonium sulphate as precipitants. More recent purification protocols use various chromatographic steps including ion exchange chromatography, hydrophobic interaction chromatography or protein A chromatography (as IgG molecules bind to protein A). All procedures used in the collection, manipulation and purification of antiserum preparations are carried out in accordance with the principles of good manufacturing practice, as discussed in the previous chapter.

Following purification, the potency of the product is determined and adjusted to a suitable strength by dilution or concentration, as appropriate. Stabilizing agents such as 0.9% NaCl or 2.25% glycine are often added, as are antimicrobial agents such as phenol or thimerosal. The antiserum preparation is sterilized by filtration and filled by aseptic technique into sterile containers which are subsequently sealed to preserve sterility. The antiserum preparation may also be freeze-dried. The antibody solution is often filled under an oxygen-free nitrogen atmosphere in order to prevent oxidative degradation of the product during long-term storage. The product should normally be stored at 2–8 °C. Under such conditions it should have a shelf-life of up to 5 years.

Immunoglobulin preparations are purified from donated bloods obtained from human volunteers. The methods of purification used are broadly similar to those used in the purification of antibodies from animal sources. Immunoglobulins purified from blood donations obtained from normal, healthy individuals usually contain a wide variety of antibody specificities. The production of such specificities was elicited over the years as the donor came into contact, either naturally or artificially, with a variety of antigens. Alternatively, high antibody titres which recognized specific antigens may be purified from donated blood obtained from individuals who have been immunized with that antigen or who have recently recovered from an illness caused by the antigenic substance. Persons recently vaccinated against hepatitis B or who have suffered from hepatitis B infection would generally exhibit high titres of anti-hepatitis B antibody.

Administration of antibody preparations may sometimes result in adverse clinical reactions. This is especially true if the antiserum preparation used is of animal origin. Adverse reactions sometimes associated with administration of animal serum include serum sickness and, in severe cases, anaphylactic shock. For this reason it is preferable to use immunoglobulin preparations of human origin whenever possible. Serum from which antibody preparations are purified should only be obtained from healthy animals or humans, and should be screened for the presence of potential pathogens prior to processing.

VACCINES

As discussed above, specific antibody preparations may be administered to individuals in order to confer immunological protection against a variety of pathogenic substances. This process is termed passive immunization as the protective antibodies are not actively generated by the body's own immune system. If, however, antigenic substances are introduced into the blood stream, either naturally or artificially, the host's immune system launches an immunological response to the challenge.

This response involves the production of antibodies that specifically recognize and bind the antigenic material that elicited their production. Antibodies are synthesized by a subset of white blood cells termed B lymphocytes. Binding of antibody to antigen should inactivate or neutralize the offending antigen. In addition, a second type of white blood cell—the T lymphocytes—may be activated, and these will also mount an immunological response. Such cells play a central role in the destruction of foreign antigenic material, such as bacteria and some viral species. After the immune system has successfully defeated the antigenic challenge, long-lived B and T cells, termed *memory cells*, remain in circulation. Memory cells are capable of recognizing the antigenic substance that elicited their initial production; if this antigenic material re-enters the body, the memory cells will trigger a full-blown immunological response. Unlike the primary response which builds up somewhat slowly, the repeat response is more rapid.

The process of vaccination is designed to exploit the natural defence mechanism conferred upon us by our immune system, and involves artificially exposing the immune system to antigenic preparations. Such processes will, by the process described above, induce active immunity against specific pathogenic organisms. If the person vaccinated subsequently comes into contact with such a pathogen, an immediate and specific immunological response is initiated. This should quickly destroy the offending organism before it has the opportunity to establish a full-blown infection.

A vaccine therefore, is a preparation of antigenic components, usually consisting of, derived from, or related to pathogens. When the vaccine is administered, it stimulates an immune reaction and thereby will confer active immunity on the recipient. This helps prevent subsequent establishment of an infection by the same or antigenically similar pathogens. Vaccines are thus used prophylactically to prevent the future occurrence of diseases which the recipient is likely to encounter. Most vaccination protocols involve administration of an initial dose followed by one or more subsequent doses over a suitable period. Administration of booster doses ensures a maximal immunological response, with associated production of high levels of circulating antibody. Vaccine production represents a significant niche of biotechnological endeavours. Most countries implement a systematic vaccination programme against key infectious diseases.

Diseases for which vaccine preparations have been developed are caused by a variety of biological entities, including viruses, rickettsiae and various microorganisms. Vaccines have been prepared that protect against other harmful microbial substances such as toxins. Perhaps the most effective method of inducing artificial immunization is to expose the recipient to a vaccine containing small quantities of the actual pathogen against which immunity is sought. This approach is rarely adopted in practice, as there is a risk that the recipient will develop the uncontrolled disease. Dead, inactivated or attenuated bacteria or viruses are

Table 5.6. Vaccine preparations most commonly used to induce active immunity. In addition to single constituent vaccines, many vaccine preparations of mixed specificity are also available: examples include measles, mumps and rubella vaccines, and vaccines for diphtheria, tetanus and pertussis

Vaccine	Vaccine preparation	Use
BCG vaccine (bacillus Calmette—Guérin)	Live attenuated strain of *Mycobacterium tuberculosis*	Active immunization against tuberculosis
Measles vaccine	Live attenuated strain of the measles virus, usually propagated in culture cells from chick embryo	Active immunization against measles
Mumps vaccine	Live attenuated strain of *Paramyxovirus parotitidis*, the causative agent of mumps—usually propagated in cultures of chick embryo cells	Active immunization against mumps
Rubella vaccine	Live attenuated strain of rubella virus, grown in cell culture	Active immunization against German measles
Poliomyelitis vaccine (Sabin vaccine—oral)	Live attenuated strain of poliomyelitis virus propagated in cell culture	Active immunization against poliomyelitis
Poliomyelitis vaccine (Salk vaccine—injection)	Inactivated poliomyelitis virus propagated in cell culture	Active immunization against poliomyelitis
Yellow fever vaccine	Attenuated strain of live yellow fever virus cultured in chick embryos	Active immunization against yellow fever
Cholera vaccine	Dead strain(s) of *Vibrio cholerae*	Active immunization against cholera
Influenza vaccines	Inactivated strains of influenza virus (individual or mixed) or suspension of influenza surface antigens (e.g. haemaglutinin and neuraminidase)	Active immunization against influenza
Pertussis vaccine	Killed strains of *Bordetella pertussis*	Active immunization against whooping cough
Plague vaccine	Formaldehyde-killed *Yersinia pestis*	Active immunization against plague
Rabies vaccine	Inactivated strain of rabies virus	Active immunization against rabies
Typhoid vaccine	Killed *Salmonella typhi*	Active immunization against typhoid fever
Typhus vaccine	Killed *Typhus rickettsiae* (*Rickettsiaprowazekii*)	Active immunization against typhus

Varicella-zoster vaccine	Live attenuated varicella virus	Active immunization against chicken pox
Diphtheria vaccine	Formaldehyde-treated toxin (e.g. a toxoid) derived from *Corynebacterium diphtheriae*	Active immunization against diphtheria
Tetanus vaccine	Formaldehyde-treated toxin (e.g. a toxoid) derived from *Clostridium tetani*	Active immunization against tetanus
Hepatitis B vaccines	Hepatitis B surface antigen isolated from plasma of carriers or produced by genetic engineering	Active immunization against hepatitis B
Haemophilus influenzae	Purified capsular polysaccharide of *Haemophilus influenzae* type B	Active immunization against *H. influenzae*—the major cause of infant meningitis
Meningococcal vaccines	Purified polysaccharide antigens obtained from one or more serotypes of *Neisseria meningitidis*	Active immunization against *N. meningitidis*—causes epidemics of meningitis
Pneumococcal vaccines	Purified polysaccharide antigens from different serotypes of *Streptococcus pneumoniae*	Active immunization against pneumococcal disease

often used. Such antigenic preparations exhibit little or no virulence but retain immunological characteristics similar to the wild-type pathogen. Organisms may be killed or inactivated by a variety of chemical or physical means. Attenuation is normally achieved by growing the organisms in an unnatural host. Safe vaccine preparations against a variety of bacterial toxins are produced by inactivating the toxin such that its toxic properties are eliminated while retaining its immunogenicity.

Many such preparations suffer from one or more disadvantages. Dead or inactivated vaccines may be significantly less immunogenic compared to their wild-type counterparts. If methods used to kill or inactivate the organisms are not consistently 100% efficient, a possibility exists that live virulent organisms may be accidently used as vaccines. There is also a danger, albeit slight, that some attenuated species might revert to the pathogenic state. Such theoretical dangers may be reduced by using a purified antigenic component of the pathogens. Surface polysaccharide and protein antigens have been purified and utilized to induce active immunity successfully in several cases. Table 5.6 lists most of the major conventional vaccine preparations commercially available.

Recent advances in vaccine development centre around recombinant DNA technologies. Such technologies allow the production and synthesis of specific protein antigens found on the pathogenic organisms, in innocuous recombinant organisms such as *E. coli*, yeast cells or mammalian cell lines. The desired protein product, which is identical to the antigen sourced from the wild-type pathogen, is then purified and employed as a vaccine. This method of vaccine production exhibits several advantages over conventional vaccine production methodologies (Table 5.7). The recombinant vaccine is extremely safe and should consist of a single antigenic constituent of the pathogen against which immunity is desired. It is thus impossible accidentally or otherwise to induce the disease state with such vaccine preparations. Furthermore, there is no possibility that the final preparation can be contaminated with additional pathogenic organisms. Administration of such a defined and structurally less complex vaccine is also less likely to induce unexpected adverse clinical reactions. Vaccine production by

Table 5.7. Advantages associated with the use of specific recombinant proteins as vaccine preparations

Clinically safe

Unlimited supply

Defined product

Less likely to cause unexpected side effects

recombinant methods also ensures a continuous and convenient supply of material from a safe source. This should help to keep production costs down and prevent accidental transmission of disease to production personnel and recipients.

The first vaccine product produced by recombinant DNA technology to be approved for human use was that of hepatitis B. Traditionally hepatitis B vaccine was prepared by purifying hepatitis B surface antigen (HBsAg) from the plasma of infected individuals. When isolated from plasma sources, HBsAg polypeptides are usually not found to be free, but are characteristically polymeric structures of 22 μm in diameter. Production of hepatitis B vaccine from such sources suffers from several drawbacks. There is a risk that the final preparation may be contaminated with active hepatitis B virions, or indeed other pathogenic viruses such as HIV. Furthermore, production of the vaccine depends upon a constant supply of plasma from infected individuals.

The hepatitis B surface antigen has been cloned and expressed in a number of recombinant systems (Table 5.8). The protein obtained from transformed yeast was the first recombinant vaccine made available for general medical use. It may be purified from yeast extract by immunoaffinity chromatography, utilising anti-HBsAg antibodies. It was approved for sale by the regulatory authorities in 1986, and is cheaper than the conventional vaccine preparations. Other hepatitis B antigens such as its core antigen (HBcAg) have also been successfully produced in recombinant systems.

Table 5.8. Some recombinant systems from which hepatitis B surface antigen has been produced

Bacterial systems	*E. coli*
Yeast systems	*S. cerevisiae*
Mammalian cell lines	Monkey kidney cells
	Mouse cells
	Human hepatoma cell line
	Monkey cell lines

A number of additional protein antigens derived from various pathogens have been produced by recombinant means (Table 5.9). Purified preparations of such recombinant antigens may well form the basis of many future vaccine preparations.

The cDNA for a surface antigen (P69) of *Bordetella pertussis* has recently been cloned and expressed at high levels in *E. coli*. Two members of the *Bordetella* family (*Bordetella pertussis* and *Bordetella parapertussis*) are the causative agents of whooping cough. Traditional pertussis vaccines are composed of killed strains of

Table 5.9. Some additional proteinaceous antigens derived from pathogens, which have been produced in recombinant systems. Many such preparations will be used therapeutically in the near future

Antigen expressed	Potential vaccine against
Bordetella pertussis surface antigen	Whooping cough
Tetanus toxin fragment C	Tetanus
Haemagglutinin	Influenza
Hepatitis A capsid proteins	Hepatitis
Poliovirus capsid proteins	Poliomyelitis
Rabies virus glycoprotein	Rabies
Surface glycoprotein D	Herpes

Bordetella pertussis. Surface antigens such as P69 purified from recombinant sources are likely to form the next generation of pertussis vaccine preparations.

Fragments of tetanus toxin have also been produced in *E. coli,* with a view to development of an alternative tetanus vaccine. Production of current tetanus vaccines involve treatment of tetanus toxin with formaldehyde, thus forming toxoid molecules. Native tetanus toxin consists of a single polypeptide chain of molecular weight 150 kDa. Proteolytic cleavage of the intact toxin yields two fragments. Fragment AB represents the 100 kDa N-terminal portion of the toxin, while fragment C represents the 50 kDa carboxyl terminal end of the toxin. Fragment C is non-toxic and is capable of inducing active immunity against tetanus in experimental animals. Recombinant fragment C preparations are likely to form the next generation of tetanus vaccines.

A number of other pathogen-derived antigens have been expressed in recombinant systems with a view to assessing their efficacy as subunit vaccines. In some instances, such recombinant proteins must be subject to certain post-translational modifications, which may best be achieved by using mammalian cells such as a Chinese hamster ovary cell line as the recombinant host.

Several other approaches have been adopted in the development of modern vaccine preparations. One such approach involves isolating or synthesizing a gene or cDNA which codes for a particular antigen obtained from a pathogen of interest. This may then be introduced into the genetic complement of a virus considered to be safe. Vaccinia virus is often used for this purpose. Recipients immunized with preparations of the live recombinant virus invariably develop immunity not only against the viral host, but also against the pathogen from which the inserted genetic information was obtained. Genes derived from hepatitis B virus and from HIV have been inserted into vaccinia viral genomes in an attempt to develop live recombinant vaccines against these pathogens.

Another research avenue entails the synthesis of specific peptide sequences identical to a portion of a protein antigen found in the pathogen. This method is attractive insofar as it results in the generation of safe vaccine preparations, free from harmful contaminants. The conformational structure adopted by any protein helps determine its level of antigenicity. In many instances treatments which destroy its native conformation will adversely affect the antigen's ability to elicit a strong immunological response. Many epitopes on the antigen's surface against which antibodies are directed are created by the specific conformational structure of the protein. Short peptide sequences may not accurately reproduce such epitopes and may be of little use in vaccinations.

Anti-idiotypic antibodies may also be successfully employed as vaccines. Upon encountering any antigen, an individual's immune system sets in motion a specific trail of immunological events. One consequence of immune stimulation relates to the production of B lymphocytes that produce antibodies which specifically recognize and bind epitopes on the foreign antigen. The antigen binding sites of such antibodies are located in their hypervariable regions. The hypervariable regions themselves contain antigenic determinants which can stimulate the production of yet another series of antibodies. The collection of antigenic determinants present on the hypervariable region are often termed *idiotype*. The second set of antibodies raised against the hypervariable region of the first set are termed *anti-idiotypic antibodies*.

Some anti-idiotypic antibodies will recognize epitopes in the primary antibody hypervariable regions which are not directly involved in binding to the foreign antigenic substance. Other anti-idiotypic antibodies, however, are raised against the specific epitopes present in the hypervariable region which do constitute the primary antibody's antigen binding site. Such anti-idiotypic antibodies carry what may be described as an internal image of the original antigen, and can mimic the antigen. They can therefore elicit the production of antibodies which will recognize and bind this original antigen.

Anti-idiotypic antibodies could be used to vaccinate against a wide range of diseases caused by various pathogens. Modern hybridoma technology (detailed later) renders possible the production of large quantities of selected antibodies. Anti-idiotypic antibodies have been used to vaccinate against hepatitis B surface antigen, poliovirus, rabies virus, *Streptococcus pneumoniae* and *Trypanosoma rhodesiense*, the causative agent of sleeping sickness. Currently, there is also great interest in the potential use of anti-idiotypic antibodies as vaccines against tumour cells. Thus far, however, the vast majority of anti-idiotypic vaccination experiments have been confined to animal studies.

Production of subunit vaccines by recombinant DNA technology is likely to have the greatest impact on vaccine technology for the foreseeable future. Anti-idiotypic antibodies could find important applications as some specialist vaccine preparations. Many antigenic determinants are non-proteinaceous in nature and thus cannot be produced by recombinant DNA technology. Anti-idiotypic antibodies can be prepared against such antigens and therefore used as effective vaccines.

Towards a vaccine for AIDS

The development of an effective vaccine against human immunodeficiency virus constitutes an intensely active area of vaccine research. HIV is the causative agent of AIDS. While the rate of transmission of this incurable disease appears to be decelerating in the Western world, its proliferation within Africa and other poorer areas of the world has reached crisis proportions. By 1992, 10 million cases of HIV infection had been reported worldwide. It is estimated that this number will have reached 40 million by the year 2000. In the USA the estimated annual cost of treating an individual infected with HIV is in the region of US $10 000. The estimated annual cost of treating a person with full-blown AIDS is in excess of $60 000. By the mid 1990s, it is expected that the cumulative cost of treating persons infected with HIV will be about $15 billion. Therapeutic substances such as azidothymidine (AZT) currently administered to AIDS patients serve only to retard the progression of the disease.

HIV is a retrovirus as its genetic material consists of RNA which, along with enzymes such as reverse transcriptase and integrase, are located in the viral core. This core material is in turn surrounded by the core protein, P24. The outer viral envelope consists of a membranous structure originally derived from the host cell membrane in which the viral particle was synthesized. Protruding from the membrane is the glycoprotein GP120. An additional glycoprotein, GP41, is found embedded in the membrane coat. GP120 and GP41 are both derived from the viral glycoprotein GP160, by proteolytic cleavage of the latter during viral replication. An additional protein, P17, is located beneath the viral envelope.

HIV is capable of binding to, and thus infecting, certain cell types. Sensitive cells must display a protein (termed CD4) on their surface in order to facilitate viral attachment. The CD4 marker is located only on selected cell types, most notably T helper cells. These cells play a critical role in the generation of cell-mediated immunity. The virus binds to CD4 via its envelope glycoprotein, GP120. Upon entry into the cell, the viral RNA is used as a template by reverse transcriptase, thus generating DNA. The viral DNA is then integrated into the host's genome, a process facilitated by the viral enzyme integrase. Once integrated, the viral genes may remain dormant for an extended period and will be replicated along with the host chromosomal DNA should the infected cells divide. However, transcription of the DNA results in the production of numerous viral particles. These particles eventually bud from the host cell membrane and hence are released into the blood stream. Particularly rapid intracellular replication may even result in lysis of the host cells.

Several different approaches continue to be evaluated with a view to developing an effective AIDS vaccine. Administration of small quantities of live wild-type virus in order to elicit a protective immunological response is obviously out of the question, as recipients would almost certainly become infected with the virus. Development of a vaccine consisting of a live attenuated strain of the virus also remains unlikely, due to the possibility of reversion to virulent form. Furthermore, it is difficult to predict what constitutes a 'safe' attenuated form of this virus, as the molecular basis by which the virus destroys the immune system is poorly understood.

Inactivated viral particles may constitute a potentially effective vaccine. Such preparations are currently being evaluated clinically. Wild-type viral strains are firstly inactivated by heat. This process causes the outer viral envelope to disintegrate. The core viral particles are then subjected to both chemical treatment and irradiation in order to ensure that no viable viral particles remain. Anti-idiotype antibodies and synthetic peptide technologies are under investigation as potential vaccine preparations. The vast majority of vaccine preparations currently undergoing clinical trials consist of specific viral antigens produced by recombinant DNA technology (Table 5.10). Somewhat mixed success has been reported in this regard. Most such antigens consist of recombinant viral GP160,

Table 5.10. Some recombinant AIDS vaccines currently undergoing clinical evaluation

Vaccine type	Developing company	US development status
GP 160	Immuno AG/National Institute of Health	Phase I clinical trials
	MicroGene Sys	Phase II clinical trials
	Wyeth-Ayerst	Phase I/II clinical trials
GP 120	Genentech	Phase I clinical trials
	Chiron/CIBA Geigy	Phase I clinical trials
P24	MicroGene Sys	Phase I clinical trials
Anti-idiotype antibody vaccine (murine monoclonal)	IDEC Pharmaceutical	Phase I clinical trials

GP120 or fragments of these molecules. These recombinant molecules are normally produced in mammalian cell lines to facilitate their post-translational glycosylation. While GP160, and GP120 which is derived from GP160, are subject to a high rate of mutation, it is considered likely that particular sequences will remain highly conserved. The presence of such preserved antigenic determinants in vaccine preparations should, in theory, help confer immunity against a wide range of HIV strains.

Several additional vaccine preparations currently being evaluated consist of recombinant core antigens such as P24 and P17. Such proteins are subject to a lesser degree of genetic drift when compared with the envelope proteins, offering the hope of providing a vaccine preparation of broad specificity. It has also been established that, as the immune function of the AIDS patient decreases, so do the circulatory levels of antibodies that recognize viral core proteins. In contrast, the level of circulatory antibodies that recognize the envelope glycoproteins remains elevated. It is therefore hoped that vaccines that specifically stimulate production of anti-core protein antibodies may be effective in suppressing the development of full-blown AIDS.

A vaccine that could successfully induce an immunological response targeted against the wild-type virus could be used as a preventive or therapeutic agent. An effective preventive vaccine should enable an immune response to be initiated against the AIDS virus immediately upon its entry into the body, thus destroying the virus before an infection can be established. The rationale behind therapeutic vaccine development is based on the fact that, subsequent to infection by HIV, the host's immune response prevents extensive viral replication for prolonged periods. This prevents progression of the disease and the patient remains clinically asymptomatic. It is hoped that vaccines used therapeutically would stimulate the immune function sufficiently to further delay development of the clinical symptoms of the disease.

Development of an effective vaccine against HIV, however, remains elusive for a number of reasons. HIV mutates at a higher rate than almost any other virus. Thus a vaccine that elicits the production of protective antibodies against one strain of the virus may not induce immunological protection against subsequent genetic variants. Furthermore, many patients suffering from AIDS exhibit high levels of circulating anti-HIV antibodies, which seemingly fail to halt the progression of the disease. It seems, therefore, that a humoral response (the production of antibodies by B lymphocytes), alone is not sufficient to combat the infection successfully.

It is likely that an effective vaccine must elicit both a strong humoral and cell-mediated immune response. T cell-mediated responses would play a crucial role in destroying cells infected with virus. Only T lymphocytes are capable of destroying cells that exhibit foreign viral antigen on their surface. Many, if not most, traditional vaccine preparations effectively stimulate only an antibody-mediated immune response.

The fact that HIV often enters a new host not as free viral particles but in the form of infected cells also renders difficult the development of an effective vaccine. Furthermore, immune surveillance may not be capable of recognizing and destroying many infected cells in which the integrated viral DNA remains quiescent. Additionally, the lack of a good animal model in which the progression of AIDS can be studied renders difficult the preclinical testing of potential vaccines. Chimpanzees probably represent the best such model system. These, however, are a protected species and are in short supply.

In addition to vaccine development, several other research strategies are being pursued with a view to defeating this viral infection. The antiviral activity of drugs known to inhibit viral enzymes such as integrase or reverse transcriptase are currently being evaluated. Many such drugs, however, are characterized by a lack of specificity and many exhibit serious side-effects.

Additional research efforts are targeted against the integrated viral DNA, in an effort to prevent activation of the viral genes. Other strategies under active consideration involve the production of recombinant soluble CD4 protein molecules. As previously discussed, viral infection is initiated by its binding to the CD4 protein found on the surface of susceptible cells. Parenteral administration of large quantities of soluble CD4 molecules may well retard or prevent viral attachment to healthy cells by acting as decoy viral receptors.

Adjuvant technology

Many antigens, when administered alone, elicit a poor immunological response. Antigen preparations are thus routinely administered in conjunction with substances termed *adjuvants*. Adjuvants function to enhance the immunological response against the injected antigenic substances by a number of means. For example, they generally enhance the immunogenicity of antigens. They also tend to protect the antigen against rapid inactivation and removal. Adjuvants may also increase antibody production by enhancing antigen presentation to a variety of immune cell types. The only adjuvant preparations currently used in humans are salts of aluminium such as aluminium hydroxide (AlH_3O_3) or aluminium phosphate (AlO_4P). Such adjuvants increase the humoral response against injected antigen as much as a hundredfold. They also exhibit few if any adverse side-effects.

A variety of even more potent adjuvant preparations are also available. These include (a) Freund's complete adjuvant and Freund's incomplete adjuvant (FCA/FIA); (b) suspensions of dead *Bordetella pertussis* cells; and (c) immunostimulatory components of mycobacterial and other cell walls. Freund's complete adjuvant consists of a mixture of dead *Mycobacterium tuberculosis* in a water and oil emulsion. The emulsifying agent normally used is Arlacel A, while the oil employed is usually mineral oil. Freund's incomplete adjuvant lacks the mycobacterial component. Freund's adjuvants have not been approved for use in humans and are restricted to use in animals due to the occurrence of side-effects upon administration. Such side-effects include severe local irritation and inflammation at the site of injection, as well as granuloma formation.

The active immunostimulatory components present in mycobacterial preparations appear to be muramyl dipeptides (MDPs), in particular *N*-acetyl muramyl L-alamyl-D-isoglutamine, as well as trehalose dimycolates (TDMs). These molecules are found in association with the mycobacterial cell wall. *N*-acetyl muramic acid is a base component of peptidoglycan, the structural polymer found in bacterial cell walls. Trehalose dimycolate consists of a molecule of trehalose to which two molecules of mycolic acid are covalently linked. Trehalose is a disaccharide consisting of two glucose residues linked via an $\alpha 1 \rightarrow 1$ glycosidic bond (Figure 5.7a). Mycolic acid is a hydroxy lipid molecule, found almost exclusively on the surface of mycobacteria (Figure 5.7b).

(a)

(b)

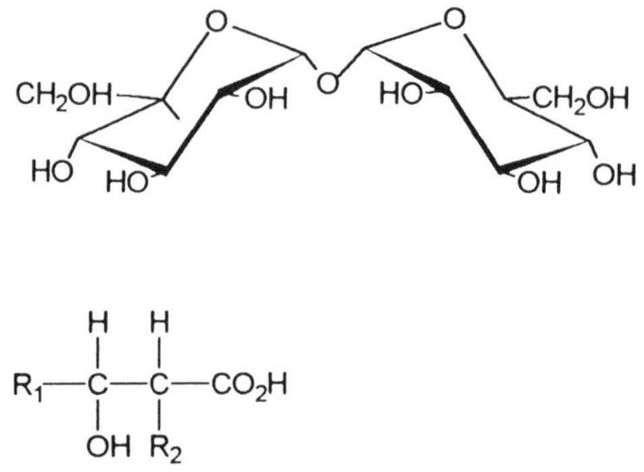

Figure 5.7. Structural formulae of (a) trehalose (α-D-glucopyranosyl-α-D-glucopyranoside) and (b) mycolic acid. R_1 and R_2 groups in mycolic acid represent long-chain aliphatic hydrocarbons

Purified extracts from mycobacterial cell walls containing such immunostimulants may also constitute valuable adjuvant material. Their administration results in marked stimulation of antibody production and an enhanced cell-mediated immune function. These stimulants are pyrogenic insofar as they potentiate release of interleukin 1 (endogenous pyrogen) from macrophages and monocytes. A variety of muramyl dipeptides have been chemically synthesized, some of which are capable of inducing a marked immunostimulatory response.

A suspension of dead *Bordetella pertussis* cells is sometimes utilized as adjuvant material. The pertussis toxin is known to stimulate antibody production. Lipopolysaccharide derived from the dead cells may also stimulate the immune function by inducing interleukin 1 secretion.

MONOCLONAL ANTIBODIES

Monoclonal antibodies are used in an ever-increasing range of biotechnological processes (Table 5.11). This is due in part to their specificity, selectivity, high antigen binding affinity and to the inexhaustible supply of identical antibody afforded by hybridoma technology. Recent advances which allow production of monoclonal antibodies in *E. coli* promise to extend even further the availability and usefulness of these highly sophisticated biological reagents.

Table 5.11. Some current applications of monoclonal antibodies

In vitro diagnostic reagents

Investigative tools for *in vivo* diagnostic purposes

Therapeutic agents

Vaccines

Ligands in immunoaffinity chromatography

Catalysts

Most antigenic substances contain numerous epitopes or regions on the surface of the antigen, which antibodies recognize. When the immune system encounters such antigens, the synthesis of a variety of antibody molecules which recognize and bind such epitopes is initiated. Each antibody type that binds to any particular epitope is produced by a clone derived from a specific B lymphocyte. If one such antibody-secreting B lymphocyte or plasma cell could be isolated and cultured successfully, antibody molecules of a single specificity—monoclonal antibodies—could be produced. Such antibody-producing cells, however, cannot be cultured in this manner over long periods.

In the mid-1970s, a technique was developed that facilitated the production of monospecific antibodies derived from a single antibody-producing cell. This process is termed *hybridoma technology*. It stems from the observation that if a population of antibody-producing cells are fused with immortal myeloma cells, a certain fraction of the resultant hybrid cells will retain the immortal characteristics of the myeloma cell while secreting large quantities of monospecific antibody.

Monoclonal antibody production initially involves the collection of lymphocytes from the spleen of a laboratory mouse immunized against the antigen of interest. A certain proportion of such lymphocytes produce antibodies which react with specific epitopes present on the antigen's surface. The lymphocytes are incubated together with mouse myeloma cells in the presence of a reagent which promotes cell fusion, such as propylene glycol. Hybrid cells derived from the fusion of myeloma cells and lymphocytes are termed hybridomas. Hybridomas may be selected from unfused cells by culture in a specific selection medium. The antibody-producing hybridomas may be separated from each other by dilution and are readily grown in tissue cultures. The resultant clones must be screened in order to determine which ones produce antibodies recognizing the antigen of interest. The most appropriate positive clones may subsequently be cultured in order to produce large quantities of monospecific antibody.

The specificity, selectivity and high binding affinity exhibited by monoclonal antibodies render them attractive biochemical tools. Moreover, once a suitable hybridoma has been identified and isolated, it may be used to produce large quantities of defined antibodies on a continuous basis. In contrast to this, the range and quantity of antibody specificities present in polyclonal antibody preparations may vary from bleed to bleed, and may not be exactly reproduced by replacement animals when the initial producing animal dies.

Antibodies are increasingly used in medicine for a variety of diagnostic and therapeutic purposes. Several dozen monoclonal antibody preparations are currently undergoing clinical evaluation. Indeed, monoclonal antibodies represent by far the largest category of biotechnologically derived products currently undergoing clinical trials. The vast majority of these monoclonal antibodies are indicated for the diagnosis and treatment of a variety of cancer types, including colorectal, breast, lung and ovarian cancer, malignant melanoma and T cell malignancies. Other monoclonal antibody preparations are undergoing assessment as potential therapeutic agents against diseases such as rheumatoid arthritis and multiple sclerosis. Yet other monoclonal preparations are being evaluated as immunosuppressive agents, and several monoclonal antibodies that specifically recognize and bind endotoxin of Gram-negative bacteria are undergoing trials to ascertain their effectiveness in treating septic shock. Septic shock is an often fatal condition which is difficult to manage by conventional therapies.

Thus far, only one monoclonal antibody, known as OKT3, has been approved for clinical use by the FDA. This antibody is used to promote a reversal of acute kidney transplant rejection. OKT3 specifically recognizes a cell surface antigen known as CD3, also referred to as the cluster of differentiation, which is associated with virtually all T cells. Binding of the antibody to CD3 can induce destruction of T cells—these cells usually mediate rejection of transplanted tissue. Treatment with OKT3 appears more effective in preventing transplant rejection than treatment with more conventional immunosuppressive drugs. Administration of OKT3, however, often induces the release of a variety of cytokines which may mediate a number of unwanted clinical responses.

Despite initial enthusiasm, the great therapeutic promise afforded by monoclonal antibody technology remains to be realized. This is particularly true in relation to the treatment of tumours. The somewhat disappointing results recorded thus far may be attributed to a number of important factors (Table 5.12). Over the past few years, however, technical advances have negated many of the initial difficulties. Such advances ensure that the future prospects of monoclonal antibodies as therapeutic agents appear bright.

Table 5.12. Factors limiting the successful application of monoclonal antibody preparations at a therapeutic level

Murine monoclonal antibodies are recognized as foreign by the host human immune system

Until recently researchers have by and large failed to generate monoclonal antibodies of human origin

Murine monoclonal antibodies when administered to humans fail to trigger a number of effector functions normally associated with the constant region of native antibody

Insufficient information exists regarding appropriate target cell surface markers which might be selectively exhibited on the surface of certain tumour types

Poor penetration of tumour mass by whole antibodies is normally observed

Monoclonal antibodies are time-consuming and expensive to produce

The immunogenic nature of murine monoclonal antibodies, when administered to humans, has perhaps presented the greatest obstacle to their successful clinical application. The administration of murine monoclonal antibodies stimulates the production of neutralizing antibodies by the host immune system. These antibodies quickly negate the therapeutic effect of the administered product. The obvious solution to such problems of immunogenicity is to develop and use monoclonal antibodies of human origin. This task has proved extremely difficult thus far. Human lymphocytes represent poor fusion partners for mouse myeloma cell lines. Upon fusion, a preferential loss of human genetic elements is often observed, and the resultant hybrids are highly unstable. Furthermore, it is often not possible to immunize human subjects using toxic or otherwise deleterious antigenic substances. Attempts to immortalize human antibody-producing cells by methods such as infection with Epstein–Barr virus have thus far proved disappointing.

Genetic engineering technologies may provide the most effective means of reducing the innate immunogenicity of murine monoclonal antibodies. Initial genetic manipulations centred around production of hybrid or chimaeric antibody molecules consisting of mouse variable regions and human constant regions. It was considered likely that such "humanized" antibodies would be significantly less immunogenic when compared to unaltered murine monoclonal antibodies. Although this is probably true, such chimaeric antibodies still retain significant portions of mouse sequence, such as the entire variable region (Figure 5.8). Such sequences can still induce substantial immunological responses.

The immunogenicity of murine monoclonal antibodies may be further reduced by "grafting" the DNA sequences coding for the antigen binding complementarity determining regions of the parent monoclonal antibody into the DNA sequences coding for a human antibody. The resultant humanized monoclonal antibody contains mouse sequences which code only for the hypervariable or antigen binding regions of the antibody (Figure 5.9). As the remainder of the molecule is entirely human, it should not elicit an antigenic response when administered to humans. Many such humanized antibodies, however, exhibit decreased antigen binding affinity when compared with the parent murine monoclonal antibody. Slight alteration of the human framework sequences which surround the hypervariable regions may restore antigen binding affinity. Although the framework regions of the variable chains do not seem to engage directly in binding antigen, they nevertheless influence such binding.

The successful production of humanized monoclonal antibodies also overcomes the inability of murine monoclonal antibodies to mediate a variety of antibody effector functions in recipient human patients. Binding of antibody to antigen facilitates the destruction and removal of the antigen by a number of means. The binding of antibody renders more efficient the process of antigen ingestion and destruction by phagocytic cells. The Fc region of bound antibody also mediates complement. Complement is composed of a number of serum proteins which, when activated, facilitate the destruction of antigen by phagocytes in addition to promoting direct lysis of invading microorganisms. The Fc regions of murine antibodies are poor mediators of such effector functions in humans. Humanized antibodies whose Fc domains are entirely human in origin do not suffer from such disadvantages.

Figure 5.8. Antibody structure (immunoglobulin G). Immunoglobulin G consists of four polypeptide chains, two heavy (identical) and two light (identical). The four polypeptide units assemble to form a Y-shaped molecule. The overall structure is stabilized by both interchain and intrachain disulphide bonds, and by noncovalent interactions. Treatment with certain proteolytic enzymes results in cleavage of the antibody at the flexible hinge (H) region, yielding two antigen binding fragments (Fab), and a constant fragment (Fc). The constant fragment (Fc) normally mediates various effector functions. Both H chains and L chains contain variable (V) regions and constant (C) regions. Variable regions contain the antigen binding site. Variable regions of antibodies of different specificity differ in amino acid sequence. Constant regions, on the other hand, do not. Structurally each H and L chain consists of a number of domains. One such domain forms the variable region of the H chain (V_H). The constant region of the H chain consists of three such domains, $C_H 1$, $C_H 2$ and $C_H 3$. L chains consist of only two domains: a variable one (V_L) and a constant one (C_L). An antibody fragment consisting of V_H and V_L domains is sometimes referred to as the F_V fragment. Each domain exhibits strikingly similar underlying architectural features, consisting of β-pleated sheets joined by loop regions. These loops in the variable domains (V_H and V_L) exhibit hypervariable sequences and form the antigen binding sites of the antibody. These hypervariable sequences are referred to as complementarity determining regions (CDR). The remaining areas of the variable domains are often termed framework regions. Framework regions exhibit reduced variability when compared to CDR sequences. Immunoglobulins are glycoproteins. The carbohydrate moiety is normally associated with the constant domains of the heavy chains. Removal of the carbohydrate groups has no effect on the antigen binding ability of the antibody but does affect various antibody effector functions $\longrightarrow$

Several novel methods have been developed which potentiate the cytocidal effect of therapeutic monoclonal antibody preparations. Most examples to date involve the coupling of radiolabels or toxins to monoclonal antibodies directed against tumour cells. The relatively large size of whole antibody molecules (the molecular weight of IgG is in the region of 150 000 Da), renders them somewhat inefficient in penetrating tumours. In this regard, radiolabelled antibody fragments which retain their antigen binding ability may be more suitable than whole antibodies in some forms of tumour therapy (Figure 5.10). Radiolabelled antibodies coupled with radioimaging techniques may also assist in the early detection of tumours.

In order to minimize toxic side effects, the toxin or radiolabel must be attached to the antibody by a stable linkage. The monoclonal antibody should also be directed against an antigenic determinant found exclusively on the tumour cell surface. Any cross-reactivity with normal human cell surface antigens would lead to unwanted destruction of healthy tissue. The detection and characterization of putative cell surface markers unique to certain cancer type remain an intensely active area of scientific research. The linkage of the cytocidal agent to the antibody must not adversely affect normal antibody functioning. Many of the first-generation antibody–cytotoxic conjugates developed yielded disappointing clinical results due to their failure to meet one or more of the above criteria.

A large number of monoclonal antibodies have been produced by hybridoma technology since its development in the mid-1970s.

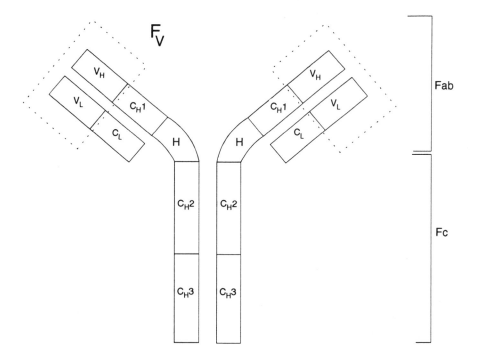

Hybridoma technology is, however, relatively complex, time-consuming and expensive. In the recent past, recombinant DNA methodologies have been used to produce monoclonal antibodies. While hybridoma production immortalizes antibody-producing cells, recombinant DNA technology may be utilized to immortalize antibody genes. This method overcomes many of the disadvantages associated with hybridoma technology. It is conceivable that in the future, monoclonal antibody production by techniques of genetic engineering will become the production method of choice. It has been suggested that such monospecific antibodies produced in bacterial systems such as *E. coli* should be termed "coliclonal antibodies".

The process of monoclonal antibody production in bacteria begins with immunization of an animal with the antigen of interest. This stimulates the proliferation of lymphocytes producing antibodies that specifically recognize various epitopes on the administered antigen. Lymphocytes from the animal may then be collected, and the cDNA coding for antibodies generated, with subsequent expression in bacteria. In this way, a library is produced containing a repertoire of antibody cDNAs. The clone producing the antibody of interest may then be selected from such a library by an appropriate antigen screening procedure.

Messenger RNA coding for antibody chains may be directly isolated from lymphocytes followed by their amplification using reverse transcriptase and the polymerase chain reaction (PCR). Alternatively, conventional cloning in bacteria of the corresponding cDNAs may be accomplished. Many initial studies concentrated on the cloning and expression in *E. coli* of cDNAs coding for antibody variable domains such as the V_H domain or the V_L domain (see Figure 5.8). Although it has been illustrated that single variable domains such as V_H are sometimes capable of binding antigen, such single domain "antibodies" may not bind antigen as efficiently as normal antibody binding fragments, which contain variable regions from both light and heavy chains. Furthermore, single domain structures such as V_H exhibit hydrophobic patches on their surface which under normal circumstances interact with V_L domains, and hence are hidden from the surrounding hydrophilic environment. Such exposed hydrophobic surfaces on single V_H domains tend to make these polypeptides somewhat "sticky".

Various strategies have been adopted in order to facilitate the production of whole antibody binding fragments in bacteria. Procedures adopted thus far include the so-called hierarchical and combinatorial approaches. The hierarchical approach involves initially isolating a clone which produces an antigen binding V_H polypeptide and subsequently coexpressing the isolated V_H cDNA with a library containing cDNAs coding for V_L domains. This library may then be screened in order to pinpoint a clone producing an

Figure 5.9. Recombinant DNA technology makes possible the production of "chimaeric" and "humanized" antibodies. Chimaeric antibodies consist of variable regions of murine origin and constant regions of human origin. Such antibodies retain significant proportions of mouse sequences, and generally remain immunogenic when administered to humans. Humanized antibodies are less immunogenic as the only sequences of murine origin they contain are those that constitute the complementarity determining regions. This first appeared in *New Scientist* magazine London, the weekly review of science and technology　　　　→

antibody fragment consisting of both V_H and V_L domains which exhibit the required specificity.

The combinational approach on the other hand involves co-expression of both heavy and light chain cDNA libraries in *E. coli*. The initial strategy involved the use of phage λ, a virus

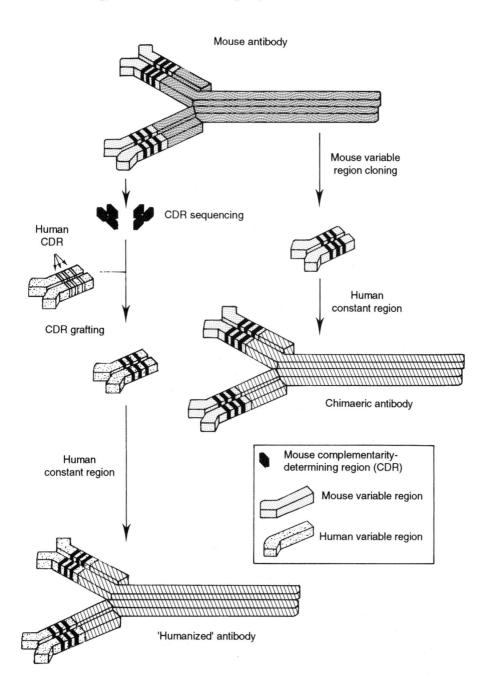

Mouse antibody

Mouse variable region cloning

CDR sequencing

Human CDR

CDR grafting

Human constant region

Chimaeric antibody

Human constant region

Mouse complementarity-determining region (CDR)

Mouse variable region

Human variable region

'Humanized' antibody

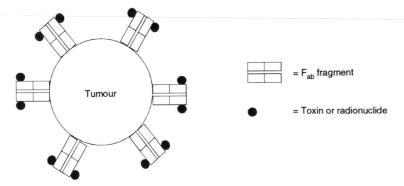

= F$_{ab}$ fragment

= Toxin or radionuclide

Figure 5.10. Cytocidal effect of antibody on a specific tumour type. Antibodies or antibody fragments which are capable of recognizing and binding specific antigenic determinants found exclusively on tumour cells are labelled with an appropriate toxin or radionuclide. Binding of antibody or antibody fragments to tumour associated markers effectively directs the toxin or radionuclide to the tumour site. This facilitates selective destruction of the tumour mass. Radionuclide labelled antibodies, coupled with radioimaging techniques, may also facilitate the detection and localization of tumours within the body

that infects *E. coli*, as vector (Figure 5.11). Complementary DNA coding for heavy and for light chains was cloned into two separate phage libraries. Cleavage of DNA from both libraries, followed by religation, allowed the generation of a combinational library which expressed both heavy and light chain variable fragments. Immunization of animals prior to isolation of lymphocyte mRNA ensures that the mRNA recovered is highly enriched in sequences specifically coding for light and heavy chain cDNAs of the antibody of desired specificity. This in turn greatly increases the likelihood that random recombination of heavy and light chains will produce whole antibody fragments capable of binding the antigen of interest.

A powerful screening and selection system has recently been introduced which facilitates the rapid identification and isolation of vector elements expressing the cDNAs coding for the antibody of interest from antibody-producing libraries. This renders the production of antibodies by methods of genetic engineering even more attractive. The system adopted involves cloning an entire repertoire of antigen binding antibody fragment cDNAs in phage libraries. Such antibody cDNAs are fused to a gene coding for a viral coat protein. Expression of the resultant antibody–coat protein hybrid in effect ensures that each phage expresses its antibody gene product on its surface. Phages containing the gene coding for the antibody of interest can be isolated quickly and easily from large libraries by passing the recombinant phage particles through a column containing immobilized antigen. Phage-expressing antibody fragments recognizing the antigen are retained on the column by immunoadsorption. After extensive washing to flush out any unbound phage, the retained phage is then eluted. The eluted phage, which contains the cDNA coding for the antibody of interest, may then be used to infect *E. coli* cells, thus allowing the production of phage antibodies. Soluble antibody fragments may also be produced by, for example, subcloning the antibody cDNA into a high expression vector and transforming a suitable host.

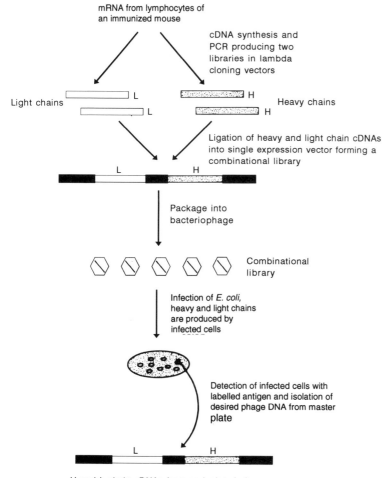

mRNA from lymphocytes of an immunized mouse

cDNA synthesis and PCR producing two libraries in lambda cloning vectors

Light chains

Heavy chains

Ligation of heavy and light chain cDNAs into single expression vector forming a combinational library

Package into bacteriophage

Combinational library

Infection of *E. coli*, heavy and light chains are produced by infected cells

Detection of infected cells with labelled antigen and isolation of desired phage DNA from master plate

H and L chain cDNAs from an isolated phage particle are cloned into a high expression system

Figure 5.11. Generation of a combinational antibody expression library in *E. coli*. Complementary DNA synthesis and PCR generate two distinct cDNA libraries, the phage DNA of which are isolated and digested, followed by religation into phage vectors. The required phage contain cDNAs for both heavy and light chain subunits of the antibodies. The desired phage are identified by screening with labelled antigen

To date most studies involve the production of antigen binding antibody fragments in prokaryotic systems. Such antibody fragments would most probably be only of limited usefulness for therapeutic applications. These fragments lack the Fc portion which mediates many antibody effector functions. This drawback could be overcome by producing whole functional antibodies as heterologous protein products in suitable recombinant systems. Ideally, the system utilized should be capable of glycosylating the recombinant protein as the native antibodies are glycoproteins.

Most antibody molecules appear to be glycosylated in their C_H2 domains. Certain antibody types exhibit additional glycosylation sites in other constant domains. Few antibody molecules appear to be glycosylated in their variable regions. The presence or absence of bound carbohydrate has no apparent influence on the antigen binding capability of the antibody. Lack of glycosylation would therefore not adversely affect the performance of monoclonal antibodies whose intended function would be that of recognizing and binding specific antigenic molecules. Most antibodies produced for industrial or diagnostic purposes would fall into this category.

In contrast to antigen binding, most antibody effector functions are reduced or abolished if normal antibody glycosylation is prevented. This is obviously of considerable clinical significance. It is thus considered likely that antibodies produced by recombinant DNA technology destined for therapeutic applications might be best expressed in mammalian cell systems capable of carrying out post-translational glycosylation reactions. Apart from influencing various effector functions, the glycosylation status of an antibody molecule may also affect other properties such as its solubility and its rate of degradation *in vivo*.

The production of monoclonal antibodies by methods of genetic engineering is likely to increase in popularity over the next number of years. The recombinant methodologies utilized in their production are already well established. Production of antibodies as heterologous protein products in systems such as *E. coli* are likely to be cheaper, less complex and less time-consuming compared with their production by hybridoma technology. Hybridoma technology, however, has proved itself to be very powerful, and many consider it likely that antibody production by recombinant methodologies will complement rather than supersede antibody production by this method.

Further reading

Books

Gosling, J.P. & Reen, D.R., (Eds) (1993). *Immunotechnology*. Portland Press, Colchester.

Smith, E.L. *et al.* (1983). *Principles of Biochemistry—Mammalian Biochemistry*, pp. 3–37. McGraw-Hill.

Articles

AIDS Medicines in Development (Drugs and Vaccines) (1992). Survey report presented by the American Pharmaceutical Manufacturers Association.

Allison, A.C. & Bayrs, N. (1987). Vaccine technology: adjuvants for increased efficacy. *Bio/Technology*, **5**, 1041–1045.

Allison, A.C. & Bayrs, N. (1987). Vaccine technology: developmental strategies. *Bio/Technology*, **5**, 1038–1040.

Biotechnology Medicines in Development (1993). Survey report presented by the American Pharmaceutical Manufacturers Association.

Bolognesi, D.P. (1990). Approaches to HIV vaccine design. *Trends in Biotechnol.*, **8**, 40.

Burton, D. (1991). Human and mouse monoclonal antibodies by repertoire cloning. *Trends in Biotechnol.*, **9**, 169–175.

Carlsson, R. & Glad, C. (1989). Monoclonal antibodies into the '90's: the all purpose tool. *Bio/Technology*, **7**, 567–573.

Carter, P. *et al.* (1992). High level *Escherichia coli* expression and production of a bivalent humanized antibody fragment. *Bio/Technology*, **10**, 163–167.

Chiswell, D. & McCafferty, J. (1992). Phage antibodies: will new "coliclonal antibodies" replace monoclonal antibodies? *Trends in Biotechnol.*, **10**, 80–84.

Cohen, J. (1993). Naked DNA points way to vaccines. *Science*, **259**, 1691–1692.

Datar, R. *et al.* (1993). Process economics of animal cell and bacterial fermentations: a case study analysis of tissue plasminogen activator. *Bio/Technology*, **11**, 349–357.

Denman, J. *et al.* (1991). Transgenic expression of a varient of human tissue type plasminogen activator in goat milk: purification and characterization of the recombinant enzyme. *Bio/Technology*, **9**, 839–842.

Ebert, K.M. *et al.* (1991). Transgenic production of a variant of human tissue type plasminogen activator in goat milk: generation of transgenic goats and analysis of expression. *Bio/Technology*, **9**, 835–838.

Finkeistein, A. & Silva, R.F. (1989). Live recombinant vaccines for poultry. *Trends in Biotechnol.*, **7**, 273–277.

Fox, J.L. (1992). FDA panel okays two factor VIII's with conditions. *Bio/Technology*, **10**, 15.

Gordon, K. *et al.* (1987). Production of human tissue plasminogen activator in transgenic mouse milk. *Bio/Technology*, **5**, 1183–1187.

Hodgson, J. (1993). Prestidigitation in the clinic. *Bio/Technology*, **11**, 159–162.

Johnston, M. & Hoth, D. (1993), Present status and future prospects for HIV therapies. *Science*, **260**, 1286–1293.

Kohler, G. & Milstein, C. (1975). Continuous culture of fused cells secreting antibody of predefined specificity. *Nature*, **256**, 495–497.

Lanzavecchia, A. (1993). Identifying strategies for immune intervention. *Science*, **260**, 937–944.

Legaz, M. *et al.* (1973). Isolation and characterization of human factor VIII (antihemophilic factor). *J. Biol. Chem.*, **248**, 3946–3955.

Markoff, A.J. *et al.* (1989). Expression of tetanus toxin fragment C in *E. coli*: its purification and potential use as a vaccine. *Bio/Technology*, **7**, 1043–1048.

Markoff, A.J. *et al.* (1990). Protective surface antigen P69 of *Bordetella pertussis*: its characterization and very high level expression in *Escherichia coli*. *Bio/Technology*, **8**, 1030.

Markwardt, F. (1991). Hirudin and derivatives as anticoagulant agents. *Thromb. Haemost.* **66**, 141–152.

McAleer, W.J. *et al.* (1984). Human hepatitis B vaccine from recombinant yeast. *Nature*, **307**, 178–180.

Mountain, A. & Adair, J.R. (1992). Engineering antibodies for therapy. In Tombs, M.P. (ed.) *Biotechnology and Genetic Engineering Reviews*, Vol. 10, pp. 1–142. Intercept.

Ogden, J. (1992). Recombinant haemoglobin in the development of red-blood-cell substitutes. *Trends in Biotechnol.*, **10**, 91–95.

Pluckthun, A. (1991). Antibody engineering: advances from the use of *Escherichia coli* expression systems. *Bio/Technology*, **9**, 545–555.

Rydel, T.J. & Tulinsky, A. (1991). Refined structure of the hirudin-thrombin complex. *J. Mol. Biol.*, **221**, 583–601.

Salk, J. *et al.* (1993). A strategy for prophylactic vaccination against HIV. *Science*, **260**, 1270–1272.

Sawyer, R.T. (1991). Thrombolytics and anticoagulants from leeches. *Bio/Technology*, **9**, 513–518.

Spalding, B.J. (1992). In hot pursuit of an HIV vaccine. *Bio/Technology*, **10**, 24–29.

Strausberg, R. & Link, R. (1990). Protein-based medical adhesives. *Trends in Biotechnol.*, **8**, 53–57.

Symons, P. *et al.* (1990). Production of correctly processed human serum albumin in transgenic plants. *Bio/Technology*, **9**, 217–221.

Thanavala, Y. (1989). Anti-idiotype vaccines. *Trends in Biotechnol.*, **7**, 62–66.

Vandegriff, K. (1992). Blood substitutes; engineering the haemoglobin molecule. In Tombs, M.P. (ed.) *Biotechnology and Genetic Engineering Reviews*, Vol. 10, pp. 403–453. Intercept.

Waldmann, T.A. (1991). Monoclonal antibodies in diagnosis and therapy. *Science*, **252**, 1657–1662.

Winter, G. & Milstein, C. (1991). Man-made antibodies. *Nature*, **349**, 293–299.

Wright, G. *et al.* (1991). High level expression of active human alpha-1-antitrypsin in the milk of transgenic sheep. *Bio/Technology*, **9**, 830–834.

Chapter 6

Therapeutic proteins: hormones, regulatory factors and enzymes

- Insulin.
  - Human insulin preparations.
- Glucagon.
- Gonadotrophins.
  - FSH, LH and hCG.
  - PMSG.
  - Superovulation and embryo transfer.
- Luteinizing hormone releasing hormone.
- Inhibin.
- Growth hormone.
- Other growth factors.
- Erythropoietin.
- Thyrotrophin.
- Corticotrophin (ACTH).
- Prolactin.
- Melanocyte stimulating hormone.
- Hormones of the posterior pituitary gland.
- Therapeutic enzymes.
  - Debriding and anti-inflammatory agents.
  - Nuclease treatment of cystic fibrosis.
  - Enzymes as digestive aids.
  - Other therapeutic enzymes.
- Cytokines.
  - Interferons.
  - Interleukins.
  - Tumour necrosis factor.
  - Colony stimulating factors.
  - Adverse clinical effects of cytokines.
- Cell adhesion factors.
- Further reading.

Administration of proteinaceous preparations forms an integral part of the management of a wide variety of disease states. Certain hormones, such as insulin, have been used clinically for many decades. Other proteins, such as interferon or somatotrophin, have been approved for clinical use within the last decade. In addition to these, there are many potential therapeutic proteins currently undergoing clinical evaluation. These include a variety of colony stimulating factors, growth factors, interferons, interleukins, tumour necrosis factors, dismutases and erythropoietins.

In the near future an explosive increase can be expected in the number of regulatory proteins approved for routine therapeutic use.

A number of factors have facilitated the rapid development of such proteinaceous therapeutic agents. Various regulatory proteins, such as growth factors and cytokines, have been characterized in detail. This fact, coupled with a greater understanding of the mechanism of action of such factors, allows their logical application as therapeutic agents.

Advances in basic understanding of how many regulatory factors function have also attracted the attention of industry.

Table 6.1. Some hormones, regulatory factors and enzymes of actual or potential therapeutic use

Protein product	Actual/intended use
Insulin	Treatment of diabetes mellitus
Glucagon	Treatment of insulin-induced hypoglycaemia
Gonadotrophins: Follicle stimulating hormone Luteinizing hormone Human chorionic gonadotrophin Pregnant mare serum gonadotrophin	Treatment of a variety of reproductive disorders in humans and animals Induction of superovulation in animals
Growth hormone	Induction of growth in children of short stature Treatment of burns, ulcers and some types of cancer Increase of milk yields in cattle (bovine growth hormone)
Additional growth factors	Various, including wound healing
Erythropoietin and additional haemopoietins	Stimulation of proliferation of a variety of blood cell types
Vasopressin	Treatment of diabetes insipidus
Oxytocin	To induce and maintain uterine contractions during childbirth
Enzymes	Various—anticoagulants, debriding agents and digestive aids
Cytokines	Various—treatment of cancer, AIDS, arthritis, asthma and other infectious diseases

Pharmaceutical companies and other commercial concerns hoping to capitalize on the applied spin-offs from such basic research have invested heavily in research and development. This serves to quicken the pace of scientific discovery and application of the scientific knowledge.

Recombinant DNA technology has had a profound influence on the development of many therapeutic protein products. This technology allows the routine large-scale production of many proteins which may be present in their natural source in only minute quantities.

This chapter details the majority of therapeutic proteins that are currently available or are the subject of clinical investigation. Some of these preparations are listed in Table 6.1.

INSULIN

Insulin is a polypeptide hormone which is produced by the beta cells of the pancreatic islets of Langerhans. The hormone was first isolated in 1921, and its amino acid sequence was determined in the late 1950s. Insulin exerts a wide variety of metabolic effects. It plays a central regulatory role in the metabolism of carbohydrates, and in the metabolism of protein and lipid. The level of secretion of insulin from the beta islet cells is primarily determined by blood glucose levels. Increases in the concentration of blood glucose induce insulin secretion which promotes the uptake of glucose by a number of tissues, particularly liver and muscle. This reduces blood glucose levels to normal values, which in turn decreases the rate of insulin release.

Failure to produce insulin results in the development of diabetes mellitus, a chronic disease characterized by an elevated level of blood glucose and by the presence of glucose in the urine. Increased rates of glycogenolysis, gluconeogenesis, fatty acid oxidation, ketone body production and urea formation result. Diabetes mellitus results in a decrease in fatty acid and protein biosynthesis. This disease is caused, in most cases, by irreversible damage of the insulin-producing beta cells. The underlying molecular events inducing such damage are poorly understood, although autoimmunity, viral infection and genetic predisposition are all believed to be contributory factors.

Diabetes mellitus may be controlled by the parenteral administration of suitable insulin preparations. Insulin was first administered to diabetic patients in 1922, just 1 year subsequent to its initial isolation. Another more rare form of diabetes, known as diabetes insipidus, is caused by a deficiency of vasopressin, a peptide hormone produced by the hypothalamus and subsequently stored in, and released from, the pituitary gland.

This hormone regulates resorption of water by the kidney. Persons suffering from diabetes insipidus may be treated by administration of vasopressin.

Insulin is initially synthesized in the pancreatic beta cells as pre-proinsulin (Figure 6.1). This molecule contains a 23 amino acid amino-terminal signal sequence, which directs the protein through the rough endoplasmic reticulum membrane to the lumen of the endoplasmic reticulum. Here, the leader sequence is removed from the preproinsulin molecule by a specific signal peptidase, yielding proinsulin.

The proinsulin molecule remains within the lumen of the endoplasmic reticulum. Small vesicles containing proinsulin subsequently bud off from the endoplasmic reticulum and fuse with membranous structures termed the Golgi apparatus. Proinsulin-containing vesicles, in turn, pinch off from the Golgi apparatus. These vesicles are often termed coated secretory granules, as they exhibit a coat composed of a protein (clathrin) on their outer surface. As they move further away from the Golgi body, these vesicles lose their clathrin coat, forming non-coated secretory granules (vesicles).

The conversion of proinsulin into insulin takes place in the coated secretory vesicles. This process involves the proteolytic cleavage of the proinsulin molecule, yielding mature insulin and the connecting peptide (C peptide). The mature insulin consists of two polypeptide chains, the A and B chains, joined by two disulphide cross-links, and has a molecular weight of 5 800 Da. The insulin A chain contains 21 amino acid residues, whereas the B chain consists of 30 amino acid residues. The A chain contains one intrachain disulphide linkage. Conversion of proinsulin to insulin generates a polypeptide sequence which originally bridged the insulin A and B chains of the proinsulin molecule. Upon the formation of mature insulin, two dipeptides are removed from either end of this bridge peptide yielding a slightly shorter peptide, the C peptide.

In addition to mature insulin, secretory granules also contain low levels of proinsulin as well as the C peptide and some proinsulin-derived amino acids. Non-coated secretory granules serve as the storage depot of insulin in the beta cells. The insulin is normally stored as a hexamer, consisting of six molecules of insulin stabilized by two atoms of zinc. Indeed, the addition of a small quantity of zinc to purified insulin results in the formation of characteristic rhombohedral crystals, with the basic crystal unit consisting of the zinc-containing insulin hexamer. The secretory granules release their contents into the blood by the process of exocytosis. Insulin is released in this manner only upon stimulation by specific secretory signals, the most significant of which is an increase in the blood glucose concentration.

Figure 6.1. Synthesis of human insulin from preproinsulin. The initial proteolytic event involves removal of a 23 amino acid signal sequence from the amino terminal end of preproinsulin, thus yielding proinsulin. Proinsulin is converted into insulin by additional proteolytic events resulting in the generation not only of insulin but also a 30 amino acid sequence (connecting peptide) and two dipeptide moieties. Mature insulin thus consists of two polypeptides, the A and B chains. The B chain contains 30 amino acid residues whereas the A chain contains 21 amino acids. The chains are covalently linked via two interchain disulphide linkages. One intrachain disulphide linkage is also present in the A chain ⟶

Insulins obtained from various different species are similar, though not identical, in their overall amino acid sequence. Rats and mice synthesize two insulin types which differ slightly from each other in amino acid sequence. The amino acid sequence of

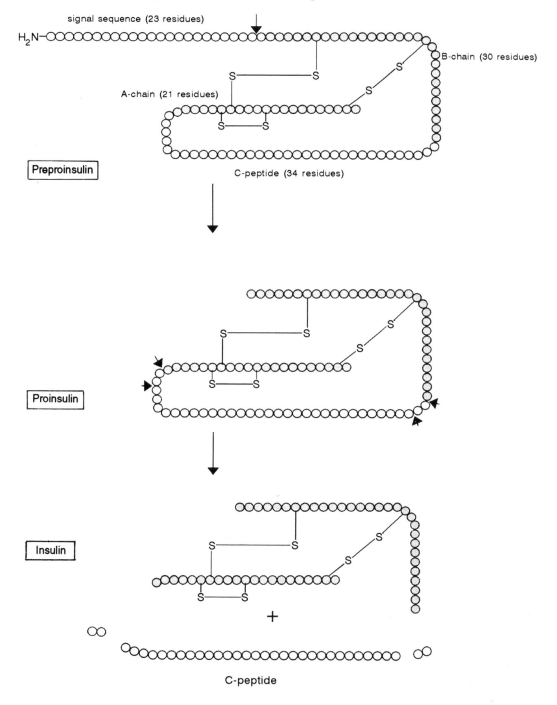

A and B chains from different species are generally very similar. Porcine insulin differs in sequence from human insulin by only one amino acid, the C-terminal amino acid of the B chain. Bovine insulin differs from the human hormone by three amino acids, while sheep insulin differs by four amino acids. The amino acid sequence of the insulin-derived C peptide, however, differs greatly from species to species.

Insulin preparations administered to diabetic patients have traditionally been obtained from the pancreatic tissue of healthy slaughterhouse animals, such as pigs or cattle. Porcine insulin may be considered to have a slight advantage over bovine preparations, as its amino acid sequence is more closely related to the native human insulin sequence. For this reason it would be marginally less immunogenic than the bovine product.

Purified insulin was first utilized therapeutically in the early 1920s. Such preparations were generally extracted from the pancreas of pig or ox by a technique employing acid–alcohol precipitation. The insulin preparations obtained were quite impure. Although insulin was first crystallized in 1926, the crystallization process was poorly understood, and researchers found it impossible to crystallize insulin consistently from crude extracts. The discovery in 1934 that zinc promoted crystallization of insulin allowed commercial producers to grow insulin crystals from crude extracts of pancreatic tissue. The crystals were isolated and subjected to recrystallization in order to purify the protein further. Such insulin preparations are often termed "conventional" insulins. Although the crystallization process resulted in significant purification of the insulin, such preparations are considered to be relatively crude by modern standards.

In addition to insulin itself, conventional insulin preparations also contain modified insulin molecules, such as arginine insulin, desamidoinsulin, insulin esters and insulin dimers. Desamido-insulins or their ethyl esters are formed during the acid–alcohol extraction of insulin. Under such conditions the amido groups of asparagine residues may be hydrolysed. Conventional insulin preparations also contain appreciable quantities of proinsulin in addition to other polypeptides such as vasoactive intestinal peptide, somatostatin and glucagon. A high molecular weight protein fraction is observed on occasions, although the recrystallization process effectively fractionates the insulin from such molecules.

Contaminants present in conventional insulin preparations can adversely affect the insulin content of the final preparation, and may also lead to clinical complications upon its administration. Many of the higher molecular weight contaminants present display protease activity. Preparations containing such unwanted activities are normally maintained in solution at acidic pH values in order to minimize proteolysis of the insulin molecule. Contaminants present are generally immunogenic. This fact is particularly relevant when one considers that many diabetic patients require insulin injections several times daily throughout their life. Repeated administrations could promote a strong immunological response, even against weak immunogens. Immunological reactions may also lead to inflammation and destruction of the tissue surrounding the site of injection.

Insulin obtained from animals, in particular from pigs, elicits only weak immunological responses when administered to humans. Nonetheless, anti-insulin antibodies are often detected in the serum of diabetic patients. The presence of such neutralizing antibodies may render necessary the administration of increased doses of insulin in order to achieve the desired biological effect. Furthermore, as antibody-bound insulin is largely resistant to the normal insulin degradative processes, such antibodies may seriously affect the activity versus time profile of the administered hormone.

In contrast to mature insulin, proinsulin contained in conventional insulin preparations is quite immunogenic. As previously discussed, the C peptide sequence of proinsulin may differ significantly from species to species.

Recrystallized insulin preparations are now usually subjected to additional chromatographic purification steps in order to reduce further the level of various contaminants traditionally associated with conventional insulin preparations. Gel filtration was the first chromatographic technique used in the further purification of insulin. Such a gel filtration step effectively separates insulin from higher molecular weight contaminants such as proteolytic enzymes, proinsulin and insulin dimers. Contaminants coeluting with the insulin fraction include desamido insulin and arginine insulin which tend to be of limited clinical significance. Other contaminants that gel filtration fails to remove from insulin preparations include a variety of pancreatic peptides of similar molecular weight to that of insulin. As insulin is eluted from gel filtration columns in one peak, preparations purified by this method are sometimes referred to as single-peak insulins.

In order to be economically viable, the gel filtration columns used to purify recrystallized insulin must be large. Some manufacturers produce a system of stack columns for this purpose (Figure 6.2). The total bed volume of the columns used is 96 litres. A sample volume of 2 litres may be applied at any one time, from which 50 g of purified insulin may be obtained. A single column run takes approximately 7 hours. Additional chromatographic steps may be used to reduce further the level of contaminants present in the purified insulin preparations. Process-scale ion-exchange chromatographic columns are often utilized for this purpose.

It has been recommended that the proinsulin content of modern insulin preparations should not exceed 10 ppm. It is also

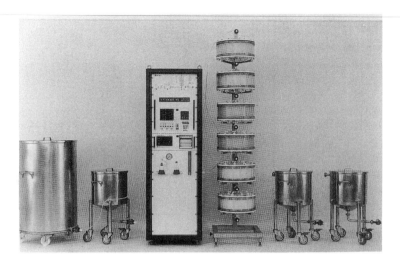

Figure 6.2. An automated system for production scale insulin purification. Its capacity is 600 g common recrystallized insulin per week. Equipment (from left to right): 300 litre stainless steel buffer tank, 50 litre conical stainless steel sample tank, control cabinet, six stack columns and two collection tanks. The columns can also be arranged in two 3-section stacks placed side by side (photograph courtesy of Pharmacia Fine Chemicals)

recommended that the content of high molecular weight protein contaminants should not exceed 10 parts per thousand or 1%.

Human insulin preparations

Since the 1970s methods have been developed that allow the production of relatively large quantities of human insulin. Some methods such as chemical synthesis of the insulin molecule, although possible, are economically impractical. Other methods have proved to be both economically viable and technically sound. These include the enzymatic modification of porcine insulin and the expression of the human insulin cDNA in recombinant microorganisms.

Human insulin was first produced by enzymatic modification of porcine insulin in the early 1970s. Human insulin differs from porcine insulin by a single amino acid. Threonine forms the carboxyl-terminus, residue 30, of the human insulin B chain, whereas an alanine residue is found in this position in the porcine molecule. Treatment of intact porcine insulin with trypsin results in the proteolytic cleavage of the B chain between residues 22 and 23 (and also between residues 29 and 30), effectively removing the carboxy-terminal octapeptide from the B chain. A synthetic octapeptide whose sequence is identical to the analogous human octapeptide may then be coupled to the trypsin treated porcine insulin. This process, outlined in Figure 6.3, effectively converts porcine insulin into human insulin. Human insulin produced by this method has been used clinically for a number of years.

Human insulin may also be produced by the application of recombinant DNA technology. Recombinant human insulin

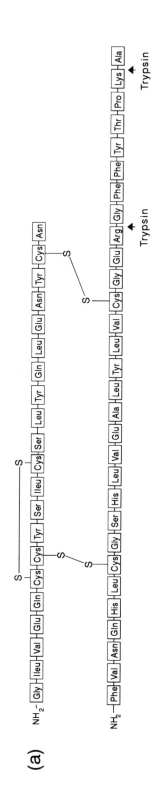

(a)

(b)

Figure 6.3. Amino acid sequence of porcine insulin (a). Trypsin cleavage sites are also indicated. Trypsin therefore effectively removes the insulin carboxyl terminus B chain octapeptide. The amino acid sequence of human insulin differs from that of porcine insulin by only one amino acid residue. Porcine insulin contains an alanine residue at position 30 of the B chain whereas human insulin contains a threonine residue at that position. Insulin exhibiting a human amino acid sequence may thus be synthesized from porcine insulin by treating the latter with trypsin, removal of the C terminus fragments generated, and by replacing this with the synthetic octapeptide shown in (b)

preparations represented the first commercial health-care product produced by this technology to be approved for widespread clinical use. The successful production of human insulin in microorganisms ensures that future insulin supplies will not depend on the availability of pancreatic tissue from slaughterhouse animals. The quantity of purified insulin obtained from the pancreas of one pig would satisfy the insulin requirement of one diabetic patient for only 3 days. Production in recombinant organisms also eliminates the risk of accidental transmission of viral disease from infected pancreatic tissue to diabetic patients.

The first insulin preparations obtained by methods of genetic engineering was produced by inserting the cDNAs coding for the insulin A and B chains into individual *E. coli* cells (strain K12). The A and B chains were then purified from the different recombinant systems and subsequently joined by disulphide bond formation. Human insulin may also be produced by expressing the nucleotide sequence coding for human proinsulin in *E. coli*. The proinsulin molecule may subsequently be converted into native insulin by the enzymatic removal of the connecting peptide. This method of production has become somewhat more popular than the method relying on the initial independent expression of the two insulin peptide chains in two separate bioreactors; this is largely due to the fact that only a single fermentation and subsequent purification scheme is necessary. Recombinant human insulin is now produced industrially on a large scale, with 50 000 litre bioreactors being used in routine production runs. Human insulin has also been produced in recombinant yeast systems.

Although the insulin produced by recombinant methods is identical to that of the native hormone, any impurities present in the recombinant preparations will be of microbial rather than animal origin. However, modern protein purification methods minimize the level of such impurities in the final product. Insulin extracts obtained from animal sources would also be heavily contaminated with non-insulin protein impurities prior to its purification. The insulin-producing beta cells of the islet of Langerhans constitute less than 1% of total pancreatic tissue.

A number of additional techniques may be employed to increase still further the purity of insulin preparations. Such techniques include hydrophobic interaction chromatography and reverse phase HPLC. These techniques, which serve to complement rather than replace existing purification techniques, have been discussed in Chapter 3. Industrial-scale purification of recombinant human insulin could thus consist of the series of steps outlined in Figure 6.4.

While the introduction of additional purification steps decreases the overall yield, they serve to increase the purity of the final

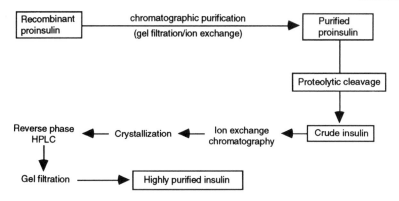

Figure 6.4. Purification of recombinant insulin produced as proinsulin in *E. coli.* Initially the proinsulin is purified chromatographically and subsequently converted into insulin by proteolytic cleavage. The insulin may be further purified by ion exchange chromatography followed by a crystallization step. The redissolved crystals can be applied to a reverse phase HPLC column. A gel filtration step is normally utilized to produce the final product

product. Purity levels approaching or exceeding 98% with an acceptable overall yield of insulin are now attainable. Reverse phase HPLC (RP-HPLC) represents a particularly powerful purification step. This chromatographic technique is capable of effectively separating the insulin molecule from a wide range of contaminants, many of which are difficult to remove by conventional chromatographic procedures. Large scale RP-HPLC columns, with bed volumes approaching 80 litres, have been built to facilitate industrial-scale insulin purification. Such columns may be used to purify several hundred grams of insulin in a single run.

Both HPLC and FPLC are also used extensively on an analytical scale during the production of modern high-purity insulin. These techniques play a significant role in in-process quality control, and in the assessment of final product purity. The speed and sensitivity with which these techniques function render them particularly suitable to such tasks. Analytical-scale RP-HPLC, for example, can distinguish between various insulin molecules that differ by a single amino acid residue. It can also detect various modified forms of insulin such as desamidoinsulin or formyl insulin, or insulin polymers.

When injected directly into the blood stream, insulin has a half-life of only minutes. Clearly such a short half-life would render the clinical management of diabetes extremely difficult. Administration by subcutaneous injection facilitates a more prolonged release of hormone from the site of injection into the blood stream. This has become the preferred method of insulin administration to diabetic patients. Insulin preparations may also be formulated in a number of ways in order to increase its duration of action. The hormone may, for example, be complexed with a protein such as protamine which further retards its release into the general circulation (protamines are a group of basic proteins found in association with nucleic acids in the sperm of certain species of fish). Alternatively, zinc may be added to the final preparation in order to promote the growth of insulin crystals in the resultant zinc–insulin suspension.

Soluble insulins are often referred to as short-acting insulins, having a duration of activity of up to 8 hours. Peak hormonal activity is observed in such cases after 3–4 hours. Long-acting insulin preparations, such as those containing protamine and/or zinc, often exhibit a duration of action of up to 36 hours.

GLUCAGON

Glucagon, like insulin, is a polypeptide hormone derived from the pancreas. It is synthesized by the alpha cells of the islets of Langerhans, and it is also synthesized by related cell types found in the gastrointestinal tract. Glucagon consists of a single polypeptide chain of 29 amino acids and has a molecular weight of approximately 3500 Da. The amino acid sequence of the hormone isolated from different animal species is almost invariably the same. As in the case of insulin, glucagon seems to be synthesized initially as proglucagon. Proglucagon is converted to glucagon by two separate proteolytic events.

Glucagon is a hyperglycaemic hormone, inducing an increase in the blood glucose concentration of recipients. In this way it opposes one of the major physiological actions of insulin. The increase in blood glucose level is promoted by stimulating an increase in the rate of gluconeogenesis and glycogen breakdown in the liver. Glucagon also promotes an increase in the rate of lipolysis and a decrease in the rate of glucose utilization by adipose tissue and muscle. It is occasionally used clinically to reverse insulin-induced hypoglycaemia in diabetic patients.

GONADOTROPHINS

Gonadotrophic hormones, as the name suggests, exert their primary effect on the male and female gonads. A number of gonadotrophins may be used clinically in the treatment of various human clinical conditions and to induce a superovulatory response in female animals.

The most commonly used gonadotrophins include follicle stimulating hormone (FSH) and luteinizing hormone (LH), which is also referred to as interstitial cell stimulating hormone (ICSH). Both LH and FSH are synthesized by the pituitary gland. Another gonadotrophin, human chorionic gonadotrophin (hCG) is produced by the placenta of pregnant women. An additional gonadotrophic hormonal preparation, menotrophin, may be isolated from the urine of postmenopausal women. Pregnant mare serum gonadotrophin (PMSG) is a gonadotrophic hormone which is found in the serum of pregnant mares.

FSH, LH and hCG

FSH, LH and hCG are glycoproteins of molecular weight 34 000, 28 500 and 36 600 Da respectively. Both LH and FSH consist of approximately 16% carbohydrate while hCG contains over 30% carbohydrate. All are dimers composed of α and β subunits. In any one animal species, the α subunit of all three is identical whereas the β subunit differs; it is thus the β subunit that confers on the particular molecule its distinct biological characteristics. The β subunit of hCG exhibits significant amino acid sequence homology with the β subunit of LH. It is therefore not surprising that hCG displays similar biological activities to LH.

In the male, FSH stimulates the production of large numbers of spermatocytes by the Sertoli cells of the seminiferous tubules. LH stimulates production of testosterone, the principal male sex hormone, by the Leydig cells of the testes. In the female, FSH and LH play critical roles in regulating the reproductive cycle.

The ovary represents the primary female reproductive organ. It produces egg cells (ova) in addition to steroid hormones. The ovary contains numerous follicles, each follicle consists of a single ovum surrounded by two layers of cells. The inner layer of cells are termed granulosa cells and these respond primarily to FSH. The granulosa cells produce oestrogens, the major female sex hormones. The outer follicular layer is composed of theca cells. These cells, which fall primarily under the hormonal influence of LH, synthesize a variety of steroids which are subsequently utilized by the granulosa cells during synthesis of oestrogen.

At the commencement of a normal menstrual cycle a group of follicles begin to mature, a process largely stimulated by FSH. Shortly thereafter a single dominant follicle emerges, and the remaining follicles regress. This first half of the menstrual cycle is often referred to as the follicular phase. The growing follicle begins to secrete increasing levels of oestrogens which ultimately trigger a surge in the concentration of LH at mid-cycle, approximately day 14 of the human female cycle. This, in turn, triggers ovulation, the release of the egg from the mature follicle, with the resultant conversion of the ruptured follicle into a structure known as the corpus luteum. The menstrual cycle now has entered its second phase, known as the luteal phase. The corpus luteum secretes oestrogens and the steroid hormone progesterone. Progesterone serves to prepare the lining of the uterus, the endometrium, for implantation of the fertilized ovum, and is required to support the growing embryo.

If fertilization does not occur, the maximum life-span of the corpus luteum in humans is normally 14 days, during which time it steadily regresses. This in turn results in a decrease in the levels of the corpus luteal hormones, oestrogen and progesterone. Withdrawal of such hormonal support results in the shedding of the endometrial tissue, which is discharged from the body during menstruation.

If follicular rupture is followed by fertilization of the released ovum, the corpus luteum does not regress and continues to produce progesterone. In the pregnant female, the corpus luteum is maintained by hCG, which is secreted by the placenta; this hormone is also found in urine, and its detection forms the basis of many pregnancy testing kits.

FSH and LH essentially regulate the reproductive process in females (Figure 6.5). In addition, FSH and LH also regulate the development and maintenance of male fertility. These hormones may be employed therapeutically to treat some forms of sterility, or other medical conditions caused by low circulatory levels of the hormones.

Figure 6.5. Changes in plasma FSH (a) and LH (b) levels during the menstrual cycle of a healthy human female

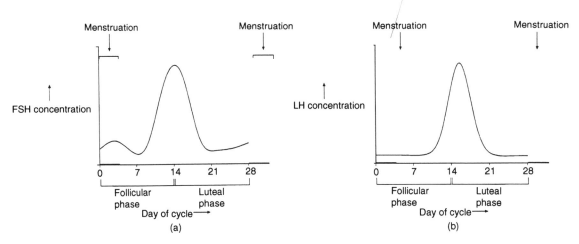

FSH may be isolated from the pituitary gland. Supply of human FSH from such sources is obviously very limited. FSH is also found in the urine of postmenopausal women. Preparations obtained from this source also contain LH activity, and are termed menotrophin. Menotrophin is purified from urine by a variety of fractionation procedures, as well as by chromatographic techniques such as ion exchange chromatography. Menotrophin preparations may be subjected to further chromatographic steps in order to remove most or all of the contaminating LH activity.

Menotrophin is often used clinically to treat a variety of reproductive complications such as anovulatory infertility, which is caused or exacerbated by low levels of circulating FSH.

Human luteinizing hormone may also be isolated from the pituitary gland. As in the case of FSH, this does not represent a commercially viable source of this hormone. As mentioned earlier, hCG exhibits many biological activities that are similar if not identical to those of LH. Since hCG can be purified relatively easily from the urine of pregnant women, hCG preparations have found widespread medical application in the treatment of conditions caused by lack of LH activity, and have been used clinically to treat infertility and delayed puberty in males. It is also administered to females together with FSH in the treatment of anovulatory infertility. In such cases, the administered FSH stimulates development of the follicle while subsequent administration of hCG initiates the mid-cycle LH surge, stimulating final maturation and subsequent rupturing of the follicle.

PMSG

Pregnant mare serum gonadotrophin, as the name suggests, is obtained from the serum of pregnant mares. It is present in serum at elevated levels between days 40 and 130 of gestation. PMSG is a gonadotrophic hormone consisting of α and β subunits and containing up to 45% carbohydrate, most of which is associated with the β subunit. The carbohydrate component is particularly high in galactose, glucosamine and sialic acid.

PMSG is a unique gonadotrophin insofar as it exhibits both FSH-like and LH-like bioactivities. Its biological specificity is conferred upon PMSG by its β subunit. PMSG is secreted by a series of small, cup-shaped outgrowths found in the horn of the pregnant uterus. These structures, which are highly enriched with PMSG, are termed endometrial cups and are found only in equines. They first become visible on day 37–40 of gestation, and reach maximum size by approximately day 70, after which they undergo steady regression. The endometrial cups are fetal rather than maternal in origin.

PMSG may be purified from serum by a procedure involving pH fractionation, alcohol precipitation, and several chromatographic steps utilizing both gel filtration and ion exchange media. This hormone fails to enjoy widespread clinical use in humans but it is used to induce a superovulatory response in certain animals.

Superovulation and embryo transfer

Gonadotrophic hormones are routinely used to induce superovulation in certain animals, most notably cattle. The aim of superovulation is to induce the simultaneous growth of multiple follicles so that several eggs are available for fertilization at the time of mating. After mating, the embryos may be collected from the mother by surgical or non-surgical techniques. These harvested embryos may then be implanted in several recipient animals which carry the offspring to term. By the use of such technologies the reproductive potential of very valuable animals, or animals with desirable genetic traits, is multiplied. The recipient animals, though less valuable than the donor, must be of reasonably high quality in order to obtain satisfactory results. Superovulation and embryo transfer are used by some farmers to speed up the genetic improvement of their herds.

Superovulation involves administration of exogenous gonadotrophins in order to stimulate follicular growth. In many agriculturally important animals, only a single follicle develops to maturity. Injection of exogenous FSH increases circulatory levels of this hormone well above its normal physiological range. This disturbs the natural hormonal balance which regulates the menstrual cycle, and can result in the development and maturation of several follicles. Gonadotrophins administered in order to induce superovulation include menotrophin, porcine pituitary FSH (P-FSH) or PMSG.

P-FSH is FSH purified from pituitary glands excised from slaughterhouse pigs. Porcine as opposed to bovine pituitaries are normally used, as the bulk of the purified product is used to superovulate cattle. The use of FSH preparations obtained from species other than the species of the intended recipient is normally encouraged, as it decreases the potential risk of transmission of species-specific diseases associated with the use of biological products obtained from slaughtered animals.

FSH is usually purified from porcine pituitary extracts by a variety of fractionation techniques, including precipitation with salts and/or alcohol. Chromatographic steps are frequently employed. Most FSH preparations available commercially contain several contaminating proteins. The most consistent superovulatory response is obtained when FSH preparations exhibiting a low LH content are used in superovulatory experiments.

Superovulatory regimens employing PMSG normally involve the administration of a single dose of this hormone. One dose is sufficient as PMSG is removed from circulation at a very slow rate. In cattle, clearance of PMSG may take up to 120 hours. The extended half-life of PMSG preparations is due to their high content of N-acetyl neuraminic acid. The long half-life of PMSG is regarded by some as detrimental to its effectiveness as a superovulatory hormone as it can lead to continued stimulation of follicular growth after ovulation, and lead to a reduced number of recoverable viable embryos. Antibodies raised against PMSG may be administered in order to mop up residual PMSG after an appropriate time period. It is now generally accepted that a greater ovulatory response is obtained by administration of FSH rather than PMSG.

Most superovulatory regimens involving FSH preparations require its administration twice daily for approximately 4 days. Regular injections of the hormone are needed to sustain multiple follicular growth, as FSH has a relatively short half-life in serum. Efforts are made to prolong the effectiveness of each injection by administering the preparation subcutaneously. This injection programme is subsequently followed by the administration of a single dose of LH, which is considered necessary to promote final maturation of the follicle and its subsequent release from the ovum.

Superovulatory responses may be variable even when FSH preparations are used to induce follicular growth. The number of viable embryos recovered from a superovulated cow can vary between 0 and 50 or over, though 5–8 embryos represent the average obtained. The variability associated with superovulatory regimens is caused by numerous factors, only some of which are clearly defined. The general health and condition of the animal influences the process, as do the exact protocols used and the experience and technical ability of the personnel involved. The ratio of FSH to LH activity in the preparations administered also seems to affect the final outcome, with preparations of FSH containing reduced LH activity appearing to be most effective. Purified recombinant FSH preparations may prove useful in this regard.

The cDNAs for the α and β subunits of bovine FSH have been cloned and expressed in a mouse epithelial cell line. Clones producing biologically active FSH were selected and their fermentation scaled up for routine production purposes. The resultant recombinant FSH may be subsequently subjected to a three-stage purification process, yielding a product which is greater than 95% pure. Such recombinant preparations would of course be devoid of contaminating LH activity.

LUTEINIZING HORMONE RELEASING HORMONE

Luteinizing hormone releasing hormone (LH-RH) is a decapeptide synthesized in the hypothalamus, and is also known as gonadotrophin releasing hormone, or gonadorelin (Figure 6.6). LH-RH secreted by the hypothalamus is conveyed directly to the pituitary gland, where it stimulates the synthesis and release of LH and FSH.

Figure 6.6. Amino acid sequence of LH-RH

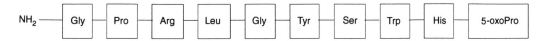

Hypothalamic regulatory factors are synthesized in minute quantities, thus their preparation from hypothalamic tissue in clinically useful quantities is impractical. Fortunately, such regulatory factors can be conveniently manufactured by direct chemical synthesis. LH-RH has been administered to both humans and animals in order to treat certain forms of subfertility or infertility.

As administration results in an increase in the secretion of gonado-trophins it improves the conception rate by enhancing normal hypothalamopituitary function. The releasing factor has also been used clinically in the treatment of certain malignant neoplasms and cryptorchidism—a condition characterized by the failure of the testes to descend into the scrotum during normal development.

In many instances analogues of LH-RH are used clinically. These generally exhibit greater potency or longer duration of action than the native releasing factor. Although LH-RH stimulates the synthesis and release of both FSH and LH, changes in the blood concentration of these two pituitary gonadotrophins do not always directly parallel each other. Other agents must exist that can selectively influence the concentration of one or other of these two hormones. One such agent currently receiving much attention is inhibin.

INHIBIN

Inhibin is a glycoprotein hormone produced by the gonads. It is synthesized in the Sertoli cells of the testes and by granulosa cells in the ovary. The hormone is a dimer, consisting of α and β subunits. Two forms of the inhibin molecule have been characterized, inhibin A and inhibin B. The α subunit is identical in both cases while their respective β subunits differ in amino acid sequence.

The synthesis of inhibin is stimulated by increased circulatory concentrations of FSH. Inhibin plays an important role in feed-back inhibition of FSH action, as it functions to selectively suppress the secretion of this hormone by the pituitary gland. Owing to its ability to decrease circulatory FSH levels, inhibin is being investigated as a potential contraceptive for both males and females.

GROWTH HORMONE

Growth hormone (GH), also known as somatotrophin, is produced by the anterior pituitary gland and is required for normal body growth and lactation. The hormone promotes general protein synthesis and also plays a role in the regulation of carbohydrate and lipid metabolism, and it promotes bodily retention of various minerals and other elements essential for normal growth.

Secretion of growth hormone from the pituitary is stimulated by a hypothalamic regulatory factor termed growth hormone releasing hormone (GHRH). Growth hormone release is inhibited by a second hypothalamic factor termed growth hormone release-inhibiting hormone or somatostatin. Somatostatin is a tetradecapeptide of molecular weight 1600 Da. It is produced primarily by the hypotha-lamus but is also synthesized by the delta cells of the pancreatic islets, and related cells of the gastrointestinal tract. Somatostatin inhibits the release of a variety of pituitary hormones in addition to growth hormone. It is also known to inhibit the release of insulin and glucagon from the pancreas.

Growth hormone initiates its anabolic effect by binding to specific cell surface receptors. The receptor consists of three distinct domains, an extracellular ligand binding domain, a transmembrane domain and an intracellular effector domain. The exact molecular mechanisms that mediate the effects of growth hormone upon binding to its receptor are poorly understood. A truncated form of the GH receptor, which approximates to the extracellular ligand binding domain, is found in serum. This serum protein is capable of binding GH and may play an important role in its clearance from the body. The binding complex consists of one molecule of GH and two molecules of the receptor-like binding protein (Figure 6.7).

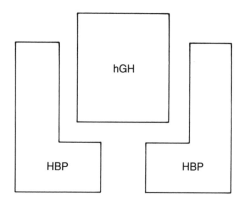

Figure 6.7. Binding of the human growth hormone (hGH) to two molecules of the hormone binding protein (HBP) occurs naturally in serum. The hormone binding protein equates to the extracellular ligand binding domain of the hGH receptor located in the plasma membrane of responsive cells. Binding of the first molecule of HBP renders more efficient the binding of the second HBP molecule to the growth hormone

Growth factor receptors have been detected in a number of tissues. The hormone exerts its primary influence on the liver, where it stimulates the synthesis of somatomedin or insulin-like growth factor 1 (IGF-1). IGF-1 directly mediates most of the growth promoting effects of GH. Other somatomedins are synthesized in the brain and in some fetal tissues. The family of somatomedins include not only IGF-1 but also IGF-2 and multiplication stimulating activity (MSA).

Because of their anabolic effects, and because the IGFs exhibit certain biological activities broadly similar to those of insulin, the somatomedins have generated considerable clinical and industrial interest. Prior to the advent of recombinant DNA technology, the somatomedins were purified from large volumes of plasma. Several commercial concerns are currently sponsoring large-scale clinical trials designed to assess the efficacy of IGF-1 in treating a number of clinical conditions.

Insufficient production of growth hormone in humans leads to dwarfism; excessive production of the hormone results in gigantism or acromegaly, a condition characterized by an increase in the size of the hands, feet and face. Growth hormone may be used clinically to treat hypopituitary dwarfism. The species-specific nature of GH makes it essential to use material from human pituitary glands.

Human growth hormone (hGH) is a polypeptide consisting of 191 amino acid residues. It contains two intrachain disulphide links and has a molecular weight in excess of 21 000 Da. Until recently, the sole source of hGH was human pituitary glands and for this reason preparations of hGH were not widely available. Furthermore, such preparations were generally not homogeneous, containing contaminants including a number of modified forms of growth hormone in addition to unrelated proteins.

The expression of the cDNA for hGH in *E. coli* facilitated the large-scale production of ample quantities of growth hormone. The recombinant product is identical in amino acid sequence to the native molecule with the exception of an additional methionine residue, due to the presence of an AUG start codon at the 5′ end of the cDNA. This methionyl-hGH exhibits identical biological activity to that of the native molecule and was first approved for clinical use in 1985.

The recombinant growth hormone may be purified to homogeneity by a series of fractionation steps including ammonium sulphate precipitation, ion-exchange chromatography on diethylaminoethyl cellulose and gel filtration chromatography on Sephacryl S-200. The ready availability of large quantities of recombinant growth hormone has accelerated investigations assessing potential applications of this hormone. Clinical evidence amassed to date suggests that the hormone may be of considerable value in treating burns, peptic ulcers, osteoporosis, and some forms of cancer. The hormone may also be used to increase the growth rate of children of short stature. Trials aimed at assessing the effectiveness of hGH in the treatment of HIV infection are currently under way.

In addition to hGH, growth hormone obtained from a number of other species has also been produced in recombinant systems. Recombinant bovine growth hormone (rbGH) may be used to increase milk yield from dairy cattle. As was the case with human growth hormone, the advent of recombinant DNA technology allowed the production of industrially significant quantities of the recombinant bovine somatotrophin. Several biotechnology companies have since developed such bovine somatotrophin preparations for use with dairy cattle.

Treatment of dairy cattle with bGH is not considered to pose a threat to the health of people consuming the milk or meat of such animals. This conclusion is based on a number of important experimental observations. Bovine growth hormone is not biologically active in humans. Furthermore, like most other proteins ingested, bGH would be degraded by the human gastrointestinal tract. Toxicity studies have shown that recombinant bGH is not orally active in rats, although this species is responsive to bGH when administered parenterally.

Administration of bGH results in elevated concentrations of IGF-1 in cows' milk. Bovine IGF-1 is identical to human IGF-1. However, toxicity studies have shown that bovine IGF-1 is inactivated when administered orally to rats. Furthermore, the concentration of IGF-1 reported to be present in the milk of cows treated with rbGH is no higher than the levels normally found in human milk.

Human somatostatin, the hypothalamic peptide which inhibits GH release from the pituitary gland, may also be used clinically. Administration of somatostatin has been found to benefit some patients suffering from gastrointestinal haemorrhage. It has also been used in the management of certain hormone-secreting tumours. Somatostatin may be obtained from hypothalamic extracts but is usually synthesized chemically. The native molecule exhibits a short half-life, and several analogues with prolonged duration of action have been developed. Such analogues include octreotide, an octapeptide which has been used to treat a number of tumour types.

OTHER GROWTH FACTORS

A number of other growth factors are currently generating considerable interest within the biopharmaceutical industry (Table 6.2). Various preparations of these factors are undergoing extensive clinical trials, and many are likely to be approved for general medical use in the near future. All of these growth factors exhibit

Table 6.2. Some growth factors and their intended applications which are currently undergoing clinical trials

Growth factor	Intended applications
Epidermal growth factor	Wound healing Skin ulcers Corneal/cataract surgery
Fibroblast growth factor	Chronic soft tissue ulcers Leg and foot ulcers Venous stasis
Insulin-like growth factor 1	Type II diabetes
Platelet-derived growth factor	Skin ulcers Diabetic ulcers

Information from the American Pharmaceutical Manufacturers Association

marked mitogenic effects by inducing cellular proliferation and division.

Epidermal growth factor (EGF) is perhaps the most extensively characterized of all growth factors. Human EGF was first isolated from urine and was termed urogastrone, owing to its ability to inhibit the secretion of gastric acid. EGF from mouse salivary glands has also been isolated and extensively characterized. EGF has been proved to exert a powerful mitogenic influence over a wide variety of cell types, including epithelial and endothelial cells and fibroblasts. The skin represents the major target tissue of EGF, where it plays an important role in growth and development of the epidermal layer. It also serves as a mitogen for a number of cell lines in culture. In addition to its presence in urine, human EGF has also been detected in serum, in the duodenum and in the salivary gland.

EGF isolated from both mouse and human sources consists of a 53 amino acid polypeptide chain, exhibiting 70% homology in amino acid sequence, and containing three intrachain disulphide linkages. The molecular weight of mouse EGF is 6041 Da, while that of human EGF is 6201 Da. EGF is initially synthesized as part of a much larger precursor protein of a molecular weight in the region of 128 000 Da. Mature EGF is released from this precursor by proteolytic cleavage. Both human and mouse EGF have been produced as heterologous proteins in recombinant bacterial systems. Owing to the large size of the mRNA coding for the EGF precursor, researchers have found it easier to synthesize a nucleotide sequence which codes for the 53 amino acids of the mature EGF molecule. The cDNA may then be expressed in *E. coli* systems.

The EGF receptor is a transmembrane protein exhibiting tyrosine kinase activity. The receptor has been discovered on the cell surface of many different cell types. Binding of EGF to its receptor induces a mitogenic response in susceptible cells by promoting increased protein and nucleic acid synthesis, by enhancing transport of certain metabolites into the cell and by increasing the rate of various intracellular metabolic activities.

Fibroblast growth factor (FGF) was first isolated from bovine brain and pituitary glands. This growth factor was found to exhibit a marked mitogenic effect on fibroblast cell lines in culture. Fibroblasts represent a particular cell type widely distributed in connective tissue. These cells produce the ground substance of such tissues in addition to the precursors of collagen and elastin fibres. It has since been demonstrated that FGF induces growth and division of numerous other cell types including osteoblasts, glial cells, smooth muscle cells and vascular endothelial cells. Its ability to induce blood vessel growth *in vivo* heightens expectations of its potential therapeutic applications.

Two types of FGF have been identified. These are termed acidic and basic FGF respectively. Two related forms of acidic FGF have been isolated. One consists of 140 amino acids while the second is a shorter version, lacking the 6 amino-terminal amino acid residues. Acidic FGFs exhibit a pI value in the region of 5–7. Basic FGF is also a single-chain polypeptide, consisting of 146 amino acid residues and having a pI value in the region of 9.5.

Both acidic and basic FGFs are capable of binding to the same receptor, although basic FGF seems to be the more potent of the two. Recombinant DNA technology has facilitated large-scale production of both human and bovine fibroblast growth factors, which has stimulated further research to elucidate the mechanism ofaction of these growth factors, and assessing their potential therapeutic uses. Both acidic and basic FGFs bind heparin. This has facilitated their purification by heparin affinity chromatography. Reverse phase HPLC has also been utilized to further purify acidic FGF.

Platelet derived growth factor (PDGF), as the name suggests, was first isolated from platelets. Subsequently, however, it has been established that this growth factor is synthesized and secreted by a number of other cell types. PDGF represents a powerful mitogen for a variety of cell types, including fibroblasts, glial cells and smooth muscle cells. This growth factor has a molecular weight of 30 000 Da, and consists of two polypeptide chains termed A and B, which are covalently linked by an interchain disulphide bond. Three isoforms of the growth factor have been isolated— the homodimers AA and BB in addition to the heterodimeric form AB. The A chain consists of 124 amino acids, whereas the B chain contains 140 amino acids.

Two distinct, although related PDGF receptors have also been identified; these have been termed α and β PDGF receptors. Both receptor types have tyrosine kinase activity. The α receptor binds all three PDGF isoforms with high affinity. The β receptor on the other hand, binds the differing PDGF isoforms with varying degrees of affinity. Binding affinity increases in the order of AA, AB, BB. PDGF seems to play an important role in the process of wound healing. A number of potential therapeutic applications of PDGF are being assessed by numerous biopharmaceutical manufacturers.

Various transforming growth factors (TGFs) which may be of potential therapeutic significance have also been identified. TGF-α exhibits similar biological activities to EGF. In addition to its presence in a variety of tumour types, TGF-α is synthesized by several normal cell types such as activated macrophages.

A distinct group of transforming growth factors termed TGF-βs have also been characterized. These factors are generally dimeric proteins consisting of two identical polypeptide chains. The two

TGF-βs which have been characterized in greatest detail are TGF-β_1 and TGF-β_2. These two factors exhibit a significant degree of amino acid sequence homology.

The TGFs exhibit a variety of biological effects, most of which relate to the modulation of growth of a number of cell types. TGF-β_1 exhibits numerous immunoregulatory effects, including inhibition of B and T cell proliferation. TGF-βs have been found in the synovial fluid of arthritic joints. Indeed, recent experiments suggest that TGF-βs may arrest the progression of arthritis. These growth factors are also the subject of intensive academic and industrial research.

Many of the growth factors thus far discovered have proved to be closely related to the polypeptide products of several known onco-genes. Transforming growth factors, as already outlined, are synthesized by a number of transformed cell types, in addition to some normal cells. Factors similar to FGF have been isolated from a number of human tumour types. The transforming protein of the simian sarcoma virus p28^{sis} shares extensive homology with PDGF. EGF shares extensive homology with TGF-α. Moreover, the EGF receptor has been implicated in the development of some tumour types. The product of the oncogene c-*erb* B exhibits a similar amino acid sequence to that of the EGF receptor.

Virtually all of the growth factors mentioned above are currently undergoing clinical evaluation in order to assess their potential as therapeutic agents. Most clinical trials concentrate on their effec-tiveness in accelerating the wound healing process and treating ulcers. An ulcer may be described as a break in the skin or a mucous membrane which fails to heal. Natural wound healing depends upon the release of growth factors which initiate the tissue repair process at the site of damage. Thrombin, factor IIa of the blood coagulation pathway, stimulates the release of α granules from blood platelets. These granules contain a variety of growth factors including FGF, PDGF and TGF. Such growth factors exert mitogenic effects on certain cell types essential for tissue repair. They also exhibit chemotactic effects, and in this way attract a number of cell types to the site of damage. Released TGF-β, for example, attracts macrophages to the area of damage. These macro-phages then release a variety of additional factors including PDGF and TGF-β. EGF is mitogenic for epithelial cells which are essential for rapid wound healing.

Clinical studies have confirmed that application of exogenous growth factors rapidly accelerates the wound healing process. Initial trials have been carried out mainly using pig skin as an animal model, as it is very similar to human skin. The encouraging results obtained seem to be reproducible in clinical trials involving humans. Furthermore it is becoming apparent that such growth factors act synergistically. Application of several distinct growth factor types to the site of damage may thus yield the most effective response.

Epidermal growth factor may find additional application in the defleecing of sheep. Administration of EGF to sheep has a transient but marked effect on the wool follicle bulb cell. This results in the development of a weakened layer in the fleece which could greatly facilitate harvesting of the wool. Large quantities of EGF would obviously be required should this become a widely adopted agricultural practice. Bulk quantities of the growth hormone could be produced by recombinant DNA technology.

ERYTHROPOIETIN

Erythropoietin is a glycoprotein hormone produced by the kidneys. This hormone stimulates the production of red blood cells, the erythrocytes, from their precursor stem cells. Human erythropoietin has a molecular weight in the region of 34 000 Da, 60% of which is carbohydrate. The hormone contains a high level of sialic acid and varying amounts of hexosamines and hexoses. Erythropoietin is found in plasma and is excreted in the urine, from which it may be purified.

Production of erythropoietin is regulated by a number of factors, most notably by tissue oxygen tension. Its plasma concentration can increase a hundredfold in highly anaemic individuals, due to insufficient oxygen supply to the tissues. Erythropoietin production can plummet as a result of certain conditions such as chronic renal failure. Such a decrease in erythropoietin levels can in turn lead to the development of secondary anaemia. The metabolic consequences of inadequate erythropoietin production may be avoided by administration of exogenous erythropoietin preparations.

Erythropoietin used clinically is produced by recombinant DNA technology. The cDNA coding for human erythropoietin was first expressed as a heterologous protein in 1984.

Erythropoietin is but one member of a large family of regulatory factors which stimulate the growth of blood cells. The various members of this family are collectively referred to as haemopoietic growth factors or haemopoietins. All blood cells are ultimately derived from a single cell type, the haemopoietic stem cells. These self-perpetuating cells are found in the bone marrow. They proliferate and differentiate under the influence of the various haemopoietic growth factors, thus yielding the range of cell types normally found circulating in plasma. The various haemopoietins associated with these differentiation pathways include not only erythropoietin, but also colony stimulating factors and various interleukins. Many of these factors have been produced in recombinant systems over the last few years and their clinical application promises to revolutionize the treatment of various blood disorders.

THYROTROPHIN

Thyrotrophin, also termed thyroid stimulating hormone or TSH, is a glycoprotein hormone produced by the anterior lobe of the pituitary. TSH is a dimer consisting of α and β subunits. The overall molecular weight of human TSH is about 28 000 Da. The α subunit of TSH is identical in sequence to the α subunit of two other glycoproteins produced by the anterior pituitary, FSH and LH. The β subunit thus confers the biological specificity upon TSH. Both the α and β subunits contain several disulphide bridges, giving TSH one of the highest sulphur content of any protein thus far characterized.

Release of TSH from the pituitary gland is controlled by two hypothalamic hormones. Thyrotrophin releasing hormone (TRH), a tripeptide, stimulates the release of thyrotrophin from the pituitary. This was the first of the hypothalamic regulatory factors to be isolated and characterized. TRH may be obtained from hypothalamic extracts or may be synthesized chemically. Several TRH analogues which exhibit varying degrees of potency have been synthesized. TRH also stimulates the release of prolactin from the pituitary gland. This releasing factor is employed clinically to assess hypothalamopituitary dysfunction.

The release of TSH from the pituitary gland is inhibited by another hypothalamic factor, growth hormone release-inhibiting hormone (somatostatin). As discussed previously, somatostatin also inhibits the release of growth hormone from the anterior pituitary gland.

Elevated circulatory levels of thyrotrophin stimulate an increase in iodine uptake by the thyroid gland and an increase in the rate of synthesis and release of the thyroid hormones, thyroxin (T_4) and triiodothyronine (T_3). Increased circulatory levels of T_3 and T_4 in turn result in a decrease in TRH and TSH secretion. Receptors that specifically bind the thyroid hormones are present in the pituitary gland and possibly in the hypothalamus. Binding of T_3 and T_4 to such receptors mediates this inhibitory effect.

Thyrotrophin has been used in the diagnosis of hypothyroidism. However, currently diagnosis of this condition relies on direct assessment of circulating levels of thyrotrophin by radioimmunoassay.

CORTICOTROPHIN (ACTH)

Corticotrophin is also referred to as adrenocorticotrophic hormone or ACTH. Corticotrophin, together with several other hormones, is synthesized in the anterior lobe of the pituitary gland (Figure 6.8). The hormone consists of a single polypeptide which in humans contains 39 amino acids and has a molecular weight of 4500 Da.

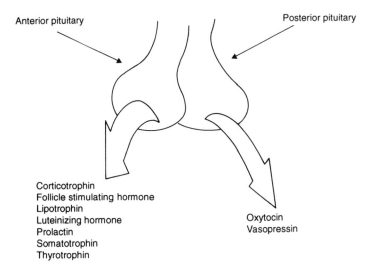

Anterior pituitary

Posterior pituitary

Figure 6.8. Hormones of the pituitary gland

Corticotrophin
Follicle stimulating hormone
Lipotrophin
Luteinizing hormone
Prolactin
Somatotrophin
Thyrotrophin

Oxytocin
Vasopressin

ACTH has been purified from a variety of animal species, and in most cases the first 24 amino acids were found to be identical in sequence in the various species. Moreover, synthetic polypeptides consisting of this 24 amino acid sequence were found to retain the full biological activity associated with ACTH. This synthetic peptide may thus be administered to humans as an alternative to the native hormone. Full-length corticotrophin purified from the pituitaries of slaughtered animals may also be used for therapeutic purposes.

Injected corticotrophin is degraded rapidly in the body. The hormone exhibits a half-life in plasma of only 15 minutes. Its therapeutic effect may be prolonged by intramuscular administration of the hormone as an ACTH–gelatin or ACTH–zinc formulation. A depot of activity is formed in this way which is released at a slow rate into the circulatory system.

Corticotrophin promotes growth of the adrenal cortex and also stimulates the synthesis and release of the adrenocortical hormones. It promotes an increase in the rate of synthesis of these steroid hormones by increasing the rate of conversion of cholesterol to pregnenolone, which represents the rate-limiting step of adrenocortical hormone synthesis.

Corticotrophin has been used clinically to induce an increase in the circulatory levels of corticosteroids, in particular cortisol and certain mineralocorticoids. Elevated levels of corticosteroids have proved beneficial in the treatment of certain medical disorders. The clinical use of corticotrophin has decreased in recent years. Most physicians now prefer to administer corticosteroids directly to the patients.

Secretion of corticotrophin from the pituitary gland is stimulated by the hypothalamic regulatory factor known as corticotrophin releasing factor (CRF). CRF is a peptide containing 41 amino acids. This factor is also known to stimulate the secretion of β-endorphin from the pituitary gland.

PROLACTIN

Prolactin is a single-chain polypeptide hormone synthesized by the anterior pituitary gland. The hormone has a molecular weight in the range 22 000–23 000 Da and contains 199 amino acids.

Prolactin exhibits a variety of biological effects. In humans this hormone, in conjunction with others such as oestrogens and glucocorticoids, stimulates the growth and development of mammary tissue. It is also responsible for the induction of lactogenesis. Binding of prolactin to its membrane receptor in the mammary tissue of nursing mothers results in an immediate increase in the concentration of cellular mRNAs coding for milk proteins such as α-lactalbumin and the caseins. Prolactin also exerts a number of effects on the gonads, where prolactin receptors have been detected in both the ovaries and testes.

Secretion of prolactin from the pituitary gland is regulated by hypothalamic factors. It is believed that inhibition of secretion is mediated by hypothalamic dopamine. A variety of factors including sleep, stress and pregnancy are known to induce secretion of prolactin. The exact molecular structure of the hypothalamic releasing factor remains to be elucidated.

MELANOCYTE STIMULATING HORMONE

Melanocyte stimulating hormone (MSH) is also known as melanotrophin. MSH is secreted by the pars intermedia of the pituitary gland. At least three forms of this peptide hormone are known to exist, α-, β- and γ-MSH. α-MSH consists of 13 amino acids and is the most potent of the melanocyte stimulating hormones. The amino acid sequence of this peptide is related to corticotrophin and identical sequences have been reported in most species examined.

The amino acid sequence of β-MSH varies slightly from species to species. It consists of 18–20 amino acids in most animals, although human β-MSH contains 22 amino acids. γ-MSH, which consists of 12 amino acids, displays no significant melanocyte stimulating activity.

MSH promotes the dispersion of melanin granules in the skin of various lower vertebrates; thus it plays an important role in adaptive colour change in such species. The physiological role of MSH in higher vertebrates including humans remains poorly understood.

Secretion of MSH from the pituitary gland is under hypothalamic control. Both hypothalamic stimulatory and inhibitory factors are known to exist. The hypothalamic releasing factor remains to be characterized. The inhibitory factor, melanostatin, is a tripeptide amide.

HORMONES OF THE POSTERIOR PITUITARY GLAND

The posterior pituitary serves as a storage and secretory organ for two hormones, oxytocin and vasopressin. Both hormones are peptides consisting of 9 amino acid residues (Figure 6.9). They are synthesized by specific hypothalamic cells and are subsequently transported in specialized vesicles along axonal pathways directly into the posterior lobe of the pituitary. Both hormones are found in association with specific carrier proteins.

Figure 6.9. Amino acid sequence of (a) oxytocin and (b) vasopressin. The similarity in amino acid sequence is striking

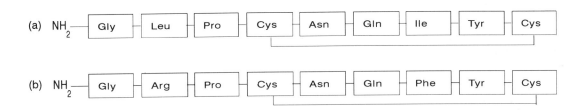

Vasopressin is also known as antidiuretic hormone (ADH). It is the major antidiuretic hormone found in the body and it exerts a direct effect on the kidney where it promotes an accelerated rate of water resorption from the distal convoluted renal tubules. It may be obtained by purification from pituitary glands or by chemical synthesis. Two forms of vasopressin are known to occur naturally. These differ in amino acid sequence by only one amino acid—residue 8 of the molecule. Arginine vasopressin (argipressin) contains an arginine residue at position 8 and has a molecular weight of 1084 Da. Vasopressin with this amino acid sequence is found in human, ovine, bovine, equine and chicken pituitaries. Lysine vasopressin, molecular weight 1056 Da, contains a lysine residue at position 8 of the vasopressin molecule. This form is found to be present in porcine pituitaries.

A deficiency in circulatory levels of vasopressin results in the onset of diabetes insipidus, a relatively rare metabolic disorder in which normal renal water resorption is impaired. Affected individuals experience a constant thirst and excrete large volumes of dilute urine. The condition may be treated successfully by the administration of vasopressin, either parenterally or as a nasal spray.

Desmopressin is an analogue of vasopressin that is often used clinically instead of vasopressin, as it exhibits greater antidiuretic activity and boasts a more prolonged period of action.

Oxytocin, like vasopressin, is a nonapeptide, containing one intrachain disulphide linkage and having a molecular weight of 1007 Da. It may be synthesized chemically or purified from the pituitary glands of slaughterhouse animals. Oxytocin induces contraction of the smooth muscle of the uterus. It is therefore used clinically to induce and maintain uterine contractions associated with labour. It is absorbed from various mucous membranes and may be administered parenterally or as a nasal spray.

Demoxytocin is a synthetic analogue of oxytocin and it is sometimes used as an alternative to the native hormone in the induction of labour, as it exhibits a higher degree of potency and enjoys a longer circulatory half-life. Oxytocin also stimulates the contraction of milk duct muscle cells, thus promoting expulsion of milk from the breast.

THERAPEUTIC ENZYMES

A variety of enzymes are used clinically in the treatment of various medical conditions (Table 6.3). Many blood coagulation factors administered therapeutically exhibit proteolytic activity. Ancrod, which is derived from the venom of the Malayan pit viper, is a serine protease which catalyses the proteolytic cleavage of microparticulate fibrin clots, and thus may be used clinically as an anticoagulant.

Table 6.3. Some enzymes which are used for therapeutic purposes

Enzyme	Therapeutic application
Ancrod (serine protease)	Anticoagulant
Tissue plasminogen activator	Thrombolytic agent
Urokinase	Thrombolytic agent
Asparaginase	Treatment of some types of cancer
Superoxide dismutase	Prevention of oxygen toxicity
Trypsin Papain Collagenase	Debriding/anti-inflammatory agents
Lactase Pepsin Pancrelipase Papain	Digestive aids

Proteolytic enzymes such as tissue plasminogen activator and urokinase are used clinically as thrombolytic agents, as they promote degradation of blood clots. These enzymes achieve their effects by activating the zymogen plasminogen, yielding the enzyme plasmin, which in turn catalyses the proteolytic degradation of fibrin strands present in clots.

All of the enzymes mentioned above which are involved in the clotting or thrombolytic process have been discussed in detail in the previous chapter.

Debriding and anti-inflammatory agents

Certain enzymes are used as debriding agents. These effectively clean open wounds by removal of foreign matter and any surrounding dead tissue. This allows for rapid healing of the wound. Enzymes such as trypsin, papain and collagenase have often been used as debriding agents. Such preparations are normally applied topically to the affected areas.

Trypsin is a proteolytic enzyme synthesized by the mammalian pancreas. It has a molecular weight of 24 000 Da and hydrolyses peptide bonds in which the carboxyl group has been contributed by either an arginine or a lysine residue.

Papain is a proteolytic enzyme isolated from the leaves and the unripe fruit of the papaya tree. It catalyses the hydrolysis of peptide bonds involving basic amino acids such as lysine, arginine or histidine. In addition to its use as a debriding agent, papain has been utilized as a meat tenderizer and for clearing beverages.

Collagenase catalyses the proteolytic destruction of collagen. Although it may be isolated from culture extracts of various animal cells, it is normally obtained from the culture supernatant of various species of *Clostridia*. Some clostridial species are pathogenic, causing diseases such as gas gangrene. The ability of such microorganisms to produce tissue-degrading enzymes such as collagenase facilitates their rapid spread throughout the body.

Administration of certain enzymes has proved beneficial in the reduction of various inflammatory responses. Such enzymes include chymotrypsin and bromelains. Chymotrypsin is a proteolytic enzyme produced in zymogen form, chymotrypsinogen, by the mammalian pancreas. Chymotrypsinogen is converted to chymotrypsin in the small intestine, where it functions to catalyse the proteolytic degradation of dietary proteins. Bromelains are plant proteases purified from the fruit or stem of the pineapple plant.

Nuclease treatment of cystic fibrosis

Cystic fibrosis (CF) represents one of the most commonly occurring genetic diseases. The frequency of occurrence varies among populations, with people of northern European extraction being most at risk. Within such populations approximately 1 in 2500 newborns are affected. It has been recognized for many years that excessive salt loss occurs in the sweat of people suffering from cystic fibrosis. More recently, the gene for this autosomal recessive disease has been identified. This gene codes for a polypeptide product which functions as a chloride channel. It thus seems that the underlying cause of cystic fibrosis may be traced to a malfunction in ion transport. Expression of the aberrant gene results in compromised function of a number of tissue types, including the pancreas and sweat glands. The major clinical sympton of CF, however, is undoubtedly the production of extremely viscous mucus in the respiratory tract, which is conducive to the establishment of recurrent bacterial infections. Such infections trigger an immune response in which large numbers of neutrophils are attracted to the site of infection. Ingestion and destruction of bacterial populations by these white blood cells result in the liberation of large amounts of DNA. This released DNA interacts with a variety of additional extracellular substances present in the infected lung, thus generating a highly viscous mucus.

In order to prevent build-up of such mucus CF patients suffering from bacterial infections are normally subjected to percussion therapy. This involves physically pounding on the patient's chest for long periods in order to dislodge the mucus, and allow the sufferer to expel it.

It was postulated some years ago that treatment of the infected lung with a DNase preparation might alleviate such respiratory symptoms, by catalysing the degradation of the extracellular DNA and hence promoting a reduction in mucus viscosity. Earlier attempts to investigate this hypothesis were hindered by lack of purified DNase preparations and lack of an effective delivery system that would target the enzyme directly to the site of infection. These difficulties have been largely overcome as recombinant human DNase is now being produced in relatively large quantities. Purified DNase preparation suitable for therapeutic administration is therefore available. This fact coupled with recent developments in aerosol technology now renders possible effective delivery of the DNase preparations into the lung by nebulization.

Clinical trials have clearly established the efficacy of the recombinant DNase preparations. This confirms earlier results which illustrated that *in vitro* incubation of DNase with mucus considerably reduced its viscosity. It thus seems likely that such recombinant DNase preparations will be utilized in the not too distant future in the clinical management of cystic fibrosis. The efficacy of such DNase preparations in the treatment of additional respiratory-related conditions such as pneumonia and chronic bronchitis is currently under active investigation.

Enzymes as digestive aids

Various enzymatic preparations may be used clinically as digestive aids. Most such enzymes are depolymerases, catalysing the enzymatic breakdown of a number of dietary components including polysaccharides, proteins and lipids. Some such enzyme preparations consist of a single enzyme which catalyses the degradation of a specific dietary substance. Others contain multiple enzymatic activities which exhibit broad digestive ability.

α-Amylase catalyses the hydrolysis of the $\alpha 1 \rightarrow 4$ glycosidic bonds which form the covalent linkage between glucose monomers in carbohydrates such as starch or glycogen. α-Amylase produced by microorganisms such as *Bacillus subtilis* or members of the *Aspergillus* family has found widespread industrial application, as discussed in Chapter 8. Amylase activity also plays an important digestive role in higher animals. The enzyme may be isolated from saliva or from pancreatic tissue and various amylase preparations have been used clinically to aid the digestion of dietary carbohydrate.

Lactase catalyses the hydrolysis of the disaccharide lactose, the principal sugar of milk, yielding glucose and galactose. Preparations of lactase find use as a digestive aid, in particular to alleviate symptoms associated with lactose intolerance. While most infants exhibit high levels of intestinal lactase activity, the adult population of many geographical regions have greatly reduced lactase activity. Such persons are unable to digest lactose, and often suffer from intestinal upset upon consumption of milk.

Various proteolytic enzymes may also be employed as digestive aids. Such enzymes include papain, bromelains and pepsin. Pepsin is secreted naturally in the stomach of most animals, where it catalyses the proteolytic degradation of dietary protein. Pepsin preparations used as digestive aids are obtained from the mucous membranes of the stomach of various slaughterhouse animals.

Some enzymatic preparations used clinically contain multiple enzymatic activities. Pancreatin, for example, is a proteinaceous preparation extracted from the pancreas. It contains amylase, protease and lipase activities and may be administered orally to patients suffering from conditions caused by deficient secretion of pancreatic enzymes. Pancrelipase is another enzymatic preparation derived from pancreatic tissue which may be employed as a digestive aid, the principal enzymatic activity of which is lipase.

Other therapeutic enzymes

Additional enzymes which are used therapeutically include asparaginase and superoxide dismutase. Asparaginase catalyses the hydrolysis of the amino acid asparagine, yielding aspartic acid and ammonia. The enzyme is a tetramer with a molecular weight in the region of 130 000 Da. It may be purified from a wide variety of microorganisms including yeast, fungi and bacteria such as *E. coli*.

All cells require asparagine to sustain normal metabolic activity. Although most cells are capable of synthesizing this amino acid, certain malignant cells lack this ability. Administration of asparaginase may thus retard or prevent the growth of such cancerous cells by reducing the level of asparagine available to them.

Asparaginase preparations used clinically are normally purified from *E. coli* or from *Erwinia chrysanthemi*. *Escherichia coli* produces two asparaginase isoenzymes of which only one is clinically effective. This enzyme is used mostly in the treatment of certain forms of childhood leukaemia.

Superoxide dismutase catalyses the conversion of the highly reactive superoxide ion (O_2^-) to hydrogen peroxide (H_2O_2). This enzyme plays an important role in all aerobic organisms, as the superoxide ion is highly reactive and capable of causing serious cellular damage. During normal metabolic activity, aerobic organisms usually reduce oxygen molecules directly to water. This reaction, as shown below, requires four electrons.

$$O_2 + 4e^- + 4H^+ \rightarrow 2H_2O$$

Incomplete oxidation of the oxygen molecule invariably generates molecules such as superoxide (O_2^-), hydrogen peroxide (H_2O_2) or hydroxyl (OH^-) ions, all of which are extremely reactive. Hydrogen peroxide and superoxide radicals cause extensive cellular damage by reacting with unsaturated fatty acids in membranes. The superoxide ion O_2^- and H_2O_2 may be generated by several cellular enzymes. Furthermore, the spontaneous oxidation of iron atoms present in various molecules such as haemoglobin also generates such reactive molecules. Two specific enzymes protect cells from the harmful effects of superoxide and hydrogen peroxide. The enzyme superoxide dismutase catalyses the conversion of superoxide to hydrogen peroxide as shown below.

$$O_2^- + O_2^- + 2H \xrightarrow{\text{superoxide dismutase}} H_2O_2 + O_2$$

Catalase then catalyses the conversion of hydrogen peroxide into oxygen and water as follows.

$$H_2O_2 + H_2O_2 \xrightarrow{\text{catalase}} 2H_2O + O_2$$

Two forms of superoxide dismutase are found in eukaryotic cells. The cytoplasmic enzyme is a dimer consisting of two identical subunits and with a molecular weight of 31 000 Da. Each enzyme subunit contains an atom of zinc and an atom of copper. Both metal atoms participate directly in the enzymatic process. The dismutase associated with mitochondria has a molecular weight of 75 000 Da and contains two atoms of manganese. Two highly homologous yet distinct superoxide dismutases are also found in aerobic prokaryotes. One form contains manganese while the other contains iron.

Superoxide dismutase isolated from bovine liver or erythrocytes has been available for a number of years and it has been used as an anti-inflammatory agent. Human superoxide dismutase may be produced as a heterologous protein in a number of recombinant systems. Such preparations are currently undergoing clinical trials in order to assess the enzyme's ability to prevent damage to tissue resulting from exposure to excessively oxygen-rich blood.

CYTOKINES

Cytokines are a large family of proteins which function in the modulation of cell-mediated immunity (Table 6.4). Cytokines that are produced by a number of cell types including lymphocytes, monocytes and macrophages. Cytokines produced specifically by lymphocytes are termed lymphokines, while those produced by monocytes are often referred to as monokines. Cytokines are synthesized and secreted in minute quantities, but they are extremely potent and exhibit wide-ranging effects on target cells. The advent of recombinant DNA technology has had a tremendous impact on cytokine research. This technology allows scientists to produce large quantities of any cytokine in recombinant systems.

Table 6.4. Cytokines of therapeutic interest

Interferons
Interleukins
Colony stimulating factor
Tumour necrosis factor

The first cytokine approved for general clinical use was interferon alpha, which gained FDA approval in 1986. Since that time various other interferon preparations and several preparations of colony stimulating factor have also come on the market. Many other cytokine preparations are currently undergoing extensive clinical evaluation.

Various different cytokines appear to be effective in treating a wide variety of human diseases. Cytokines have been indicated for use in the management of various cancer types and adjuvants to cancer and AIDS therapy (Table 6.5). They have also been indicated in the treatment of rheumatoid arthritis, multiple sclerosis, asthma, allergies and various infectious diseases. It is confidently expected that many additional cytokine preparations will soon become available to the medical community.

Table 6.5. Some cytokines currently approved, or undergoing clinical trials, for the treatment of AIDS or AIDS-related complex

Cytokine	Developing companies
Interferons alpha (several types)	Schering-Plough Hoffmann-La Roche Interferon Sciences Burroughs Wellcome
Granulocyte-macrophage colony stimulating factor (GM-CSF)	Sandoz Genetics Institute Schering-Plough Immunex
Macrophage colony stimulating factor (M-CSF)	Genetics Institute
Interleukin 2	Chiron
Interleukin 3	Sandoz

Data from the American Pharmaceutical Manufacturers' Association.

Interferons

The first cytokine to be identified and studied was interferon (IFN), which was discovered in 1957. It has since been demonstrated that all vertebrates produce a variety of interferons. These proteins are species-specific and exhibit a wide range of biological activities. Most mammals produce at least three types of interferon: interferon alpha (IFN-α), interferon beta (IFN-β) and interferon gamma (IFN-γ). Humans produce several types of IFN-α, but only one IFN-β and one IFN-γ.

IFN-α and IFN-β both appear to bind the same cell surface receptor and are both acid stable. These are collectively termed type I interferons. IFN-γ is acid-labile and binds to a distinct receptor; this interferon has been termed type II interferon. Interferons exhibit a wide range of biological activities (Table 6.6).

Table 6.6. Some of the biological activities exhibited by interferons

Resistance to viral attack in a broad range of cell types

Modulation of the rate of growth of a variety of cell types

Key regulatory role in the immune response

Upon becoming infected by a virus, many cells synthesize and release interferons. The interferons subsequently induce other cells to become resistant to viral attack. In this way, interferon plays a broad protective role against a wide range of viruses. In order to confer resistance against viral infection, susceptible cells must encounter the interferon prior to becoming exposed to the viruses. Interferon preparations have, however, proved beneficial in the treatment of a variety of conditions caused by established viral infections. Interferon alpha has been found to be effective in the treatment of hepatitis B infections and also in the treatment of genital warts, caused by papillomavirus.

In addition to viruses, several other agents are capable of inducing cells to produce interferon. Such additional agents include double-stranded RNA, various mitogens, a variety of bacteria and a number of other cytokines.

Interferons influence the growth of a variety of cells. IFN-α and IFN-β are both capable of inhibiting the growth of a range of cell types—both normal and malignant. Such interferon preparations have been used successfully in the treatment of hairy cell leukaemia, AIDS-related Kaposi's sarcoma, lymphoma and melanomas.

Interferons are also capable of modulating the immune response. Various IFNs induce expression of major histocompatibility complex (MHC) genes. IFNs may also induce an increase in the number of cell surface receptors on a variety of cell types for other cytokines. Many interferon preparations stimulate enhanced cytotoxic activity of various immune cells such as macrophages, lymphocytes and natural killer cells.

Interferon alpha. Interferon alpha is also known as leukocyte interferon or lymphoblastoid interferon, as it was initially isolated from these cell types (Table 6.7). At least 16 different types of human IFN-α are known to exist. Although these proteins exhibit significant sequence homologies, they are different in their primary structure and are encoded for by a family of related genes. Most human IFN-αs are single-chain polypeptides consisting of 165–166

Table 6.7. Human interferons

Interferon type	Additional name	Cell type from which it was initially isolated
Interferon alpha (at least 16 distinct but related types exist)	Leukocyte interferon or lymphoblastoid interferon	Leukocytes or lymphoblastoid cells
Interferon β	Fibroblast interferon	Fibroblasts
Interferon γ	Immune interferon	Activated T lymphocytes

amino acids. Although many IFN-αs are devoid of carbohydrate side-chains, several are glycoproteins, exhibiting varying degrees of glycosylation. The molecular weights of the IFN-αs may range from 16 000 Da to in excess of 26 000 Da, depending upon the carbohydrate content of the molecule. The enzymatic or chemical removal of the carbohydrate of such glycosylated interferons seems to have little adverse effect on their bioactivity. Most of the recombinant IFN-αs are produced in *E. coli* and, thus, are devoid of any post-translational glycosylation modifications.

Beta and gamma interferons. Human IFN-β, also termed fibroblast interferon, was traditionally derived from fibroblasts which had been exposed to certain viruses or to polynucleotides, and it may now also be produced by recombinant DNA technology. Human IFN-β was the first interferon molecule to be purified. It is a glycoprotein of 166 amino acids and has a molecular weight in the region of 20 000 Da.

Interferon gamma, often referred to as immune interferon, is produced largely by activated T-lymphocytes and seems to be one of the more important immunomodulatory interferons. Human IFN-γ is a single-chain glycoprotein consisting of 143 amino acids of a molecular weight ranging from 15 000 Da to 25 000 Da, depending upon the degree of glycosylation. As previously mentioned, IFN-γ differs significantly from other interferons in terms of its physicochemical properties and it binds a distinct receptor type.

Production and purification of interferons. Interferon preparations may now be produced by culturing appropriate cell types or by recombinant DNA technology. Up to the 1970s, the majority of interferons available for clinical use were sourced from human leukocytes (white blood cells), obtained directly from transfusion blood supplies. This leukocyte interferon consisted of a variety of related IFN-α molecules. The final IFN preparation obtained was approximately 1% pure, and vast quantities of transfused blood were required to produce minute amounts of product. Although the potential clinical applications of IFN were recognized at that time, the scarcity of these cytokines rendered impossible their widespread clinical application.

In the late 1970s, bulk quantities of interferons were first produced by mammalian cell culture. A specific strain of a human lymphoblastoid cell line termed the Namalwa cell line was most often employed in this regard. After induction by Sendai virus, this cell line produces appreciable quantities of leukocyte (α) interferon. Industrial-scale production of Namalwa cell leukocyte interferon is undertaken in large culture vessels, many of which have a capacity in excess of 8000 litres. Highly purified interferon preparations may be obtained from this source. The purified product has been shown to consist of at least eight distinct molecular species of IFN-α.

Interferons are also produced in large quantities by recombinant DNA methodologies. Most recombinant preparations were first developed in the early 1980s. Several recombinant human IFN-αs, in addition to IFN-β and IFN-γ have been produced in *E. coli*, yeast and in some eukaryotic cells such as cultured monkey cells and Chinese hamster ovary cells. Most of the interferon preparations currently undergoing clinical trials has been produced in *E. coli*. Owing to the inability of *E. coli* to carry out post-translational modifications, these recombinant products are not glycosylated. This fact seems to have little adverse effect on the medical efficacy of the products.

A wide variety of chromatographic techniques have been used in the purification of various interferon preparations. Such techniques include affinity chromatography with a variety of immobilized ligands such as lectins, reactive dyes, concanavalin A and phenyl groups, in addition to anti-interferon monoclonal antibodies. Other techniques used have included ion-exchange, gel filtration and mineral chelate chromatography. HPLC has proven to be a particularly powerful technique in the preparation of modern, high-purity interferons. The introduction of sensitive and convenient immunoassays also facilitated the rapid purification of interferon preparations. Initial interferon assays relied on monitoring their antiviral activity in cell culture and such bioassays were time-consuming and complex.

Most of the clinical applications of interferons to date centre around the treatment of various tumour types, in addition to their use as antiviral agents. Studies indicate that interferon preparations may be beneficial in the treatment of a variety of other conditions, including asthma, multiple sclerosis and arthritis. A greater understanding of the molecular mechanism of action of interferons, and how they interact with each other, will no doubt allow a more logical application of these powerful agents in the treatment of diseases. Interferon preparations currently produced and their intended clinical applications are summarized in Table 6.8.

The interleukins

The interleukins represent yet another family of cytokines. At least ten different interleukins have thus far been characterized and many of these preparations are currently the subject of clinical investigation. Some interleukins such as interleukin 1 and interleukin 2 are already employed in the treatment of a number of medical conditions.

Table 6.8. Interferon preparations currently available or undergoing clinical evaluation

Product name	Company	Intended applications	US development status (as of 1992)
Actimmune (IFN-γ1b)	Genentech	Management of chronic granulomatous disease	Approved 1990
Alferon N (IFN-αn3)	Interferon Sciences	Genital warts	Approved 1989
Intron A (IFN-α2b)	Schering-Plough	Hairy cell leukaemia Genital warts AIDS related Kaposi's sarcoma Non-A, non-B hepatitis	Approved 1986 Approved 1988 Approved 1988 Approved 1991
Roferon-A (IFN-α2a)	Hoffmann-La Roche	Hairy cell leukaemia AIDS related Kaposi's sarcoma	Approved 1986 Approved 1988
Actimmune (IFN-γ1b)	Genentech	Small-cell lung cancer Atopic dermatitis Trauma-related infections Renal cell carcinoma Asthma and allergies	Phase III clinical trials Phase II clinical trials Phase II clinical trials
Alferon LDO (IFN-αn3)	Interferon Sciences	AIDS and ARC	Phase I/II clinical trials
Betaseron (IFN-β)	Berlex Laboratories	Multiple sclerosis Cancer	Phase III clinical trials Phase I/II clinical trials
Immuneron (IFN-γ)	Biogen	Rheumatoid arthritis Venereal warts	Phase II/III clinical trials Phase II clinical trials
Interferon gamma (IFN-γ)	Amgen	Cancer Infectious diseases	Phase II clinical trials
Intron A (IFN-α2b)	Schering-Plough	Various cancer types Chronic hepatitis B Delta hepatitis Acute hepatitis B Chronic myelogenous leukaemia	Application submitted Application submitted Application submitted Phase III clinical trials Phase III clinical trials
R-Frone (IFN-β)	Serono Laboratories	Various malignancies	Phase I clinical trials
Roferon A (IFN-α2a	Hoffman-La Roche	Colorectal cancer (in combination with W/5 flourouracil) Hepatitis Chronic myelogenous leukaemia AIDS and ARC	Phase II clinical trials Undergoing clinical trials

Data from the American Pharmaceutical Manufacturers Association.

The interleukins display a wide range of biological activities. Most function as regulators of the immune response, and are synthesized by a variety of cell types. Secreted interleukins subsequently modulate immunological and other activities by binding to specific surface receptors present on a variety of cell types. Binding of the interleukin to its receptor initiates cellular responses via intracellular messenger systems which remain to be characterized. Due to the similarities in their overall modes of action, the interleukins are often referred to as the hormones of the immune system. Various other cytokines also function to regulate the immune response.

Interleukin 2. Interleukin 2 (IL-2) is perhaps the best characterized of all the interleukins. This molecule plays a pivotal role in normal immunological functioning, and is likely to become one of the most widely used biopharmaceutical products of the 1990s. IL-2, also known as T cell growth factor, is a single-chain glycosylated polypeptide of molecular weight 15 500 Da. The protein consists of 133 amino acids arranged in four major α-helical regions. These are aligned in antiparallel fashion such that their hydrophilic regions are directed towards the surrounding aqueous environment.

Analysis of the crystalline structure of IL-2 preparations reveals that the molecule is devoid of any β-conformational sequences. Interleukin 2 contains one intrachain disulphide linkage which is essential for normal biological activity. The molecule has an isoelectric point between 6.5 and 6.8. The human gene coding for IL-2 is comprised of four exons separated by three introns.

Interleukin 2 is synthesized and secreted by T lymphocytes upon their activation, either by an antigen or a mitogen. It then stimulates the further growth and differentiation of activated T and B lymphocytes. It also potentiates the activity of natural killer cells and monocytes. In this way, this cytokine plays a central role in normal humoral and cell-mediated immunity.

Together with antigen, IL-2 is one of the primary molecular effectors capable of inducing an antigen-specific immune response, as proposed initially by MacFarlane-Burnet in his clonal selection theory. MacFarlane-Burnet rightly predicted that an extensive repertoire of lymphocytes was present naturally in the body, one member of which would specifically recognize virtually any antigen that the body might encounter. The presentation of antigen to such a specific lymphocyte would induce the clonal expansion of that particular cell, and in this way the immune system would mount an immediate and specific response against the offending antigen.

When an antigen gains entry into the body some of the antigenic material is invariably ingested by macrophages. These cells then "present" fragments of the ingested material on their surface, hence the origin of the term "antigen presenting cells". A small proportion of circulating T lymphocytes recognize and bind presented antigen via a specific cell surface receptor. These antigen specific cells then undergo selective clonal expansion, while other circulating lymphocytes, which do not recognize the antigen, remain quiescent. This growth and division of such T cells is dependent upon the presence, not only of antigen, but also of interleukin 1 (IL-1). IL-1 is produced by activated macrophages. Activated T helper cells, in turn, produce IL-2 which stimulates further T cell differentiation. Together with antigen and interleukin 6, IL-2 also co-stimulates the proliferation of B lymphocytes capable of producing antibodies which bind selectively to the antigen. Activated B and T cells also begin to express IL-2 receptors on their surface. Quiescent lymphocytes are normally devoid of such receptors.

In contrast to most lymphocytes, natural killer (NK) cells constitutively express IL-2 receptors on their surface. Binding of IL-2 to such receptors induces the immediate proliferation of NK cells. IL-2 also stimulates increased production and secretion of various NK cell products such as a variety of additional cytokines, including tumour necrosis factor, IFN-γ and colony stimulating factors. These cytokines serve to further potentiate the overall immunological response. NK cells comprise approximately 10% of circulating lymphocytes. These cells are believed to play a particularly important role in immunological reactions to cancer cells and to virally infected cells.

The IL-2 receptor has recently been characterized in detail. The high affinity receptor consists of two distinct transmembrane polypeptide chains which interact with each other non-covalently. The smaller of the two receptor molecules, the α subunit, has a molecular weight of 55 000 Da and consists of three segments: a large extracellular domain capable of binding IL-2, a transmembrane domain and an intracellular domain consisting of 13 amino acids. The larger polypeptide component, the β subunit, of the IL-2 receptor has a molecular weight of 75 000 Da. This polypeptide also contains extracellular, transmembrane and intracellular domains. In this case, however, the intracellular domain is relatively large, consisting of 286 amino acids. It is this portion of the molecule that initiates the intracellular response upon binding of IL-2.

The exact mechanism by which signal transduction is accomplished has not been elucidated. The IL-2 receptor seems to be devoid of enzymatic activity; however, various serine- and tyrosine-specific kinases have been implicated in IL-2 signalling.

The IL-2 molecule contains two distinct receptor binding sites on its surface. One site recognizes and binds the α receptor subunit while the other binds the β subunit, as illustrated in Figure 6.10.

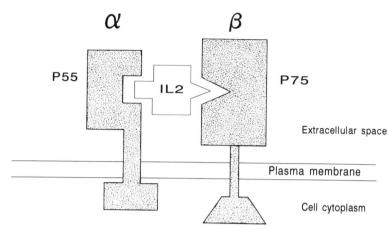

Figure 6.10. The interleukin 2 receptor

Various studies have revealed substantial differences in the affinity and kinetics of IL-2 binding to the two distinct receptor subunits. IL-2 binds the α (55 kDa) subunit rapidly with association/dissociation rates almost comparable to diffusional values. In contrast, IL-2 binds the β receptor subunit (75 kDa) with greater affinity and with decreased association/dissociation rates. The combination of these two binding characteristics in the intact heterodimeric structure yields receptors of very high affinity.

Interleukin 2 is secreted in small quantities by activated T helper cells. Several tumour cell lines, such as the Jurkat leukaemia line, produce increased quantities of this cytokine. Cell lines such as these provided much of the IL-2 for its initial characterization. Such a source would, however, be totally inadequate in terms of producing clinically significant quantities of this interleukin. Recombinant DNA technology now facilitates the large-scale production of virtually unlimited amounts of interleukin 2. Both the native human IL-2 gene and IL-2 cDNA have been used in recombinant protein production. The resultant protein product exhibits a range of biological activities identical to those of the native IL-2 molecule. Although IL-2, along with many other lymphokines produced naturally, is normally glycosylated, most recombinant products are devoid of carbohydrate side-chains. Lack of glycosylation seems to have little, if any, adverse effect on the biological activity of such recombinant products. Furthermore, it is somewhat serendipitous that lymphokine-containing inclusion bodies in recombinant systems are readily solubilized and that the denatured lymphokines readily renature, yielding fully active molecules. Quite a number of such recombinant IL-2 products are currently used clinically or are undergoing extensive clinical trials.

Interleukin-2 has been used for a number of years in the treatment of some forms of cancer. More recently its therapeutic potential in the treatment of a variety of infectious diseases, including AIDS, has been realized. The effectiveness of IL-2 in treating such conditions lies in its ability not only to enhance B and T cell responses, but also in its ability to mobilize natural killer cells. These cells are capable of inducing lysis of damaged, virally infected or cancerous cells and they also synthesize and release a cascade of cytokines, which further potentiates the immune response. Experiments in animal models carried out in the early 1980s illustrated that lung tumours regressed if NK cells were removed from the body, activated by incubation with IL-2 and then reintroduced into the body along with additional IL-2.

More recently, the apparent effectiveness of IL-2 in the treatment of a variety of infectious diseases, in particular those resistant to conventional treatment, has further heightened medical interest in this cytokine. Observations relating to its effectiveness in the treatment of AIDS and lepromatous leprosy have generated particular interest in this regard.

As IL-2 plays a pivotal role in the clonal expansion of both activated T and B lymphocytes, this cytokine may also prove to be useful as a vaccine adjuvant. A large proportion of conventional adjuvants are unsuitable for use in humans as they promote a variety of unacceptable side-effects. Unfortunately, these are invariably the most potent adjuvants known. Many such adjuvants potentiate the immune response by promoting the activation of macrophages. These activated cells release IL-1, which in turn induces the production of a variety of other cytokines, primarily IL-2. IL-2 then orchestrates the subsequent immunological response. Direct administration of IL-2 seems to be one possible approach to the development of future vaccine adjuvants.

A molecular understanding of IL-2 and its mode of action also allows researchers to logically manipulate IL-2-mediated immune responses. This knowledge has already facilitated the development of a number of ingenious methods which may be used to clinically induce selective immunosuppression. Such treatments could have important consequences for the survival of allografts or in the treatment of autoimmune conditions. Allografts (tissue or organs grafted or transplanted from a donor to a recipient) result in the immune system of the recipient perceiving the transplant organ as being non-self, due to the presence of foreign antigens. An immunological response against the transplant is initiated, leading to transplant rejection. Thus far, the sole means of preventing rejection is to induce broad immunosuppression in the recipient by administering drugs such as glucocorticoids and cyclosporin. These drugs function by inhibiting the production of IL-2.

Selective immunosuppression may be induced by a number of means. IL-2 receptors are expressed solely on activated B and T lymphocytes. The remaining unactivated B and T lymphocytes are devoid of such receptors. Administration of IL-2–toxin conjugates will thus lead to selective destruction of activated lymphocytes. Alternatively, monoclonal antibody preparations that react specifically with IL-2 receptors but fail to trigger an intracellular response may be administered. Administration of such antibodies will block IL-2 binding and hence prevent IL-2 mediated proliferation of the activated lymphocytes. Soluble IL-2 receptors have also been developed. Parenteral administration of such polypeptides limits or prevents normal IL-2 functioning as they compete with native IL-2 receptors for binding of the cytokine. Other strategies employed in promoting clinical immunosuppression include the development of IL-2 analogues that bind to the IL-2 receptor, but fail to trigger an intracellular response. Selective immuno-enhancement may, of course, be induced by the exogenous administration of IL-2 and IL-2 analogues may also be developed that exhibit increased immunostimulatory activity.

Recent progress in our understanding of immune functioning illustrates just how efficiently this system carries out its protective role. A greater appreciation of how immune surveillance is regulated at a molecular level now allows researchers to intervene in this natural process, and increase or decrease its effectiveness as required.

Interleukin 1. Although IL-2 is perhaps the best-known and most extensively characterized interleukin, several other members of this family hold considerable therapeutic promise. Interleukin 1 has been assessed clinically in the treatment of several medical conditions, most notably cancer, with some successes.

Interleukin 1 is produced by a number of cell types, predominantly by phagocytic cells such as macrophages and monocytes. There are two distinct forms IL-1, termed IL-1α and IL-1β. Although these interleukins are the products of two distinct genes and exhibit only limited homology, they bind to the same receptor and induce identical biological responses. Both IL-1α and IL-1β are initially synthesized as precursor molecules with molecular weights of 31 000 Da. These precursors are subsequently processed proteolytically, yielding active biological molecules with molecular weights of 17 500 Da. Human IL-1α has an isoelectric point of 5.5, while that of IL-1β is close to 7.0.

Two molecular forms of the IL-1 receptor have been identified. These have been termed type I and type II receptors. The type I receptor is found to be present mainly on the surface of T cells and fibroblasts. The type II receptor is present on B lymphocytes. Both IL-1α and IL-1β bind to both receptors.

IL-1 exhibits a wide range of biological activities (Table 6.9). Such activities include promotion of the proliferation of a wide variety of cell types, induction of fever, stimulation of wound healing and regulation of certain inflammatory events. IL-1 therefore has also been termed *leukocyte activating factor*, *lymphocyte activating factor*, *T cell replacing factor* and *endogenous pyrogen*.

Table 6.9. Selected biological activities of IL-1

Promotes proliferation of
　thymocytes
　fibroblasts
　haemopoietic cells
　lymphocytes

Promotes wound healing

Induces hepatic acute phase protein synthesis

Stimulates release of prostaglandin and collagenase from synovial cells

Stimulates release of additional cytokines from selected cells

Induces a fever response

Induces sleep

IL-1 promotes the release of IL-2 from activated T lymphocytes, and many biological activities attributed to IL-1 may be mediated, at least in part, by IL-2. IL-1 has also been implicated in the progression of a number of disease states. It has been isolated, for example, from the synovial fluid of arthritic patients.

IL-1 preparations were initially obtained from the supernatant fluid of macrophages activated by incubation with bacterial lipopolysaccharide. Although small quantities of IL-1 could be purified in this way, its large-scale production was impractical.

The cloning and expression of the IL-1 cDNA in *E. coli* has facilitated the large-scale production of biologically active IL-1. This, in turn, has rendered practical the widespread medical application of this cytokine. Large-scale clinical trials designed to assess the effectiveness of IL-1 in treating a variety of cancers in addition to promoting wound healing are currently under way.

A soluble form of the IL-1 receptor has also been expressed in recombinant bacterial systems. Parenterally administered soluble receptor should compete with native receptor for binding to IL-1 and, in this way, lessen its biological effects. This would probably benefit patients suffering from a number of conditions such as septic shock, where many of the adverse clinical symptoms observed are due to the synthesis and release of inappropriately high concentrations of IL-1.

Other interleukins. A number of additional interleukin preparations are undergoing clinical evaluation (Table 6.10). Interleukin-3 (IL-3) is one such example. This cytokine plays a central role in promoting the differentiation of certain cells in the bone marrow, thus forming mature leukocytes and other cells. IL-3 is also known as multi-potential colony stimulating factor.

IL-3 is a glycoprotein consisting of 133 amino acids. It is produced predominantly by activated T lymphocytes and thus may be termed a lymphokine. Biologically active IL-3 has been produced in a number of recombinant systems, including recombinant bacterial, yeast and mammalian cells. IL-3 produced as a heterologous protein product in the bacterium *Bacillus licheniformis* first entered clinical trials in 1989. The lymphokine was purified from this source by a variety of chromatographic techniques, including hydrophobic

Table 6.10. Some interleukin preparations currently undergoing clinical trials

Product name	Developing company	Applications
PEG (IL-2)	Cetus	AIDS
Proleukin (IL-2)	Cetus	Renal cell carcinoma Cancer Kaposi's sarcoma
Recombinant IL-1α (human)	Immunex	Prevention of chemotherapy or radiotherapy-induced bone marrow suppression Cancer immunotherapy
Recombinant IL-1β (human)	Immunex Syntex	Prevention of chemotherapy or radiotherapy-induced bone marrow suppression Treatment of melanoma Immunotherapy Wound healing
Recombinant IL-2 (human)	Amgen	Cancer immunotherapy
Recombinant IL-2 (human)	Hoffman-La Roche Immunex	Cancer immunotherapy in combination with other drugs
Recombinant IL-3 (human)	Hoechst-Roussel Immunex	Bone marrow failure Platelet deficiencies Autologous bone marrow transplantation Adjuvant to chemotherapy Peripheral stem cell transplant
Recombinant IL-4 (human)	Schering-Plough	Treatment of immunodeficiency diseases Cancer therapy Vaccine adjuvant
Recombinant IL-4 (human)	Sterling Drug Immunex	Cancer immunomodulator

Data from the 1991 report of the American Pharmaceutical Manufacturers Association.

interaction chromatography, two ion-exchange chromatography steps, and a gel filtration step. IL-3 exhibits significant therapeutic potential in the treatment of a number of medical conditions, including bone marrow failure or bone marrow transplants.

Interleukin 4 (IL-4) is synthesized by T helper lymphocytes. This glycoprotein, of molecular weight 20 000 Da, is also termed B cell growth factor. IL-4 exhibits a broad range of biological activities, including stimulating the proliferation of a number of cell types such as B and T lymphocytes, thymocytes and mast cells. IL-4 promotes enhanced secretion of IgG and IgE from stimulated B cells and also enhances antigen presentation by monocytes.

Most IL-4 responsive cells have a specific receptor on their cell surface. The receptor has a molecular weight of 140 000 Da. A soluble form of this receptor has been produced by recombinant DNA methods. This soluble receptor consists of the extracellular ligand binding domain of the native receptor molecule. Studies using animal models have shown that administration of such soluble receptor fragments can block many of the biological activities of IL-4.

IL-4 is required *in vivo* to stimulate production of IgE, which plays a central role in the development of a variety of hypersensitivity reactions. Thus IL-4 antagonists such as the soluble receptor may find clinical application in the control and treatment of a variety of IgE-induced allergic reactions.

Interleukin 6 (IL-6) is another cytokine that may well prove to be therapeutically beneficial. This glycoprotein is synthesized by macrophages and fibroblasts. IL-6, together with IL-2, plays an important role in promoting differentiation and proliferation of activated B lymphocytes. IL-6 seems to induce the expression of various acute phase proteins in hepatocytes, and this cytokine has been linked with various autoimmune diseases. The IL-6 receptor appears to be distributed on a wide variety of cell types.

One of the more recently named interleukins is interleukin 10. IL-10 had previously been termed cytokine synthesis inhibitory factor (CSIF) and is produced mainly by activated monocytes. As its alternative name suggests, this interleukin strongly inhibits the synthesis of a number of other monocytic cytokines, including IL-1α, IL-1β, IL-6, IL-8, TNFα and granulocyte-macrophage colony stimulating factor. As many of these cytokines have been implicated in a variety of inflammatory reactions, IL-10 may play an important regulatory role in inflammatory responses. IL-10 also seems to exhibit autoregulatory activity as it strongly downregulates its own production.

Human IL-10 exhibits extensive sequence homology at the amino acid level with a specific open reading frame present in the genome of the Epstein–Barr virus. The protein product of this viral gene, which has been termed viral-IL-10, exhibits many biological properties similar to that of human IL-10.

In addition to monocytes, IL-10 has been shown to react with a variety of other cell types, including T and B lymphocytes, thymocytes, macrophages and mast cells.

Another recently described interleukin which holds considerable therapeutic promise is interleukin 12. IL-12 was originally termed natural killer cell stimulatory factor and is produced by macrophages and B lymphocytes. It is known to stimulate INF-γ release from T lymphocytes and natural killer cells. IL-12 also induces differentiation of certain T helper cells from uncommitted T cells. In this way, it is believed that IL-12 plays a vital role in orchestrating a cell-mediated immune response. Many microbial and other infections primarily trigger such a cell-mediated response, thus administration of IL-12 may prove particularly beneficial in combating such infections. Furthermore, inclusion of IL-12 as an adjuvant may prove beneficial in the production of many vaccines.

Tumour necrosis factor

Tumour necrosis factor (TNF) is another member of the cytokine family of regulatory proteins. Two forms of TNF are now recognized, TNFα and TNFβ. Although both proteins bind the same receptors and elicit broadly similar biological responses, they are distinct molecules and share less than 30% homology. The original protein termed tumour necrosis factor, although still referred to as TNF, is more properly termed TNFα; it is also known as cachectin. TNFβ is also referred to as lymphotoxin.

The discovery of TNFα stems from observations made at the beginning of the twentieth century by an American surgeon called William Coley. Coley observed that the tumours of some cancer patients regressed or disappeared after the patients had suffered a severe bacterial infection. Coley therefore attempted to treat cancer patients by administering live bacteria to his patients. This approach suffered from several disadvantages, not least of which was the inability to control the ensuing infection in those pre-antibiotic days. In an effort to surmount such difficulties, Coley subsequently developed a vaccine consisting of dead bacterial cells in suspension, which became known as Coley's toxins. Some clinical successes were recorded when cancer patients were treated with such toxins. Consistent results were never attained, however, and this method of treatment fell from medical fashion.

The active component of Coley's toxins was later shown to be a complex biomolecule termed lipopolysaccharide. This molecule is found in association with the outer membrane of Gram-negative bacteria. Lipopolysaccharide contains both lipid and polysaccharide components, and is also referred to as endotoxin.

Lipopolysaccharide itself is devoid of any anti-tumour activity. The serum of animals injected with lipopolysaccharide was found to contain a factor toxic to cancerous cells, and this factor, produced by specific cells in response to lipopolysaccharide, was termed tumour necrosis factor (necrosis refers to cell death). Lipopolysaccharide represents the most potent known stimulant of TNFα production.

Tumour necrosis factor α is synthesized by a wide variety of cell types, most notably activated macrophages, monocytes, certain T lymphocytes and NK cells, in addition to brain and liver cells. Although lipopolysaccharide is the most potent inducer of TNFα synthesis, various other agents such as some viruses, fungi and parasites also stimulate the synthesis and release of this cytokine. Furthermore, TNFα may act in an autocrine manner, stimulating its own production.

Native human TNFα is a homotrimer, consisting of three identical polypeptide subunits tightly associated about a threefold axis of symmetry. This arrangement resembles the assembly of protein subunits in many viral capsids. The individual polypeptide subunits of human TNFα are non-glycosylated and consist of 157 amino acids. The molecule has a molecular weight of 17 300 Da and contains one intrachain disulphide linkage.

X-ray crystallographic studies of the TNFα monomer have shown that the molecule is elongated and exhibits extensive β-conformational secondary structure. Much of its amino acid sequence forms two β-pleated sheets, each of which contains five antiparallel β strands.

Human TNFα is synthesized initially as a 233 amino acid precursor molecule. Proteolytic cleavage of a signal sequence releases native TNFα. TNFα may also exist in a 26 000 Da membrane-bound form.

TNFα induces its biological effects by binding specific receptors present on the surface of susceptible cells. Two distinct TNFα receptors have been identified. One receptor (TNF-R55) has a molecular weight of 55 000 Da, whereas the second receptor (TNF-R75) has a molecular weight in the region of 75 000 Da. These two distinct receptor types show no more than 25% sequence homology. THF-R55 is present on a wide range of cells, whereas the distribution of the type II receptor is more limited. Both are transmembrane glycoproteins with an extracellular binding domain, a hydrophobic transmembrane domain and an intracellular effector domain.

The exact molecular mechanisms by which TNFα induces its biological effects remain to be determined. Binding of this cytokine to its receptor seems to trigger a variety of events mediated by

G-proteins in addition to the activation of adenylate cyclase, phospholipase A_2 and protein kinases. The exact biological actions induced by TNFα may vary from cell type to cell type. Other factors, such as the presence of additional cytokines, further modulate the observed molecular effects attributed to TNFα action on sensitive cells.

Initial clinical interest in TNFα stemmed from the observation that this cytokine showed selective cytotoxicity towards transformed cells *in vitro*. It was also shown that the presence of interferons potentiated this effect.

The TNFα gene has been cloned and inserted in a variety of recombinant expression systems, both bacterial and eukaryotic. The resultant availability of large quantities of purified, biologically active TNF has facilitated clinical evaluation in a number of diseases, most notably cancer. Many such trials, using TNF either alone or in combination with interferons, yielded disappointing results. This may be due, at least in part, to the dosage regimen employed. Large quantities of TNFα may not be administered to patients owing to its toxic—if not lethal—side-effects. Clinical studies continue, however, and new discoveries such as the fact that lithium ions may potentiate TNF antitumour activities ensure sustained scientific interest.

TNF seems to play a central role in several other physiological processes such as inflammation, the immune response and septic shock. Septic shock is a serious medical condition usually associated with Gram-negative bacterial infection. Lipopolysaccharide present in the outer membrane of such Gram-negative bacteria play a central role in the development of this condition. Lipopolysaccharide, as previously mentioned, is the most potent known inducer of TNF synthesis, and it has been demonstrated clinically that administration of exogenous TNFα can induce symptoms identical to those seen in patients suffering from septic shock. Furthermore, it has been shown that pretreatment with anti-TNFα antibodies exerts a protective effect on animals subsequently challenged with potentially lethal doses of lipopolysaccharide.

Prolonged production of inappropriately elevated levels of TNFα has also been implicated in the development of cachexia, the wasting syndrome often associated with chronic parasitic or other infections, and with cancer.

TNFα modulates the immune response by a number of means. It stimulates the production of a variety of other cytokines, in particular interferons IL-1 and IL-6, and various colony stimulating factors. It also promotes synthesis of platelet-derived growth factor. TNFα exhibits many biological activities similar to those of IL-1. It also stimulates the proliferation of a variety of T lymphocytes and plays a role in antibody production by B lymphocytes. TNFα enhances the activity of various phagocytic cells such as macrophages. It has also been implicated in the development of certain autoimmune disorders, including that of rheumatoid arthritis. In addition, it has been shown that administration of anti-TNFα antibodies may decrease or prevent rejection of grafted or transplanted tissues.

TFNβ, previously known as lymphotoxin, was actually discovered prior to TNFα, but at that time it was not recognized as a TNF. TNFβ is produced by T lymphocytes and is a glycoprotein consisting of 171 amino acids. Although TNFα and TNFβ exhibit only limited amino acid sequence homology, they both bind to the same receptors and exhibit broadly similar biological activities. The TNFβ cDNA has been cloned and expressed in recombinant *E. coli*. Its resultant availability in large quantities allows its biochemical characteristics and clinical potential to be more fully assessed.

Colony stimulating factors

The serum of healthy individuals normally contains three distinct types of leukocytes (white blood cell): lymphocytes, granulocytes and monocytes. Lymphocytes may be subdivided into B and T cells, which function to promote antibody-mediated and cell-mediated immunity respectively.

Granulocytes may be further categorized as basophils, neutrophils and eosinophils. These cells are capable of detecting and destroying bacteria and other foreign particles. Some may also play a role in allergic responses. Monocytes are also capable of engulfing and destroying bacteria and other foreign particles, in addition to tissue debris.

As previously discussed, all circulating blood cells are ultimately derived from a single cell type, the haemopoietic stem cells. These cells are located in the bone marrow. A variety of haemopoietic growth factors stimulate the proliferation and differentiation of these stem cells, ultimately yielding this variety of blood cells (Figure 6.11). Erythropoietin, for example, stimulates the formation of erythrocytes from stem cells; colony stimulating factors, on the other hand, promote the proliferation and differentiation of

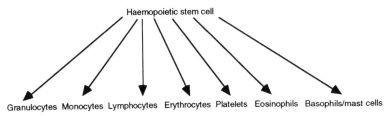

Figure 6.11. Various blood cell types derived from haemopoietic stem cells

certain precursor cells derived from stem cells, yielding mature granulocytes and macrophages. These two cell types play important roles in defending the body against a wide range of pathogens. Any factor that decreases the circulating level of these cell types renders an individual extremely susceptible to infectious disease.

The term *colony stimulating factor* reflects the ability of these substances to promote the growth *in vitro* of various leukocytes in clumps or colonies. A number of different colony stimulating factors have been identified and characterized (Table 6.11). These include granulocyte-macrophage colony stimulating factor (GM-CSF), granulocyte colony stimulating factor (G-CSF), macrophage colony stimulating factor (M-CSF) and multipotential colony stimulating factor (multi-CSF). The latter is in fact identical to IL-3.

Table 6.11. Colony stimulating factors. Individual CSFs exhibit a broad range of molecular weights depending upon their degree of glycosylation

Name	Composition	Molecular weight (Da)
GM-CSF	Single-chain glycoprotein	$\leq 30\,000$
G-CSF	Single-chain glycoprotein	$\leq 20\,000$
M-CSF	Dimeric glycoprotein	$\leq 90\,000$
Multi-CSF	Single chain glycoprotein	$\leq 30\,000$

Colony stimulating factors are produced by a variety of cell types, including lymphocytes and fibroblasts. The various CSFs exhibit little amino acid sequence homology and receptors for all four subtypes have been identified on responsive cells. All the receptors are transmembrane glycoproteins having an extracellular ligand binding domain, a transmembrane segment and an intracellular effector domain. The intracellular domain of M-CSF displays tyrosine kinase activity. The mechanisms by which the other CSFs generate an intracellular signal in response to ligand binding is as yet poorly understood. It is possible, however, that binding of ligand to all four receptor types initiates a number of intracellular signalling sequences.

The advent of recombinant DNA technology has facilitated the large-scale production of colony stimulating factors. In many instances post-translational glycosylation is not required to sustain biological activity. CSFs produced in prokaryotic expression systems may thus be used clinically. Recombinant CSFs have also been produced in mammalian expression systems which are capable of carrying out post-translational modifications, especially glycosylation.

Administration of colony stimulating factors normally promote a significant increase in the serum level of macrophages and granulocytes. Many clinical studies have successfully illustrated that administration of such cytokines can greatly augment the resistance of susceptible patients to infection. CSFs employed clinically thus far are normally administered to patients whose circulatory levels of leukocytes are diminished, due to diseases such as AIDS, or due to therapeutic regimens such as irradiation or chemotherapy associated with cancer treatments. In many cases, accelerated granulocyte-macrophage production and enhanced leukocyte activity have been recorded. Various colony stimulating factors are currently undergoing clinical evaluation, as illustrated in Table 6.12.

Table 6.12. Some colony stimulating factor preparations approved for sale or currently undergoing clinical trials

Product name	Company	Application
Leukine (GM-CSF)	Immunex	Autologous bone marrow transplantation
Prokine (GM-CSF)	Hoechst-Roussel	Chemotherapy-induced neutropenia
Neupogen (G-CSF)	Amgen	AIDS, leukaemia, aplastic anaemia
Leucomax (GM-CSF)	Sandoz Schering-Plough Genetics Institute	Treatment of low blood cell counts
Macrolin (M-CSF)	Cetus	Cancer, fungal disease

Data from the American Pharmaceutical Manufacturers Association.

Adverse clinical effects of cytokines

Since the 1980s there has been an explosive increase in our understanding of immune function. Intense academic and applied research has facilitated the speedy application of this knowledge at a clinical level. Both established and future potential applications of cytokines have aroused great scientific interest, and these drugs are among the most exciting of the biopharmaceutical products under development.

It has been shown that most cytokines also mediate a range of undesirable, if not toxic, side-effects. Many of the adverse clinical symptoms associated with infection or other disease states are most probably caused by cytokines released naturally in the body.

Administration of interferons often induces various flu-like symptoms in humans. Recipients suffer from loss of appetite, aches and pains, shivering and fatigue. Many of these symptoms are reproduced in clinical studies involving the administration of a variety of other cytokines to human patients.

Interleukin 1 is also known as endogenous pyrogen because of its ability to induce fever. This interleukin, both directly, and indirectly via the induction of a cascade of additional cytokines, exhibits a variety of additional undesirable side-effects and has been linked to a variety of pathological states, including arthritis.

Various side-effects are associated with administration of IL-2. These include general malaise, fever, chills, nausea and hypertension. Many of these toxic effects however, are most probably attributable to IL-2-induced release of TNFα.

Quite a number of undesirable clinical side-effects are directly attributable to release of TNFα. As previously outlined, this cytokine has been implicated in certain autoimmune diseases, in inflammatory reactions and has also been shown to mediate cachexia and septic shock. Such numerous toxic effects have thus far curbed much of the potential clinical promise of this cytokine. Indeed, a considerable research effort has been directed towards developing antibodies, soluble receptor fragments and other drugs capable of neutralizing the activity of TNFα.

Development of drugs capable of modifying cytokine behaviour is likely to form an important part of future cytokine research. Such drugs are assured of a ready market as it is becoming more apparent that inappropriate production, overproduction or prolonged production of these powerful regulatory molecules may well be implicated in a wide range of disease states.

CELL ADHESION FACTORS

Another group of biomolecules which may well prove to be of enormous therapeutic value are cell adhesion factors. These factors are required to sustain normal embryonic development, and in the adult they play a vital role in a variety of processes such as the immune response, wound healing and inflammation. In addition, adhesion factors appear to play a central role in the development of a variety of disease states, including osteoporosis and cancer metastasis. Recent research has proven fruitful in providing an increased understanding of these factors, and elucidating their functions within the body.

Cell adhesion factors are generally expressed on the surface of many cell types, and promote adhesive interactions, either with the extracellular matrix or with other cells. Thus far, four distinct groupings of adhesion factors have been identified: integrins, selectins, cadherins and some members of the immunoglobulin superfamily.

Integrins are generally heterodimeric proteins, and are expressed on the surface of a wide range of cell types. While some integrins promote cell–cell adhesion, most seem to mediate cell–matrix interactions during cell adhesion to basement membranes.

Selectins are monomeric, and are found mainly in association with various blood cells and endothelial cells. Each selectin contains a calcium-dependent lectin domain through which its adhesive properties are mediated by binding various carbohydrate groups. Selectins appear to play an important role in promoting interactions between various blood cell types.

Cadherins are a family of adhesion molecules present on the surface of many cell types that largely mediate adhesion between similar cells. Some members of the immunoglobulin superfamily are expressed on the surface of certain cell types, most noticeably lymphocytes, and play a role in cell–cell interactions.

The integrin family of adhesion factors have thus far attracted the most attention from the biotechnology industry. Most integrins seem to recognize a specific tripeptide sequence, arginine-glycine-aspartate, on their target protein receptors. Some synthetic peptides containing this tripeptide sequence are capable of binding integrins and form the basis of new therapeutic approaches currently undergoing clinical evaluation. One gel containing such a peptide has been found effective in promoting wound healing. Application of the gel to slow healing wounds or ulcers appears to accelerate the wound-healing process. This may well be due to the localization and retention of certain wound-healing cell types in the affected area by virtue of their binding via surface integrins to the applied peptide.

Osteoporosis, the progressive degeneration of bone tissue resulting in brittle bones, represents one disease condition whose clinical progression may well be retarded by intervention using integrin therapy. It appears that certain cell types that promote bone damage actually attach to the bone via surface integrins.

Many additional therapeutic applications of cell adhesion factors, or molecules capable of interfering with the adhesive functioning of such factors, may become apparent as scientific information relating to these molecular species accumulates. For example, the adhesion molecules on the surface of metastatic cancer cells are known to differ from those displayed on non-metastatic cancer cells. Adhesion factor therapy may thus one day play a role in prevention of the spread of cancer cells.

FURTHER READING

Books

British Pharmacopoeia (1993). HMSO, London.

Dieleman, S. *et al.* (eds) (1992). *Clinical Trends and Basic Research in Animal Reproduction*. Elsevier, Amsterdam.

Martindale, The Extra Pharmacopoeia, 29th edn (1989). Pharmaceutical Press, London.

United States Pharmacopoeia XXII (1990). United States Pharmacopoeial Convention.

Articles

Allen, G. (1982). Structure and properties of human interferon-α from Namalway lymphoblastoid cells. *Biochem. J.*, **207**, 397–408.

Allen, G. et al. (1987). Synthesis and cloning of a gene coding for a fusion protein containing mouse epidermal growth factor. *J. Biotechnol.*, **5**, 93–114.

Allen, G. & Fantez, K. (1980). A family of structural genes for human lymphoblastoid (leukocyte-type) interferon. *Nature*, **287**, 408–411.

Balkwill, F. (1988). The body's protein weapons. *New Scientist*, June 16th issue.

Bienz-Tadmor, B. (1992). Biopharmaceuticals and conventional drugs: clinical success rates. *Bio/Technology*, **10**, 521–525.

Biotechnology Medicines in Development (1991). *Annual Survey Report*. Pharmaceutical Manufacturers Association, Washington, DC.

Blundell, T.L. et al. (1972). Insulin: the structure in the crystal and its reflection in chemistry and biology. *Adv. in Protein Chem.*, **26**, 279–402.

Boland, M.P. et al. (1991). Alternative gonadotrophins for superovulation in cattle. *Theriogenology*, **35**, 5–17.

Camussi, G. et al. (1991). The molecular action of tumor necrosis factor-α. *Eur. J. Biochem.*, **202**, 3–14.

Chappel, S. et al. (1988). Bovine FSH produced by recombinant DNA technology. *Theriogenology*, **29**, 235.

Chien, K. (1993). Molecular advances in cardiovascular biology. *Science*, **260**, 916–917.

Closset, J. et al. (1983). Purification of the 22,000- and 20,000-mol.wt. forms of human somatotropin and characterization of their binding to liver and mammary binding sites. *Biochem. J.*, **214**, 885–892.

Closset, J. & Hennen, G. (1978). Porcine follitropin, isolation and characterization of the native hormone and its α and β subunits. *Eur. J. Biochem.*, **86**, 105–113.

Collins, F. (1992). Cystic fibrosis: molecular biology and therapeutic implications. *Science*, **256**, 774–779.

Cosman, D. et al. (1990). A new cytokine receptor superfamily. *Trends Biochem. Sci.*, **15**, 7, 265–270.

DeJong, F. (1988). Inhibin. *Physiol. Rev.*, **68**, 555–595.

deVos, A. et al. (1992). Human growth hormone and extracellular domain of its receptor; crystal structure of the complex. *Science*, **255**, 306–312.

Edgington, S. (1993). Nuclease therapeutics in the clinic. *Bio/Technology*, **11**, 580–582.

Fiers, W. (1991). Tumor necrosis factor. *FEBS Lett.*, **285**, 2, 199–212.

Fuh, G., et al. (1993). Rational design of potent antagonists to the human growth hormone receptor. *Science*, **256**, 1677–1680.

Furman, T. et al. (1987). Recombinant human insulin-like growth factor II expressed in *E. coli*. *Bio/Technology*, **5**, 1047–1056.

Golde, D. & Gasson, J. (1988). Hormones that stimulate the growth of blood cells. *Scientific American*, **259**, 1, 34–42.

Gonzalez, A. et al. (1990). Superovulation of beef heifers with follitrophin: a new FSF preparation containing reduced LH activity. *Theriogenology*, **33**, 519–529.

Gray, P. et al. (1982). Expression of human immune interferon cDNA in *E. coli*, and monkey cells. *Nature*, **295**, 503–508.

Hynes, R. & Lander, A. (1992). Contact and adhesive specificities in the associations, migrations, and targeting of cells and axons. *Cell*, **68**, 303–322.

Johnson, I.S. (1983). Human insulin from recombinant DNA technology. *Science*, **219**, 632–637.

Juskevich, J. & Guyer, C. (1990). Bovine growth hormone: human food safety evaluation. *Science*, **149**, 875–884.

Kaplan, G. *et al.* (1992). Rational immunotherapy with interleukin-2. *Bio/Technology*, **10**, 157–162.

Kerr, I. & Stark, G. (1991). The control of interferon inducible gene expression. *FEBS Lett.*, **285**, 2, 194–198.

Konrad, M. (1989). Immunogenicity of proteins administered to humans for therapeutic purposes. *Trends in Biotechnol.*, **7**, 175–179.

Krimpenfort, P. *et al.* (1991). Generation of transgenic dairy cattle using *in vitro* embryo production. *Bio/Technology*, **9**, 844–847.

Kroeff, E. *et al.* (1989). Production-scale purification of biosynthetic human insulin by reverse-phase high performance liquid chromatography. *J. Chromatog.*, **461**, 45–61.

Kronheim, S. *et al.* (1986). Purification and characterization of human interleukin 1 expressed in *E. coli. Bio/Technology*, **4**, 1078–1082.

Kurokawa, T. *et al.* (1987). Cloning and expression of cDNA encoding human basic fibroblast growth factor. *Febs Lett.*, **213**, 189–194.

Lanzavecchia, A. (1993). Identifying strategies for immune intervention. *Science*, **260**, 937–944.

Lee-Huang, S. (1984). Cloning and expression of human erythropoietin cDNA in *E. coli. Proc. Natl Acad. Sci. USA*, **81**, 2708–2712.

Lindsell, C. *et al.* (1986). Variability in FSH:LH ratios among batches of commercially available gonadotrophins. *Theriogenology*, **25**, 167.

Liu, D. (1992). Glycoprotein pharmaceuticals: scientific and regulatory considerations and the US orphan drug act. *Trends in Biotechnol.*, **10**, 114–120.

Looney, C. *et al.* (1988). Superovulation of donor cows with bovine follicle-stimulating hormone (bFSH) produced by recombinant DNA technology. *Theriogenology*, **29**, 271.

Malefyt, R. *et al.* (1991). Interleukin 10 (IL-10) inhibits cytokine synthesis by human monocytes: an autoregulatory role of IL-10 produced by monocytes. *J. Exp. Med.*, **174**, 1209–1220.

Maliszewski, C. & Fanslow, W. (1990). Soluble receptors for IL-1 and IL-4: biological activity and therapeutic potential. *Tibtech*, **8**, 324–329.

March, C. *et al.* (1985). Cloning sequence, and expression of two distinct human interleukin 1 cDNA's. *Nature*, **315**, 641.

Metcalf, D. (1991). Control of granulocytes and macrophages: molecular, cellular and clinical aspects. *Science*, **254**, 529–533.

Moore, W. & Ward, D. (1980). Pregnant mare serum gonadotrophin. *J. Biol. Chem.*, **255**, 6923–6929.

Nagata, S. *et al.* (1980). Synthesis in *E. coli* of a polypeptide with human leukocyte interferon activity. *Nature*, **284**, 316–320.

Nicol, S. & Smith, M.L. (1960). Amino acid sequence of human insulin. *Nature*, **187**, 483–485.

Orci, L. *et al.* (1988). The insulin factory. *Scientific American*, **259**, 3, 50–61.

Old, L. (1988). Tumor Necrosis Factor. *Scientific American*, **258**, 5, 41–49.

Olson, K. *et al.* (1981). Purified human growth hormone from *E. coli* is biologically active. *Nature*, **293**, 408–411.

Packard, B. (1989). Molecular-level understanding of cytokines and tumour therapy. *Trends in Biotechnol.*, **7**, 78–79.

Pestka, S. *et al.* (1987). Interferons and their actions. *Ann. Rev. Biochem.*, **56**, 727–777.

Pisa, P. *et al.* (1992). Selective expression of interleukin 10, interferon γ and granulocyte-macrophage colony stimulating factor in ovarian cancer biopsies. *Proc. Natl Acad. Sci. USA*, **89**, 7708–7712.

Robinson, C. (1990). Polypeptide growth factors—a growth area for biotechnology. *Trends in Biotechnol.*, **8**, 59–60.

Ruttenberg, J. (1972). Human insulin: facile synthesis by modification of porcine insulin. *Science*, **177**, 623–626.

Sairam, M. (1979). Studies on pituitary follitropin, an improved procedure for the isolation of highly potent ovine hormone. *Arch. Biochem. Biophys.*, **194**, 63–70.

Scorer, C. *et al.* (1991). Amino acid misincorporation during high level expression of mouse epidermal growth factor in *E. coli. Nucl. Acids Res.*, **19**, 3511–3516.

Scott, D.A. (1934). Crystalline insulin. *Biochem. J.*, **28**, II, 1592–1602.

Scott, P. (1993). IL-12: initiation cytokine for cell mediated immunity. *Science*, **260**, 496–497.

Smith, K. (1989). Interleukin futures. *Bio/Technology*, **7**, 661–667.

Smith, K. (1988). Interleukin 2, inception, impact and implications. *Science*, **240**, 1169–1176.

Smith, K. (1990). Interleukin 2. *Scientific American*, **262**, 3, 26–33.

Sofer, G. (1986). Current applications of chromatography in biotechnology. *Bio/Technology*, **4**, 712–715.

Sprang, S. (1990). The divergent receptors for TNF. *Trends Biochem. Sci.*, **15**, 10, 366–368.

Springer, T. (1990). Adhesion receptors of the immune system. *Nature*, **346**, 425–434.

Taniguchi, T. *et al.* (1980). Expression of the human fibroblast interferon gene in *E. coli. Proc. Natl Acad. Sci. USA*, **77**, 9, 5230–5233.

tenDijke, P. & Iwata, K. (1989). Growth factors for wound healing. *Bio/Technology*, **7**, 793–798.

Thomas, K. & Gimenez-Gallego, G. (1986). Fibroblast growth factors: broad spectrum mitogens with potent angiogenic activity. *Trends Biochem. Sci.*, **11**, 81–84.

Thornbeck, G. *et al.* (1992). Involvement of endogenous tumor necrosis factor α and transforming growth factor β during induction of collagen type II arthritis in mice. *Proc. Natl Acad. Sci. USA*, **89**, 7375–7379.

Travis, J. (1993). Biotech gets a grip on cell adhesion. *Science*, **260**, 906–908.

Van Brund, J. (1989). Lymphokine receptors as therapeutics? *Bio/Technology*, **7**, 668–669.

Van Leen, R. *et al.* (1991). Production of human interleukin 3 using industrial microorganisms. *Bio/Technology*, **9**, 47–52.

Waldmann, T. (1989). The multi-subunit interleukin 2 receptor. *Ann. Rev. Biochem.*, **58**, 875–911.

Weaver, J. *et al.* (1988). Production of recombinant human CSF-1 in an inducible mammalian expression system. *Bio/Technology*, **6**, 287–290.

Zurawski, G. (1991). Analysing lymphokine-receptor interactions of IL-1 and IL-2 by recombinant DNA technology. *Tibtech*, **9**, 250–257.

Chapter 7

Proteins for diagnostic purposes

- Enzymes as diagnostic reagents.
  - Assay of blood glucose.
  - Assay of blood cholesterol and triglyceride.
  - Assay of blood urea and uric acid.
  - Immobilized enzymes as diagnostic reagents.
  - Immobilized enzyme electrodes.
- Antibodies as diagnostic tools.
  - Radioimmunoassay.
  - Enzyme immunoassay.
  - Enzyme-linked immunosorbent assay (ELISA).
  - Enzymes used in EIAs.
  - Trends in immunoassay development.
  - Immunological assays for HIV.
  - Latex based and other immunological assays.
  - Membrane bound diagnostic systems.
  - *In vivo* applications of antibodies as diagnostic aids.
- Further reading.

Many proteins are currently used for a variety of diagnostic purposes. An ever-increasing understanding of normal and abnormal metabolic activity allows clinicians to link changes in the concentration of various biomolecules to disease states or to impending events. Sensitive and specific diagnostic assays capable of detecting and quantifying many marker molecules are now available. These diagnostic systems greatly assist medical practitioners in the accurate diagnosis or prediction of medical abnormalities, thus allowing them to formulate the most appropriate therapeutic responses.

Clinical chemistry is the scientific discipline charged with detecting, monitoring and quantifying a broad variety of "marker" substances present in biological samples. A wide range of biomolecules are of potentially significant diagnostic value. Such substances include low molecular weight molecules such as urea, glucose, cholesterol or steroid hormones. Many substances of higher molecular weight, such as specific proteins which may be released from damaged tissue or whose normal concentration is altered due to a particular metabolic aberration, are also of diagnostic value (Table 7.1). Additional diagnostic tests have been developed which detect supramolecular assemblies including serum lipoproteins, viruses or microorganisms.

It is important to note that most disease conditions are dynamic rather than static in nature. The value of many diagnostic results is often closely correlated not only with the sensitivity and specificity of the test, but also with the speed with which results may be obtained. Indeed, the progression of many disease states is often monitored by performing repeat tests on samples obtained from the patient at appropriate time intervals.

A large number of diagnostic systems have been designed to facilitate their automation. The use of automatic multisample analysers in any clinical laboratory increases the speed, efficiency, throughput and economy with which diagnostic assays are carried out.

Blood and urine are the most commonly assayed biological samples, although other bodily products, such as faeces, saliva or sweat, may also be used.

Proteins most often used as analytical tools for diagnostic purposes include a variety of enzymes and antibody preparations. Enzymes are used largely to detect and quantify various medically significant metabolites present in biological samples. Antibodies are normally used to detect and quantify the presence of specific antigens in such samples. In a reversal of this procedure, antigens may be employed to detect the presence of specific antibodies in serum samples. This latter approach is illustrated by the use of specific antigens associated with the AIDS virus in order to detect the presence of anti-HIV antibodies in blood or serum samples. Immunoassays using antibodies are normally used in the detection of larger molecular weight compounds such as proteins.

A wide range of alternative techniques also form the basis for numerous diagnostic tests. Such techniques include electrophoresis, conventional chromatography, HPLC, isoelectric focusing and chromatofocusing, all of which are routinely used in modern clinical chemistry laboratories. More recently a variety of diagnostic tests based upon DNA technology have been introduced in the clinical arena. Such techniques may be used to detect a variety of genetic abnormalities which may serve as markers, indicating the presence of a number of genetic disorders such as Duchenne muscular dystrophy or cystic fibrosis. DNA tests are characterized by an excellent degree of specificity, sensitivity and rapidity. Such tests are poised to make a significant impact in the future of laboratory medicine.

Table 7.1. Some proteins of diagnostic significance. Diseases potentially associated with an increase or decrease in the concentration of such proteins are also listed

Protein	Diseases associated with increased concentration of protein	Diseases associated with decreased concentration of protein
Albumin	Blood–brain barrier damage	Malnutrition; liver cirrhosis; severe burns; inflammation; renal disease
Alpha-1-acid glycoprotein	Inflammatory conditions; malignant neoplasms	Severe hepatic damage; nephrotic syndrome
Alpha-1-antitrypsin	Acute hepatic diseases; active cirrhosis; inflammatory disease	Genetic deficiency; chronic pulmonary disease
Alpha-2-macroglobulin	Nephrotic syndrome; oral contraceptive use; diabetes	Disseminated intravascular coagulation; stress
Apolipoprotein A1	Atherosclerosis; hyperlipidaemias	Atherosclerosis; Tangier disease
Apolipoprotein B		
C-Reactive protein	Inflammation; tumours; tissue destruction; active rheumatoid arthritis	
Ceruloplasmin	Inflammatory conditions; oral contraceptive use	Wilson's disease; malnutrition
Complement C3	Inflammatory disease	Autoimmune disease; lupus; chronic hepatitis distress; neonatal respiratory tissue injury
Complement C4	Acute inflammatory disease; bacterial infections	Acute glomerular nephritis; chronic hepatitis; autoimmune disease; lupus
Haptoglobin	Inflammation	Haemolytic anaemia; sickle-cell anaemia; hepatic disease
Immunoglobulin A	IgA myeloma; cirrhosis; chronic liver disease; chronic infections	Immune deficiency; non-IgA myeloma
Immunoglobulin G	Chronic infections: IgA myeloma; liver disease; multiple sclerosis; meningitis	Immune deficiency
Immunoglobulin M	IgM myeloma; liver disease; acute hepatitis	Immune deficiency; non-IgM myeloma
Kappa (light chains)	Monoclonal gammopathies; Bence Jones light chain disease	Monoclonal gammopathies

Lambda (light chains)	Monoclonal gammopathies; Bence Jones light chain disease	Monoclonal gammopathies
Microalbumin	Early stage renal disease; diabetes; hypertension	
Prealbumin		Malnutrition; liver disease; acute inflammation
Properdin factor-B		Autoimmune diseases; sickle-cell disease; bacterial diseases
Rheumatoid factor	Rheumatoid arthritis; mixed connective tissue disease; viral and bacterial disorders	
Transferrin	Iron deficiency; pregnancy; acute hepatitis; oral contraceptive use	Inflammation

Data kindly supplied by Beckman Diagnostic Systems Group.

In keeping with the central theme of the book this chapter concentrates on the use of proteins as diagnostic reagents. Particular attention is given to the use of both enzymes and antibodies. This chapter reviews many of the more commonly used diagnostic strategies which depend upon protein reagents, and gives examples that have found wide application in clinical chemistry laboratories.

ENZYMES AS DIAGNOSTIC REAGENTS

A variety of enzyme preparations have been used as diagnostic reagents for many years. Enzymes may be used to detect and estimate the levels of specific analytes present in biological samples. In certain situations, enzymes may also be used to catalytically remove specific compounds present in biological samples, which may interfere with the assay of a particular metabolite. Enzymes are also widely used as labels in enzyme immunoassay (EIA) systems. Their high degree of selectivity coupled with their catalytic efficiency render many enzymatic preparations ideal diagnostic reagents.

The majority of enzymes initially used as diagnostic reagents were obtained from plant and animal sources. Specific examples include peroxidase from horseradish and cholesterol esterase from porcine pancreas. Table 7.2 lists some such enzymes which are used as diagnostic reagents. Some enzymes (for example hexokinase) are obtained from yeast and other microorganisms).

Microorganisms produce a large array of enzymes with a wide range of functions and capabilities. In addition to being an alternative source of enzymes, many microbial species harbour novel enzymes with potential applications in assaying biological substances.

Many microorganisms produce enzymes of superior stability compared with analogous enzymes derived from animal sources. Particular attention has been focused on enzymes produced by a variety of thermophilic bacteria, as such enzyme preparations exhibit impressive stability characteristics when stored under suboptimal conditions. Such characteristics are attractive in diagnostic reagents, because reagent stability and consequent extended shelf-life is of paramount importance in the development of any diagnostic assay.

Many enzymes are produced in large quantities by microorganisms. Fermentation technology makes possible the economical production of such enzymes. Moreover, recent advances in DNA technology facilitates the large-scale production of virtually any

Table 7.2. Some enzymes used directly or indirectly as diagnostic reagents. The source of enzymes and their likely applications are listed

Enzyme	Source	Application
Acetyl cholinesterase	Bovine erythrocytes	Analysis of organophosphorus compounds such as pesticides
Alcohol dehydrogenase	Yeast	Determination of alcohol levels in biological fluids
Alkaline phosphatase	Calf intestine and kidney	Conjugation to antibodies allows its use as an indicator in ELISA systems
Arginase	Beef liver	Determination of L-arginine levels in plasma and urine
Ascorbate oxidase	*Cucurbita* species	Determination of ascorbic acid levels; eliminates interference by ascorbic acid
Cholesterol esterase	Pig pancreas	Determination of serum cholesterol
Creatine kinase	Rabbit muscle, beef heart, pig heart	Diagnosis of cardiac and skeletal malfunction
Glucose-6-phosphate dehydrogenase	Yeast, *Leuconostoc mesenteroides*	Determination of glucose and ATP in conjunction with hexokinase
Glucose oxidase	*Aspergillus niger*	Determination of glucose in biological samples in conjunction with peroxidase; a marker for ELISA systems
Glutamate dehydrogenase	Beef liver	Determination of blood urea nitrogen in conjunction with urease
Glycerol kinase	*Candida mycoderma*	Determination of triglyceride levels in blood in conjunction with lipase
Glycerol-3-phosphate dehydrogenase	Rabbit muscle	Determination of serum triglycerides
Hexokinase	Yeast	Determination of glucose in body fluids
Peroxidase	Horseradish	Indicator enzyme for reactions in which peroxide is produced
Phosphoenolpyruvate carboxylase	Maize leaves	Determination of CO_2 in body fluids
Urease	Jack bean	Determination of blood urea nitrogen; marker enzyme for ELISA systems
Uricase	Porcine liver	Determination of uric acid
Xanthine oxidase	Buttermilk	Determination of xanthine and hypoxanthine in biological fluids

Data kindly supplied by Biozyme Laboratories Ltd.

enzyme in recombinant systems. This technology is likely to have a major impact in the diagnostic arena.

Most enzyme preparations used for diagnostic purposes have been subjected to some form of purification. Unlike protein preparations destined for parenteral administration, usually it is not necessary to purify diagnostic reagents to homogeneity. However, it is of crucial importance to ensure that the purification procedure employed removes any proteins or other molecules present in the initial preparation which might interfere with the assay or lead to erroneous results.

Most enzymatic preparations used in the detection of various molecules of diagnostic significance are free in solution. However, for some applications, the enzyme used is immobilized or is in a reagent strip format. Examples of these latter formats are detailed later.

Enzymes used in the detection and quantitation of specific metabolites normally use the metabolite in question as a substrate. To be of diagnostic value, changes in the concentration of one or other of the coreactants, cofactors or products of the reaction must be readily monitored. Two of the most commonly employed enzyme types in diagnostic systems are dehydrogenases and oxidases. In the case of dehydrogenases, progression of the reaction may be followed by monitoring the conversion of NADH to NAD^+ or vice versa, as illustrated below (S, reaction substrate; P, reaction product).

$$S + NAD^+ \xrightarrow{\text{dehydrogenase}} P + NADH + H^+$$

Formation or disappearance of NADH may be readily monitored spectrophotometrically, as NADH absorbs light strongly at 340 nm, whereas NAD^+ does not absorb at this wavelength. Progression of the reaction may also be followed by fluorescence or luminescence. From the above generalized example, it is clear that one molecule of NADH is formed for each molecule of substrate S (the metabolite whose concentration is to be determined) converted. Thus, in such a system, the total quantity of NADH formed reflects the quantity of substrate present in the sample analysed.

When considering this example, it becomes evident that all assay reagents should be present in excess, such that it is the concentration of substrate which dictates the final quantity of NADH formed. If insufficient NAD^+ were present, for example, the reaction would cease once all the cofactor was converted to NADH, although significant quantities of the substrate, S, might still remain. In such circumstances, the quantity of substrate present in the sample analysed would be underestimated.

Oxidases, which yield hydrogen peroxide as a reaction product, are also commonly used in diagnostic systems. However, in most such instances none of the final products of this primary reaction (e.g. P or H_2O_2 in the example below) are readily quantified. A second enzymatic step is therefore included in the assay system. The second enzyme utilizes one of the products of the initial enzymatic step as one of its substrates, H_2O_2 in the generalized example illustrated below. Unlike the primary enzymatic step, one or other of the products of this second reaction is easily measured by use of an appropriate assay. In the example cited, the second enzyme utilizes the H_2O_2 produced in the initial reaction, together with a second colourless substance, as substrates, yielding H_2O and a coloured dye as products. The quantity of dye produced can be determined by monitoring the change in absorbance at an appropriate wavelength. As long as all the reagents present in the assay system are in excess, the quantity of dye produced by such a coupled assay is directly related to the quantity of substrate present in the sample analysed.

$$S + \tfrac{1}{2}O_2 + H_2O \xrightarrow{\text{oxidase}} P + H_2O_2$$

$$H_2O_2 + \text{colourless substance} \xrightarrow[\text{e.g. peroxidase}]{\text{second enzyme}} \text{dye (coloured)} + H_2O$$

The concepts outlined in the generalized examples discussed above are illustrated more clearly by the specific examples as discussed below. Several alternative enzymatic methods have been developed to assay many biological substances of diagnostic interest.

Assay of blood glucose

The concentration of glucose present in blood is an important diagnostic marker for several disease states, especially diabetes mellitus. Blood glucose determinations constitute one of the most common assays carried out in clinical chemistry laboratories. Most such determinations are based upon specific assays, of which there are several available. A system based on glucose oxidase was one of the first enzymatic analytical procedures developed. The reaction principle upon which this coupled assay system is based is as follows:

$$\text{Glucose} + O_2 + H_2O \xrightarrow{\text{glucose oxidase}} \text{gluconic acid} + H_2O_2$$

$$2H_2O_2 + 4\text{-aminophenazone} + \text{phenol} \xrightarrow{\text{peroxidase}}$$
$$\text{quinoneimine} + 4H_2O$$

In this coupled assay, the quantity of glucose in the initial sample is directly related to the quantity of quinoneimine formed. Unlike the other reaction products, quinoneimine is a red-violet dye whose concentration can be determined by measuring its absorbance at 500 nm.

In the above example the reagents present in the assay system consist of the enzymes glucose oxidase and peroxidase, and the reagents phenol and 4-aminophenazone. A specified volume of this reagent cocktail would be incubated with a specific volume of the serum sample to be assayed. The reaction would be allowed to proceed for an appropriate period, typically 5–10 minutes, at an appropriate temperature, typically 20–37 °C, until all the glucose present in the serum sample had been enzymatically converted to gluconic acid. Several additional assays in which serum samples have been replaced with standard glucose solutions are normally run concurrently. The glucose concentration present in the serum sample can be calculated by comparison of the unknown with standard absorbency values.

A second assay system commonly used to determine blood glucose levels employs the enzymes hexokinase and glucose-6-phosphate dehydrogenase (G6P-DH). The reaction principle is outlined below:

$$\text{Glucose} + \text{ATP} \xrightarrow{\text{hexokinase}} \text{glucose 6-phosphate} + \text{ADP}$$

$$\text{Glucose 6-phosphate} + \text{NAD}^+ \xrightarrow{\text{G6P-DH}}$$

$$\text{gluconate 6-phosphate} + \text{NADH} + \text{H}^+$$

The concentration of glucose present in the sample is related to the quantity of NADH formed by the coupled assay. This may be easily monitored by measuring the absorbance at 340 nm.

The above assay systems are capable of rapidly, accurately and conveniently determining the concentration of blood glucose present in any sample tested. Such a result, however, determines the concentration of glucose present at the time the blood sample was taken.

Diagnostic tests that monitor average long-term blood glucose concentrations are also available. Such tests are invariably based on assessing the level of glycosylated haemoglobin present in a serum sample. Elevated blood glucose concentrations promote the glycosylation of haemoglobin molecules by covalent linkage of a glucose residue to the terminal valine residue of the haemoglobin β chain. This reaction may occur progressively in the erythrocyte throughout its normal 120 day life-span. Thus, the level of glycosylated haemoglobin present in a blood sample reflects blood glucose levels over a period of several weeks. This assay is invaluable in assessing the long-term effectiveness of therapeutic approaches as applied to diabetic patients. Separation of glycosylated haemoglobin from native haemoglobin is normally achieved by electrophoretic or chromatographic procedures.

Determination of the presence of glucose in urine also constitutes an important diagnostic test. Such a system is often used in the initial detection of diabetes, as discussed later.

Assay of blood cholesterol and triglyceride

An association between atherosclerosis and elevated plasma cholesterol levels has been established. This association relates in particular to levels of cholesterol associated with low-density lipoproteins (LDL-cholesterol). The condition of atherosclerosis is responsible for the deaths of 1 in every 2 individuals in most Western societies.

Owing to their largely hydrophobic nature, both cholesterol and triglycerides are transported in blood in the form of soluble complexes termed lipoproteins (Figure 7.1). The outer layer of lipoprotein particles consists largely of phospholipids and proteins termed apoproteins. The hydrophilic region of these molecules are oriented towards the surrounding aqueous environment. The internal portion of such lipoproteins is composed predominantly of triglycerides and cholesterol covalently linked to long-chain fatty acids via ester bonds, cholesterol-esters.

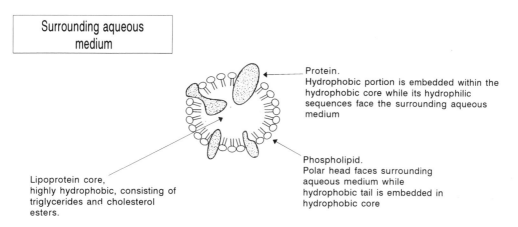

Based upon differences in physicochemical properties, lipoproteins may be classified as high-density lipoproteins (HDL), low-density lipoproteins (LDL) or very low-density lipoproteins (VLDL) (Table 7.3). Chylomicrons, another group of lipoproteins, are found in plasma, particularly after the ingestion of a lipid rich meal. Chylomicrons are the largest lipoproteins and consist almost exclusively of triglycerides; they serve to transport dietary fats.

Figure 7.1. Generalized structure of plasma lipoprotein

Table 7.3. Approximate content of triglycerides and cholesterol in various lipo-proteins

	Chylomicrons	VLDL	LDL	HDL
Triglycerides (%)	85–90	50–60	10	3–4
Cholesterol (%)	2–4	15	45	17–18

Increases in the circulatory concentration of LDL-cholesterol in particular is linked to the development of atherosclerosis. Elevated levels of HDL-cholesterol, on the other hand, appear to exert a protective effect against the development of fatty deposits on the inner walls of arteries characteristic of atherosclerosis. This may be due to the fact that HDL functions to transport cholesterol from peripheral tissues to the liver where it may be converted to bile acids.

Several diagnostic tests capable of measuring serum concentrations of cholesterol are available. Total serum cholesterol levels may be estimated by employing the three-step coupled enzymatic system illustrated below:

$$\text{Cholesterol ester} + H_2O \xrightarrow{\text{cholesterol esterase}} \text{cholesterol} + \text{fatty acids}$$

$$\text{Cholesterol} + O_2 \xrightarrow{\text{cholesterol oxidase}} \text{4-cholestenone} + H_2O_2$$

$$2H_2O_2 + \text{phenol} + \text{4-aminoantipyrine} \xrightarrow{\text{peroxidase}}$$

$$\text{quinoneimine} + 4H_2O$$

In the initial reaction, cholesterol esterase catalyses the hydrolytic cleavage of cholesterol esters, yielding free cholesterol which may then be oxidized by cholesterol oxidase, and the resultant H_2O_2 can be quantified by the peroxidase system. The end product, quinoneimine, is a red dye which may be easily quantified by measuring its absorbance at 500 nm.

Total serum HDL-cholesterol may be specifically quantified using the above assay method. In this case it is necessary firstly to remove the other lipoproteins from the serum sample. This may be achieved by the addition of phosphotungstic acid and magnesium ions, which promote the precipitation of LDL, VLDL and chylomicrons. Following centrifugation, the cholesterol content of the HDL-containing supernatant may be assayed by the above procedure.

Elevated levels of serum triglycerides have been linked to atherosclerosis and coronary artery disease. Various enzymatic systems have been developed which facilitate appraisal of triglyceride concentrations in serum. One commonly used system, based upon four sequential enzymatic reactions, is illustrated as follows:

$$\text{Triglycerides} + H_2O \xrightarrow{\text{lipases}} \text{glycerol} + \text{fatty acids}$$

$$\text{Glycerol} + ATP \xrightarrow{\text{glycerol kinase}} \text{glycerol 3-phosphate} + ADP$$

$$\text{Glycerol 3-phosphate} + O_2 \xrightarrow[\text{oxidase}]{\text{glycerol-3-phosphate}}$$

$$\text{dihydroxyacetone phosphate} + H_2O_2$$

$$H_2O_2 + \text{4-aminoantipyrine} + \text{chlorphenol} \xrightarrow{\text{peroxidase}}$$

$$\text{quinoneimine dye} + HCl + 2H_2O$$

The concentration of triglycerides in the serum sample provided is proportional to the quantity of dye produced. Quinoneimine levels are determined by measuring the absorbance at 500 nm.

A variety of additional enzymatic systems, such illustrated below, have been developed to quantify triglyceride levels in biological samples.

$$\text{Triglycerides} + H_2O \xrightarrow{\text{lipases}} \text{glycerol} + \text{fatty acids}$$

$$\text{Glycerol} + ATP \xrightarrow{\text{glycerol kinase}} \text{glycerol 3-phosphate} + ADP$$

$$ADP + \text{phosphoenolpyruvate} \xrightarrow{\text{pyruvate kinase}} ATP + \text{pyruvate}$$

$$\text{Pyruvate} + NADH + H^+ \xrightarrow{\text{lactate dehydrogenase}} \text{lactate} + NAD^+$$

Assay of blood urea and uric acid

Determination of urea and uric acid is carried out frequently in most clinical laboratories. Elevated levels of these metabolites are often indicative of a variety of metabolic disorders, including diseases of the kidney. Most methods used to estimate urea include the enzyme urease, which catalyses the hydrolytic cleavage of urea as shown below.

$$\text{Urea} + H_2O \xrightarrow{\text{urease}} \text{ammonia} + CO_2$$

A variety of methods may be used to detect and quantify the amount of ammonia formed by this reaction. One such method relies upon the chemical detection of the ammonia formed. Ammonium ions react with phenol and hypochlorite in the presence of sodium nitroprusside, forming a blue coloured complex which absorbs light strongly at 640 nm.

An alternative enzymatic method used to quantify the ammonia produced uses the enzyme glutamate dehydrogenase, as follows:

$$\text{Urea} + H_2O \xrightarrow{\text{urease}} 2NH_4{}^+ + CO_2$$

$$2\alpha\text{-ketogluturate} + 2\,\text{NADH} + 2\,\text{NH}_4{}^+ \xrightarrow[\text{dehydrogenase}]{\text{glutamate}}$$

$$2\,\text{glutamate} + 2\,\text{NAD}^+ + 2\,H_2O$$

The NAD^+ formed by the second reaction may be monitored, and is proportional to the quantity of urea present in the initial sample.

Uric acid may also be quantified by a number of enzymatic systems. One of the more common methods uses a coupled assay system consisting of uricase and peroxidase, as shown below. The end product, the red dye quinoneimine, has an absorbance maximally at 500 nm.

$$\text{Uric acid} + O_2 + 2\,H_2O \xrightarrow{\text{uricase}} \text{allantoin} + CO_2 + H_2O_2$$

$$2\,H_2O_2 + \text{4-aminophenazone}$$

$$+ \text{3,5-dichlor-2-hydroxybenzene sulfonic acid} \xrightarrow{\text{peroxidase}}$$

$$\text{quinoneimine} + 4H_2O$$

An alternative method for the quantitation of uric acid is as follows:

$$\text{Uric acid} + O_2 + 2H_2O \xrightarrow{\text{uricase}} \text{allantoin} + H_2O_2 + CO_2$$

$$H_2O_2 + \text{ethanol} \xrightarrow{\text{catalase}} \text{acetaldehyde} + 2H_2O$$

$$\text{Acetaldehyde} + \text{NADP}^+ \xrightarrow[\text{dehydrogenase}]{\text{aldehyde}} \text{acetate} + \text{NADPH} + H^+$$

The increase in absorbance at 340 nm due to the formation of NADPH is proportional to the quantity of uric acid present in the sample.

Immobilized enzymes as diagnostic reagents

There has been a steady increase in the development and use of diagnostic reagents in immobilized form. Enzymes immobilized on reagent strips are used for selected diagnostic purposes in clinical laboratories, directly by doctors and indeed by patients themselves.

Among the most commonly employed reagent strips are those designed to detect glucose in body fluids. The wafer-thin plastic strips contain a narrow pad at one end, to which the sample is applied. The sample material diffuses inwards into the body of the pad, which houses the appropriate diagnostic enzyme, in addition to any ancillary reagents required.

Most such systems are designed so that one of the products of the final enzymatic reaction is a chromogen and thus coloured. Thus, most systems are based on visual examination of any colour development in the pad after the sample is applied. The intensity of the colour produced reflects the concentration of the metabolite in the sample applied; such kits may, therefore, be used in a semiquantative fashion.

The pad present on reagent strips used to detect glucose usually contains immobilized glucose oxidase and peroxidase. The peroxidase reaction produces a coloured dye as one of its products. Immobilized enzymes are also used in a variety of enzyme electrodes.

Immobilized enzyme electrodes

A wide variety of immobilized enzymes have been used in the construction of "enzyme electrodes". Such electrodes are capable of rapidly and accurately determining the concentration of selected analytes present in a given test solution. Many such enzyme electrodes have been used for diagnostic purposes. Table 7.4 summarizes some of the metabolites of diagnostic significance for which specific enzyme electrodes exist.

The principles adopted in the design and construction of any enzyme electrode are basically straightforward (Figure 7.2). The

Table 7.4. Analytes of diagnostic importance for which specific quantitative enzyme electrodes have been developed. The enzymes used in these test systems are also listed

Analyte	Enzyme used in its detection
Glucose	Glucose oxidase
Urea	Urease
Uric acid	Uricase
Amino acids	L-Amino acid oxidase
Cholesterol	Cholesterol oxidase
Alcohols	Alcohol oxidase
Penicillin	Penicillinase

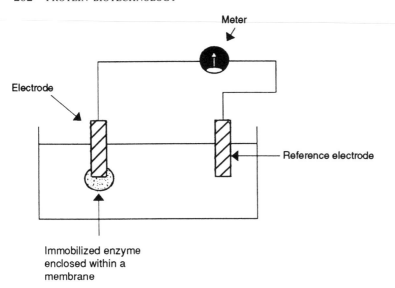

Figure 7.2. Diagrammatic representation of an enzyme electrode

immobilized enzyme is contained within a membrane which surrounds the chosen electrode. The pore size of the outer membrane which faces the surrounding aqueous environment of the test sample must be sufficiently large to allow free penetration of all the substrates necessary for the immobilized enzyme reaction: one of these substrates should be the metabolite to be quantified. In order to protect the immobilized enzyme, the outer membrane must exclude entry of proteins, particulate matter and whole cells. The inner membrane, between the immobilized enzyme and the electrode surface, must in no way interfere with the diffusion of reaction substrates or products; these molecules must freely come into contact with the electrode surface.

Several electrode types may be used in constructing the enzyme electrode. Electrodes that specifically detect gases such as oxygen, carbon dioxide and ammonia, and various ionic species, are all commercially available. The electrode chosen will depend upon what products are produced by the enzymatic reaction. If, for example, oxygen is evolved, then the oxygen electrode may be used. The quantity of oxygen molecules impinging on the electrode surface is then converted into some form of detectable signal, such as a flow of electrical current. This signal can then be monitored. Readings obtained when the electrode is immersed in solutions containing known concentrations of the analyte allow a standard curve to be constructed.

In the construction of an enzyme electrode an enzyme is normally chosen that (a) utilizes as substrate the metabolite whose concentration is to be determined, and (b) has additional substrates or reaction products that may be detected and quantified by the actual electrode.

When the enzyme electrode is immersed in the test sample, the analyte whose concentration is sought diffuses through the outer membrane, comes into contact with the immobilized enzyme and hence is catalytically transformed. The reaction products then pass freely through the inner membrane and impinge on the electrode surface, generating a signal. The magnitude of this signal is proportional to the concentration of substrate present in the sample tested. The exact value may be calculated by reference to the standard curve. Heat generated during enzymic reactions may also be monitored.

Enzyme electrodes and other detection systems using immobilized enzymes are often both economical and popular, for a number of reasons. Immobilization ensures that the enzyme is reusable and often increases the stability of the enzyme itself, thus effectively prolonging the electrode's working life. Many such enzyme electrodes may be used repeatedly for weeks or months.

Enzymes may be immobilized by a number of methods, discussed more fully in Chapter 9. Briefly, immobilization generally involves either chemically coupling the enzyme to a solid support or physically entrapping the enzyme within a polymeric substance such as polyacrylamide. In some instances, it may be unnecessary to use a purified enzyme preparation in the construction of an enzyme electrode. In many instances however, it may be advisable to subject the initial enzyme preparation to some degree of purification in order to remove inhibitory or other undesirable contaminants such as proteases from the final preparation.

ANTIBODIES AS DIAGNOSTIC TOOLS

Enzymes are often used for diagnostic purposes because of their catalytic properties and their specificity. Another group of proteins, antibodies, are also frequently used as diagnostic reagents because they too exhibit extreme specificity in their recognition of a particular ligand, the antigen that stimulated their production. Antibody preparations are often used in the detection and quantitation of a wide variety of specific antigens, many of which are of considerable diagnostic significance. It must always be borne in mind that the immunological activity of a substance does not necessarily equate to its biological activity. This is particularly true with regard to protein antigens. Many proteins may retain their immunological identity, even under conditions that promote a reduction of their biological activity. In certain circumstances, antigen molecules may be used as diagnostic reagents in order to detect and quantify the presence of specific antibody species in serum samples.

Assays that employ antibodies to detect and quantify antigens are generally termed *immunoassays*. The substance of interest is used as an antigen and injected into animals in order to elicit the production of antibodies against that particular molecule. If the antigen is too small to elicit an immunological response, it may first be attached to a large carrier molecule to render it antigenic. Either monoclonal or polyclonal antibody preparations may be used in immunoassay systems.

Antibody molecules have no inherent characteristics which would facilitate their direct detection in immunoassay systems. A second important step in developing a successful immunoassay therefore involves the incorporation of a suitable marker. The marker serves to facilitate the rapid detection of antibody–antigen binding. Immunoassay systems that use radioactive labels as marker systems are termed radioimmunoassays (RIA), whereas systems using enzymes are termed enzyme immunoassays (EIA).

Radioimmunoassay

Radioimmunoassays (RIA) were among the first immunoassay systems to be developed. Such systems have subsequently been successfully used in the quantitation of a large variety of bio-molecules present in the body at very low concentrations. RIA technology has found widespread application both in basic research and for many routine analytical applications.

Two basic reagents are required when initially developing an RIA system: (a) antibodies that have been raised against the antigen of interest, and (b) highly purified antigen that has been labelled with a radioactive tag such as iodine 125. In all assay systems the concentration of antibody added must be limiting, while the concentration of radiolabelled antigen must be in excess.

In a typical radioimmunoassay, appropriate volumes of antibody, labelled antigen and sample supplied for assay are incubated together. As the antibody concentration is limiting, the antigen (Ag) present in the sample supplied will compete with the labelled antigen (Ag*) for binding to antibody (Ab).

$$Ag + Ag^* + Ab \text{ (limiting)} \rightleftharpoons AgAb + Ag^*Ab + Ag + Ag^*$$

In all such systems, the quantities of both antibody and labelled antigen used remain constant. Thus, when equilibrium is reached, the greater the quantity of antigen present in the assay sample, the less labelled antigen will be bound by antibody.

The next step in the assay involves separation of bound from free antigen. Separation is normally achieved by using techniques which precipitate the antibody–antigen complex while leaving unbound antigen, both labelled and unlabelled, in solution. Physical separation may then be achieved by decanting the supernatant from the assay tube, such that the precipitate remains. In practice, precipitation of antigen–antibody complex from solution is most often achieved by adding to the reaction mixture a second antibody, which has been raised against the first or primary antigen-binding antibody.

Upon separation of free from bound antigen, the level of radioactivity associated with bound antigen is normally assessed. As previously mentioned, the more unlabelled antigen present in the assay sample, the lower will be the proportion of labelled antigen bound and hence the lower the radioactive count obtained.

A standard curve relating antigen concentration to levels of radioactivity bound may be constructed by assaying several samples containing known concentrations of unlabelled antigen. The concentration of antigen present in the samples may thus be calculated by reference to the standard curve.

Enzyme immunoassay

The successful introduction of radioimmunoassay has revolutionized many areas of clinical and other biological sciences. However, there are a number of disadvantages associated with the use of radioactive elements in such systems. Disadvantages include:

- The need for radiological protection.
- The generation of radioactive waste
- The short shelf-life of some assays due to the short half-life of some radioactive isotype used.
- The necessity for expensive analytical equipment, and often dedicated areas in which to perform the assays.

Such disadvantages have led to the development of immunoassay systems using alternative labels. Some of these systems use fluorescent or chemiluminescent tags. However, immunoassay systems using enzymes as marker substances have proved to be particularly popular. Enzyme immunoassay (EIA) systems take advantage of the extreme specificity and affinity with which antibodies bind the antigens that stimulated their initial production, coupled to the catalytic efficiency of enzymes which facilitates straightforward detection and quantitation. In general many EIAs exhibit similar sensitivities to RIAs but are free from most of the disadvantages associated with RIAs.

EIAs are sometimes described as heterogeneous or homogeneous systems. Heterogeneous assays involve the separation of bound and free label (as is the case with standard RIAs). Certain EIAs have been developed in which the activity of the enzyme label is significantly altered by binding of antibody to antigen. In such cases, there is no requirement to separate free from bound; assay systems of this design are termed homogeneous systems.

Enzyme immunoassays were first introduced in the early 1970s. In most such systems the antibody was immobilized on a solid surface, such as on the internal walls of the wells in a microtitre plate (Figure 7.3). As in the case of classical RIAs, the immobilized antibody was incubated with a known amount of labelled antigen, in addition to unknown quantities of antigen present in the samples being assayed. As the level of antibody present is limiting, labelled and unlabelled antigen compete with each other for binding. The greater the quantity of unlabelled antigen present in the assay sample, the less labelled antigen will be retained by the immobilized antibody. After allowing antibody–antigen binding to reach equilibrium, unbound antigen is removed by a washing step. The amount of enzyme-labelled antigen retained is assayed for enzymatic activity.

A variant of this EIA involves immobilizing the antigen and employing an antibody–enzyme conjugate. The principle involved

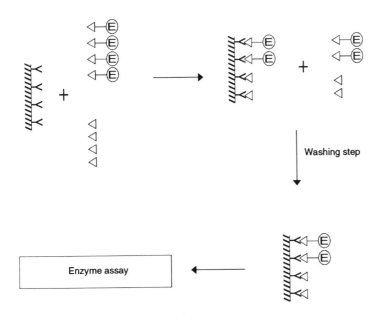

Figure 7.3. Principle of competitive solid-phase EIA. The bound enzymatic activity is inversely proportional to the quantity of unlabelled antigen present in the sample assayed

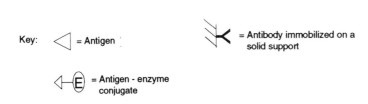

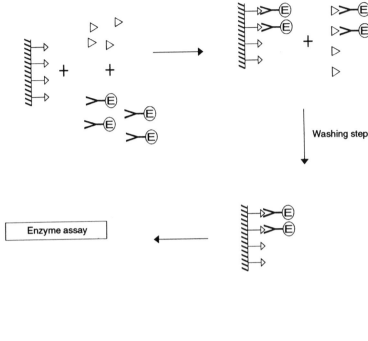

Figure 7.4. Enzyme immunoassay system using immobilized antigen. The higher the level of antigen present in the sample for assay, the lower the level of enzyme–antibody conjugate retained on the immobilized phase, and hence the lower the level of enzymatic activity recorded

Washing step

Enzyme assay

Key: = Antibody - enzyme conjugate

= Antigen

= Immobilized antigen

is outlined in Figure 7.4. In this case, the immobilized antigen and the free antigen present in the assay sample compete for binding to a limited amount of enzyme-labelled antibody. The more free antigen is present in the assay sample, the less enzyme–antibody conjugate will bind the immobilized antigen. A washing step removes all unbound material, and subsequent enzyme assay allows accurate estimation of the quantity of bound enzyme.

A standard curve of antigen concentration versus bound enzymatic activity may be constructed by assaying several samples containing known quantities of antigen. Readings for unknowns may therefore be quantified by reference to this standard curve.

Enzyme linked immunosorbant assay (ELISA)

Since the introduction of EIA over 20 years ago, many variations on the enzyme immunoassay concept have been designed. One of the most popular EIA systems in use is that of the enzyme-linked

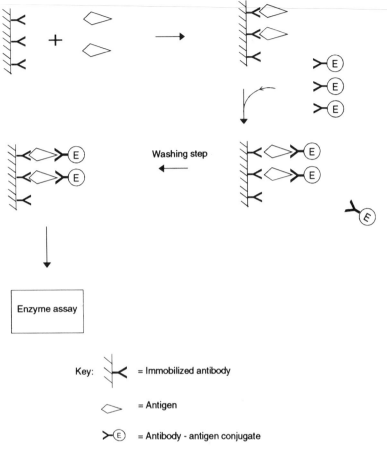

Figure 7.5. Principle of non-competitive ELISA

Washing step

Enzyme assay

Key: ⊢≺ = Immobilized antibody

◇ = Antigen

≻Ⓔ = Antibody - antigen conjugate

immunosorbent assay (ELISA). The principle upon which the ELISA system is based is illustrated in Figure 7.5. In this form it is also often referred to as the "double antibody sandwich technique".

In the basic ELISA system, antibodies raised against the antigen of interest are adsorbed onto a solid surface—again, usually the internal walls of microtitre plate wells. The sample to be assayed is then incubated in the wells. Antigen present will bind to the immobilized antibodies. After an appropriate time, which allows antibody–antigen binding to reach equilibrium, the wells are washed.

A preparation containing a second antibody labelled with a tag, which also recognizes the antigen, is then added. If monoclonal antibodies are used, this second monoclonal antibody must recognize an epitope on the antigen surface which differs from the epitope recognized by the primary or immobilized monoclonal antibody. The second antibody will also bind to the retained antigen and the enzyme label is conjugated to this second antibody.

Subsequent to a further washing step, to remove any unbound antibody–enzyme conjugate, the activity of the enzyme retained is assayed. The activity recorded is proportional to the quantity of antigen present in the sample assayed. A series of standard antigen concentrations may be assayed to allow construction of a standard curve. The standard curve facilitates calculation of antigen quantities present in "unknown" samples.

Enzymes used in EIAs

A wide variety of enzymes have been used as markers in various ELISA systems. Suitable enzymes may be chosen on the basis of a number of criteria. Other than catalytic properties, one obvious criterion is that the activity of the enzyme chosen be easily monitored. Many of the enzymes used produce a coloured product which may be easily monitored by colorimetric or other appropriate methods. Enzymes most often used as labels include alkaline phosphatase and horseradish peroxidase, in addition to β-galactosidase, glucose oxidase and urease. All such enzymes utilize a suitable chromogenic substrate.

Alkaline phosphatase isolated from calf intestine was one of the first enzymes to be used in ELISA systems. The substrate normally used is para-nitrophenylphosphate (PNPP). PNPP is enzymatically hydrolysed by alkaline phosphatase, releasing inorganic phosphate and para-nitrophenol (PNP), which is yellow and absorbs light at 405 nm. The substrate used with β-galactosidase is normally O-nitrophenyl β-D-galactopyronoside, with colour development being measured at 420 nm. Even more sensitive product detection may be obtained for such enzymatic systems if a fluorogenic substrate is used.

The covalent coupling or conjugation of the chosen enzyme to the second antibody used in ELISA systems may be achieved by a number of chemical methods. Perhaps the simplest of such methods involves chemical conjugation with glutaraldehyde (Figure 7.6), which is a homobifunctional reagent—its two reactive groups which link two proteins together are identical. Glutaraldehyde reacts irreversibly with the ϵ-amino group of lysine, forming covalent linkages, as illustrated diagrammatically in Figure 7.6.

The glutaraldehyde method is popular due to its simplicity and is inexpensive. All conjugation methods, however, are likely to result

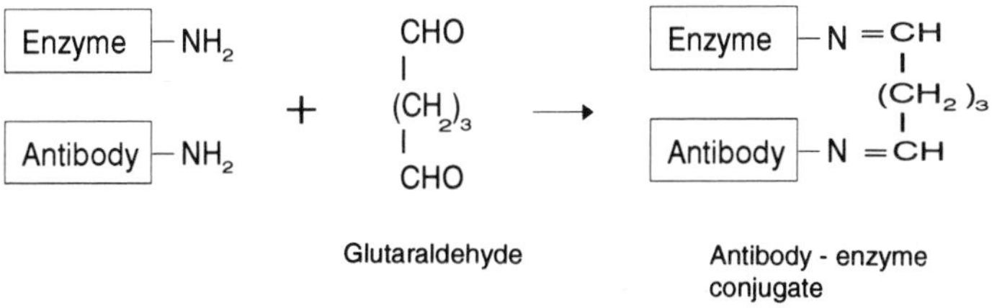

Glutaraldehyde

Antibody - enzyme conjugate

Figure 7.6. Chemical coupling of enzyme to antibody species using glutaraldehyde. Amino groups shown on the free enzyme and free antibody are ϵ amino groups of lysine residues. The reactants are incubated for a period of 1 hour or more. Excess free lysine may be added for efficient termination of the coupling procedure. Unreacted glutaraldehyde may be removed by dialysis. The procedure will inevitably yield a proportion of antibody–antibody and enzyme–enzyme conjugates in addition to the enzyme–antibody conjugate

in the inactivation of a proportion of coupled enzyme molecules and indeed antibodies.

Trends in immunoassay development

Several definite trends have been observed in the design of many of the more modern immunoassays. An increasing proportion of such immunoassays have incorporated non-isotopic detection systems. The incorporation of enzymatic detection systems in immunoassays is becoming particularly popular. Horseradish peroxidase is the enzyme most often used in such systems. Its popularity reflects its extremely high catalytic efficiency, its relatively small size and the availability of a variety of specific and sensitive assay systems by which peroxidase activity may be monitored. Alkaline phosphatase is probably the next most popular enzyme used as a label in EIAs.

Increasing numbers of new assays are also based on solid phase systems. This normally renders separation of free from bound antigen quite straightforward. The proportion of new immunoassay systems using monoclonal antibodies is increasing steadily. Monoclonal antibodies can exhibit enhanced specificity compared with polyclonal preparations. Monoclonal antibody technology also affords a continuous supply of antibodies of defined specificity.

Modern immunological assays are also becoming more user friendly. In the clinical laboratory, this is being achieved by the design and installation of automatic and semiautomatic analysers, and by the modification of immunoassay methodology to facilitate its applicability to such automated systems. User-friendly assay kits have also been developed which personnel with no scientific training may use outside the laboratory. Typical examples of such immunoassay systems are the pregnancy detection kits which are sold over the counter for home use. Such kits are discussed at the end of this chapter.

Although an increasing proportion of newly developed immunoassays are of the enzyme immunoassay type, RIA systems continue to play a significant role in modern clinical and research laboratories. A wide range of biological molecules—large and small, protein and non-protein—are, and will continue to be, assayed by RIA systems. RIA systems designed to detect and quantify molecules such as serum erythropoietin and serum apolipoproteins are among examples of recently developed immunoassays which rely on radiolabel detection.

Immunological assays for HIV

Among the more recently developed immunological assays is an assay designed to detect persons infected by the HIV virus.

Worldwide sales of such diagnostic kits exceeded US $100 million by 1990. Many kits initially manufactured were used largely to screen donated blood or blood products for the presence of HIV. More recently, an increasing proportion of such kits are being used to screen personnel at high risk of acquiring the disease, in addition to people who must demonstrate that they are HIV-negative in order to emigrate to certain countries or be considered for various types of employment.

Assay systems have been developed that are capable of recognizing either HIV-1, the major causative agent of AIDS in the industrialized world, or HIV-2, the major causative virus in Africa. Thus far, however, the majority of diagnostic systems on the market are those that specifically detect HIV-1.

The host's immune function launches an immunological reaction upon infection with HIV. This is characterized, in part, by the presence of anti-HIV antibodies in the serum. The infected individual generally remains asymptomatic and the disease may progress no further for several years before the development of AIDS-related complex and, subsequently, full-blown AIDS. The occurrence of anti-HIV antibodies in the serum of all individuals infected by the virus forms the basis for the vast majority of HIV diagnostic systems. Most such systems are of the enzyme immunoassay type, with the majority of these being ELISAs.

In such assays HIV antigens are immobilized on a supporting surface, such as in the wells of microtitre plates. The serum sample to be analysed is then incubated in these wells, allowing any anti-HIV antibodies present to bind the immobilized antigen. After washing, a second antibody, binding to bound human antibody, is added. A suitable enzyme label, such as horseradish peroxidase, is conjugated to this second antibody. Unbound conjugate is subsequently removed by washing and any retained enzymatic activity is detected by a suitable assay system. The activity retained is proportional to the quantity of anti-HIV antibody present in the serum sample assayed. This is summarized diagrammatically in Figure 7.7.

Initial ELISA systems used immobilized antigen which had been partially purified from HIV particles grown in tissue culture. Most modern ELISA systems, however, use specific, highly purified HIV antigens which are produced by recombinant DNA technology. Genes coding for HIV proteins such as P24 (the core protein), GP41, GP120 and GP160 (envelope proteins) have all been produced in recombinant systems.

Production of antigen by recombinant methods not only eliminates the hazards of working with the actual virus itself, but also ensures the production of a defined viral antigen. This, in turn, has enhanced the sensitivity and specificity of the ELISA and minimized the occurrence of false positive reactions.

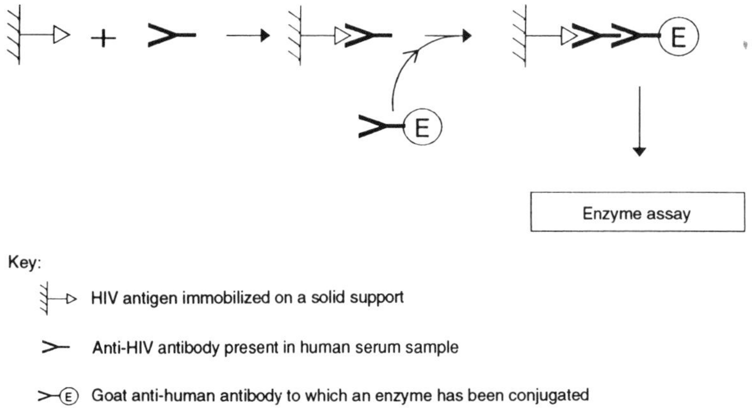

Figure 7.7. ELISA system designed to detect the presence of anti-HIV-1 antibody present in an assay sample

Key:

⊢▷ HIV antigen immobilized on a solid support

≻ Anti-HIV antibody present in human serum sample

≻ⓔ Goat anti-human antibody to which an enzyme has been conjugated

Examples of additional immunoassay systems recently developed to detect a variety of clinically significant biological molecules are described in many of the articles listed at the end of this chapter.

Latex-based and other immunoassay systems

A variety of other immunoassay systems have been developed in addition to the classical RIAs and EIAs. Latex-based and membrane-based immunoassays represent two such alternative systems.

Uniform spherical particles of the polymeric polystyrene-based substance latex have found widespread application in the development of a variety of immunologically based diagnostic tests. Most latexes are manufactured from polystyrene, and particle sizes used generally range between 0.1 μm and 1.1 μm; individual latex particles therefore are not visible to the unaided human eye. Incubation of latex with a protein solution leads to absorption of the protein molecules onto the surface of the latex particle. Alternatively, various reactive groups may be initially introduced into the latex. This permits the covalent linkage of proteins to such particles.

Latex particles coated with antibody can be utilized to detect the presence of a specific antigen in a biological sample. Conversely, latex particles coated with a specific antigen may be used to detect the presence of antibodies which recognize that specific antigen. Latex-based diagnostic systems rely on the inherent specificity of an antibody–antigen reaction, and the bifunctional nature of antibody binding, to promote agglutination of the latex particles. Some diagnostic kits are available in which the latex particles have been replaced by erythrocytes.

As illustrated in Figure 7.8, when latex particles coated with antibody raised against a specific antigen are incubated with a biological sample containing that antigen, the ensuing antigen–antibody reaction results in the formation of large aggregates of latex particles, i.e. the latex agglutinates. The antigen effectively

Figure 7.8. Latex agglutination assay system. In (a) latex particles have been coated with antibody. The presence of appropriate antigen in the sample tested results in agglutination of the latex particles. The antigen present effectively acts as a bridge between adjacent latex particles. In (b) the latex particles have been coated with antigen. In this case, the presence of appropriate antibody in the assay sample results in agglutination of the latex, i.e. the antibody acts as a bridge between adjacent latex particles. Agglutination can easily be detected visually as illustrated in (c). In this case the well on the left-hand side of the plastic card contains unagglutinated latex particles, whereas the well to the right contains agglutinated latex ⟶

acts as a bridge between adjacent latex particles. This process of agglutination is evident as the newly aggregated particles are clearly seen by the naked eye.

In practice, such assay systems are initiated by mixing a sample of antibody or antigen-coated latex with the biological sample to be analysed. If the sample contains the appropriate antigen or antibody, the latex will agglutinate. The agglutination process is perceived visually as the transformation of a homogeneous milk-like latex mixture into a more granular form. The agglutinated latex particles are similar in appearance to discrete grains of sea sand.

(a) Latex particles coated with antibody + Antigen present in biological sample → Agglutination of latex particles

(b) Latex particles coated with antigen + Antibody present in biological sample → Agglutination of latex particles

(c)

Latex-based assays are popular for a number of reasons, most notably the speed with which results are obtained and as little or no specialized equipment is required. Latex assays are normally carried out on plastic-coated cards, which have a series of slight circular indentations on their surface. The latex and the biological sample to be assayed are incubated together in one such surface indentation. The card is gently rocked back and forth to ensure continued mixing of the latex and the sample. Agglutination, should it occur, will be visible within 3–5 minutes, in contrast to a typical EIA which requires a minimum of 2 hours to yield results.

Latex tests are primarily qualitative. Semiquantitative estimations may be obtained by assaying a series of dilutions of the antigen-containing sample until agglutination no longer occurs. The minimum concentration of antigen required to promote agglutination in such a system is then determined by assaying a series of dilutions of an antigen standard. The presence of interfering substances, particularly in undiluted blood or urine samples, may yield false positive results in some cases.

Numerous latex-based assays have been developed and many are extensively used. Examples of useful latex-based diagnostic tests include those designed to detect pregnancy or various infectious agents and those that detect various serum factors associated with arthritis.

Shortly after implantation of a fertilized egg, the human placenta begins to synthesize and secrete human chorionic gonadotrophin (hCG); hCG exhibits many biological activities similar to those of luteinizing hormone (LH). It functions primarily to maintain the corpus luteum during the first weeks of pregnancy. Human chorionic gonadotrophin is not normally synthesized in healthy, non-pregnant females; however, it is found in both the serum and urine of pregnant women, and therefore represents an ideal diagnostic marker for pregnancy. The fact that this hormone may be detected in the urine of pregnant females makes its assay all the more desirable, as urine samples may be conveniently obtained.

Many of the initial hCG-based pregnancy detection systems were based on bioassays. This involved parenteral administration of urine samples to immature female rabbits or rats. Growth and ripening of the ovarian follicles in such animals indicated the presence of hCG, and hence pregnancy. Modern pregnancy diagnostic systems are virtually all based upon immunological detection of hCG in urine samples. Both RIA and ELISA systems are available, in addition to the latex-based particle agglutination assays.

Most latex-based pregnancy detection systems use latex coated with anti-hCG antibodies. The presence of hCG in the urine sample

Figure 7.9. Indirect latex-based pregnancy detection systems. (a) Series of events occurring when a urine sample obtained from a pregnant female is tested; (b) Series of events occurring when a urine sample from a non-pregnant female is tested $\longrightarrow$

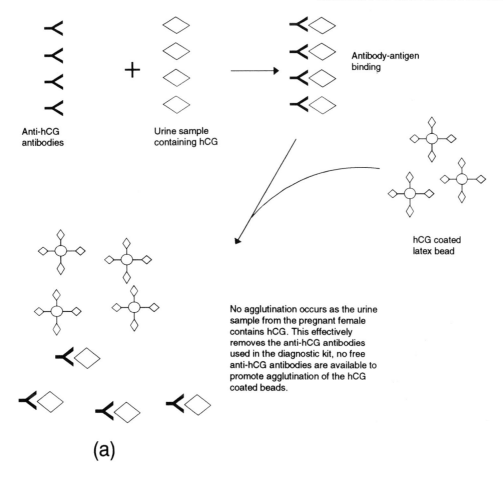

Anti-hCG
antibodies

Urine sample
containing hCG

Antibody-antigen
binding

hCG coated
latex bead

No agglutination occurs as the urine
sample from the pregnant female
contains hCG. This effectively
removes the anti-hCG antibodies
used in the diagnostic kit, no free
anti-hCG antibodies are available to
promote agglutination of the hCG
coated beads.

(a)

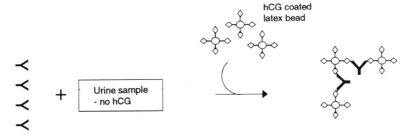

anti-hCG antibody

Urine sample
- no hCG

hCG coated
latex bead

Agglutination of the latex particles occurs as the
non-pregnant female's urine contains no hCG
The anti-hCG antibodies therefore bind to the hCG
coated latex beads and agglutination results

(b)

thus promotes agglutination of the latex beads. Such systems may be termed direct assay systems. The principle involved is illustrated in Figure 7.8a.

Alternative indirect latex-based pregnancy detection kits use latex particles coated with hCG. The three components are incubated in these indirect assay systems. Initially, anti-hCG antibody and a sample of urine are incubated together. The latex particles are then added and should the sample be from a pregnant female, the hCG present in the urine will bind the anti-hCG antibody. Thus all the free hCG and anti-hCG antibody are mopped up. In such cases, no agglutination will be noted upon addition of the latex beads. This is illustrated diagrammatically in Figure 7.9a. If, on the other hand, the sample assayed by this indirect system is from a non-pregnant female, there will be no hCG in the urine sample supplied, and the binding sites of the anti-hCG antibodies mixed initially with the urine sample will remain empty. In this case agglutination will be observed upon the addition of the hCG-coated latex beads (Figure 7.9b).

The presence of a variety of infectious agents may also be indicated by a number of latex-based assays. Many such systems employ latex particles coated with antibody which recognize antigens associated with the infectious agent. Some assays, however, employ latex particles coated directly with antigen. Agglutination of such particles indicate the presence of anti-antigen antibodies in the serum sample analysed. This indicates that the individual tested has come in contact with the infectious agent in question and has mounted an immunological response to it.

Examples of such latex-based assays which are available commercially include systems that detect the causative agents of hepatitis B, syphilis and AIDS. Most hepatitis B latex agglutination tests are based upon incubation of blood samples with latex particles coated with antibody raised against hepatitis B surface antigen. Agglutination indicates the presence of hepatitis B particles in the sample.

Some latex-based syphilis tests utilize latex particles coated with antigens purified from *Treponema pallidum*, the syphilitic causative agent. Agglutination will occur if the blood sample assayed contains antibodies to this microorganism. The presence of such antibodies in the serum sample indicates that the sample donor has come into contact with *Treponema pallidum*. Most latex-based HIV tests involve the use of latex particles coated with recombinant HIV-1 antigens.

Several latex-based diagnostic tests also find widespread application in rheumatology. Such systems include those designed to detect the presence of rheumatoid factor and C-reactive protein. Rheumatoid factors are autoantibodies which specifically bind to human IgG. Their presence in serum is usually indicative of rheumatoid arthritis. C-reactive protein is a serum protein whose concentration increases several hundredfold subsequent to acute infections.

Membrane-bound diagnostic systems

Antibodies immobilized on membranes such as nitrocellulose may also be used as diagnostic reagents. Thoughtful design of such systems, in particular those designed to detect pregnancy, has contributed greatly to their ease of use. This point is illustrated in the generalized membrane-based hCG detection system outlined below. Such systems have become popular and are usually sold "over the counter" by pharmacists.

In such systems, two thin lines of antibody are sprayed in the shape of a cross on the surface of the nitrocellulose membrane (Figure 7.10). This process is achieved by specialized industrial spraying equipment. The antibody applied along one line specifically binds hCG. The antibody applied along the second line binds alkaline phosphatase. At this stage, both these antibody lines are invisible to the human eye.

Figure 7.10. Membrane-bound pregnancy (hCG) detection system

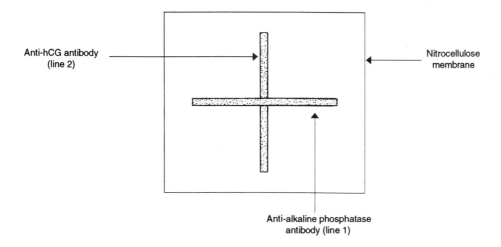

Anti-hCG antibody (line 2)

Nitrocellulose membrane

Anti-alkaline phosphatase antibody (line 1)

Another component used in these systems is a freeze-dried preparation of anti-hCG antibody conjugated to alkaline phosphatase (AP). The urine sample obtained for testing is used to reconstitute this antibody–enzyme conjugate. The reconstituted conjugate is then allowed to come into contact with the membrane surface.

A urine sample from a non-pregnant female will contain no hCG. Therefore, the hCG binding sites of the antibody–enzyme conjugate reconstituted by the urine sample remain unoccupied. When this sample comes into contact with the membrane surface

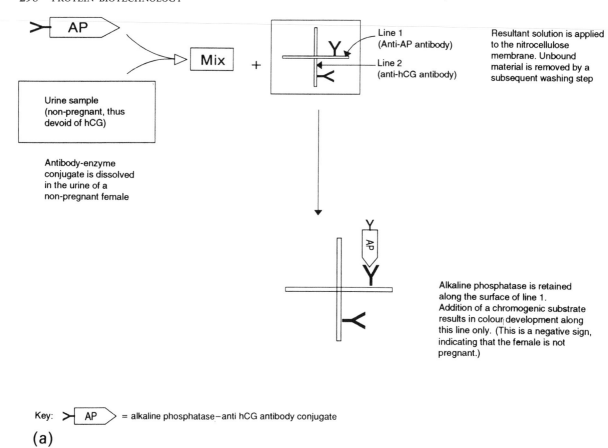

Resultant solution is applied
to the nitrocellulose
membrane. Unbound
material is removed by a
subsequent washing step

Line 1
(Anti-AP antibody)

Line 2
(anti-hCG antibody)

Urine sample
(non-pregnant, thus
devoid of hCG)

Antibody-enzyme
conjugate is dissolved
in the urine of a
non-pregnant female

Alkaline phosphatase is retained
along the surface of line 1.
Addition of a chromogenic substrate
results in colour development along
this line only. (This is a negative sign,
indicating that the female is not
pregnant.)

Key: >─| AP > = alkaline phosphatase–anti hCG antibody conjugate

(a)

the conjugate is bound, via the AP, to the anti-AP antibody line
(line 1 in Figure 7.10). Nothing binds to the anti-hCG antibody line
(line 2 in Figure 7.10), as no hCG is present in the urine.

After the membrane has been rinsed in order to remove unbound
material, it is briefly immersed in a solution containing an alkaline
phosphatase chromogenic substrate such as PNPP. Colour
development is therefore witnessed all along the anti-AP antibody
line due to the presence of bound AP (Figure 7.11a). No colour
development occurs along line 2 as no hCG is present. Thus a minus
sign is produced indicating that the female is not pregnant.

On the other hand, the urine of a pregnant female does contain
hCG and this hCG is bound by the antibody–enzyme conjugate
upon its reconstitution with the urine sample. In this case, the
conjugate is bound by both antibody lines upon coming in contact
with the membrane surface. The conjugate binds line 1 via the AP
moiety. Binding to line 2 (the anti-hCG antibody line) occurs
because the hCG acts as a bridge as illustrated in Figure 7.10b.
In this case, immersion of the membrane in a developing solution

Figure 7.11. Principle of membrane bound
hCG (pregnancy) detection kit. (a) Events
occurring if female is not pregnant; (b)
events occurring if female is pregnant

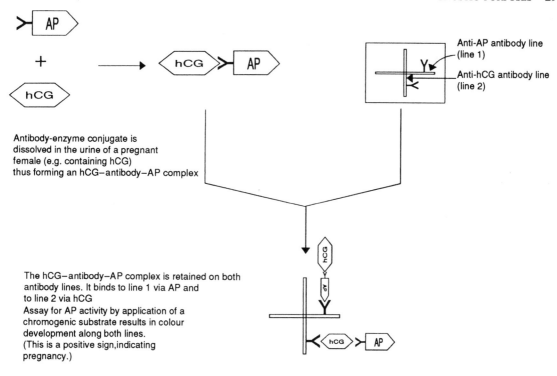

Antibody-enzyme conjugate is
dissolved in the urine of a pregnant
female (e.g. containing hCG)
thus forming an hCG–antibody–AP complex

The hCG–antibody–AP complex is retained on both
antibody lines. It binds to line 1 via AP and
to line 2 via hCG
Assay for AP activity by application of a
chromogenic substrate results in colour
development along both lines.
(This is a positive sign, indicating
pregnancy.)

Key: = alkaline phosphatase-anti-hCG antibody complex

        = hCG

(b)

results in colour development along both lines, so that a plus sign is
obtained, indicating pregnancy.

In vivo applications of antibodies as diagnostic aids

Monoclonal antibodies directed against specific antigen bearing
tumours have also been used *in vivo* to detect and localize such
tumours. Antibodies used must bind to a specific antigen which is
expressed only by the tumour tissue. Most such antibodies are
firstly conjugated to an appropriate radionuclide such that the
tumour can be detected by scanning for radioactivity. Obviously,

such antibody conjugates must be highly purified prior to their parenteral administration. Recent advances in identification of surface antigens produced specifically by abnormal cells ensure a bright future for such *in vivo* diagnostic systems.

FURTHER READING

Books

Gosling, J.P. & Reen, D.R. (eds) (1993). *Immunotechnology*. Portland Press, Colchester.

Tijssen, P. (1985). *Practice and Theory of Enzyme Immunoassays*. Elsevier, Amsterdam.

Articles

Albers, J. *et al.* (1990). The unique lipoprotein (a): properties and immunochemical measurement. *Clin. Chem.*, **36**, 12, 2019–2026.

Atkinson, T. (1983). The potential of microbial enzymes as diagnostic reagents. *Phil. Trans. R. Soc.* (London), **B300**, 399–410.

Bahl, O. (1969). Human chorionic gonadotrophin, purification and physicochemical properties. *J. Biol. Chem.*, **244**, 3, 567–574.

Beltz, G. *et al.* (1989). Development of assays to detect HIV-1, HIV-2 and HTLV-1 antibodies using recombinant antigens. In Albertini A. *et al.* (eds) *Molecular Probes: Technology and Medical Applications*, pp 131–142. Raven Press, New York.

Deftos, L. (1991). Bone protein and peptide assays in the diagnosis and management of skeletal disease. *Clin. Chem.*, **37**, 7, 1143–1148.

Diamandis, E. & Christopoulos, T. (1991). The Biotin-(strept)Avidin system: principles and applications in biotechnology. *Clin. Chem.*, **37**, 5, 625–636.

Freedman, D. *et al.* (1986). The relation of apolipoproteins A-1 and B in children to parental myocardial infarction. *New Engl. J. Med.*, **315**, 721–726.

Gosling, J. (1990). A decade of development in immunoassay methodology. *Clin. Chem.*, **36**, 8, 1408–1427.

Gottfried, T. & Urnovitz, H. (1990). HIV-1 testing: product development strategies. *Trends in Biotechnol.*, **8**, 35–40.

Klotzsch, S. & McNamara, J. (1990). Triglyceride measurements: a review of methods and interferences. *Clin. Chem.*, **326**, 9, 1605–1613.

Labeur, C. *et al.* (1990). Immunological assays of apolipoproteins in plasma: methods and instrumentation. *Clin. Chem.*, **36**, 4, 591–597.

Moskowitz, S. *et al.* (1983). Prealbumin as a biochemical marker of nutritional adequacy in premature infants. *J. Paediatr.*, **102**, 5, 749–753.

Price, C. (1983). Enzymes as reagents in clinical chemistry. *Phil. Trans. R. Soc.* (London), **B300**, 411–422.

Schlageter, M. (1990). Radioimmunoassay of erythropoietin: analytical performance and clinical use in haematology. *Clin. Chem.*, **26**, 10, 1731–1735.

Smith, T. *et al.* (1992). What's up Doc? *Technology Ireland*, Sept., 22–24.

Waldmann, T. (1991). Monoclonal antibodies in disease and therapy. *Science*, **252**, 1657–1662.

Whicher, J. & Evans, S. (1990). Cytokines in disease. *Clin. Chem.*, **36,** 7, 1269–1281.

Wu, A. (1989). Creatine kinase isoforms in ischemic heart disease. *Clin. Chem.*, **35,** 1, 7–13.

Chapter 8

Polymer-degrading enzymes of industrial significance

- Sources of industrial enzymes.
  - Thermophilic microorganisms as enzyme producers.
- Sales value of bulk enzymes.
- Industrial applications of bulk enzymes.
- Carbohydrases.
  - Amylases.
  - Glucose isomerase.
  - Industrial importance of starch conversion.
  - Additional sugar degrading enzymes.
  - Lignocellulose-degrading enzymes.
  - Pectin and pectic enzymes.
- Proteolytic enzymes.
  - Proteolytic enzymes of medical significance.
  - Proteases used in the brewing and baking industries.
  - Proteases used in the meat and leather industries.
  - Proteases incorporated into detergent products.
  - Proteases used in cheese manufacture.
- Lipases.
- Safety aspects.
- Further reading.

In the preceding chapters emphasis has been placed on aspects of biotechnology relating to the production and utilization of proteins of therapeutic and diagnostic significance. Such proteins are normally produced in small quantities and are generally highly purified. As with many health-care products, they are often expensive and economic considerations are not of paramount importance in their production and marketing.

This chapter focuses on a different group of protein products, the "industrial" or "bulk" enzymes. This group includes amylases, cellulases, lignocellulose-degrading enzymes, pectinases, proteases and lipases. The majority of these enzymes may be described as hydrolytic depolymerases. In contrast to enzymes used for therapeutic or diagnostic purposes, such industrial enzymes are produced in large quantities, of the order of thousands to hundreds of thousands of kilograms annually, and are normally processed only to a very limited degree. Furthermore, in most instances economic considerations such as production costs are of critical importance to their commercial applications (Table 8.1).

Table 8.1. Comparison of some attributes of industrial and medically important proteins

Bioactive industrial proteins	Medical/diagnostic proteins
Produced in large quantities	Produced in small quantities
Crude or partially purified	Extensively purified
Economic considerations critical	Economic considerations are of secondary importance to functional excellence of product
Function: mostly enzymes	Function: various (hormones, growth factors, cytokines, other regulatory factors, blood factors, vaccines)
Source: mainly microbial	Source: mainly human or animal and recombinant products
Generally extracellular	May be intracellular or extracellular

SOURCES OF INDUSTRIAL ENZYMES

The majority of industrial enzymes are produced by microorganisms. The producer strains are usually members of a family of microbes classified as GRAS (generally recognized as safe). Such bulk enzymes are produced primarily by bacteria and fungi, most notably by members of the genera *Bacillus* and *Aspergillus* (Table 8.2.)

Bacillus strains have traditionally been utilized in the production of industrially important enzymes for a number of reasons: with the exception of the *Bacillus cereus* group, they are all GRAS listed, and are widely distributed in nature. Furthermore, they may be easily cultured in relatively inexpensive media. Members of the genus *Bacillus* are capable of producing large quantities of many desirable enzymatic activities, most of which are secreted directly

Table 8.2. Some industrially important enzymes, their sources and applications

Enzyme	Source	Application
α-Amylase	*Bacillus amyloliquefaciens* *B. licheniformis* *B. subtilis* *Aspergillus oryzae*	Hydrolyses $\alpha 1 \rightarrow 4$ linkages in starch. Used to liquefy starch and reduce its viscosity
β-Amylase	*Bacillus polymyxa* *B. circulans* Barley	Enzymatic degradation of starch, yielding the disaccharide maltose
Glucoamylase (amyloglucosidase)	*Aspergillus niger* *Rhizopus* spp.	Hydrolysis of starch, yielding dextrose
Pullulanase	*Bacillus* spp. *Aerobacter aerogenes* *Klebsiella* spp.	Debranching of starch by hydrolysis of $\alpha 1 \rightarrow 6$ glycosidic linkages
Glucose isomerase	*Bacillus coagulans* *B. stearothermophilus* *Streptomyces* spp. *Arthrobacter*	Production of high-fructose syrup by conversion of glucose to fructose
β-Galactosidase (lactase)	*Bacillus coagulans* *Streptomyces* spp. *Saccharomyces* spp. *Aspergillus* spp.	Hydrolysis of milk lactose, yielding glucose and galactose
Invertase (sucrase)	*Saccharomyces* spp.	Hydrolysis of sucrose, yielding glucose and fructose
Cellulase and hemicellulase	*Trichoderma* spp. *Sporotrichum cellulophilum* *Actinomyces* spp. *Aromonas* spp. *Aspergillus niger*	Enzymatic hydrolysis of cellulose-containing material
Pectinases	*Aspergillus niger* *Fusarium* spp.	Enzymatic hydrolysis of pectin
Proteases	*Bacillus amyloliquefaciens* *B. subtilis* *Streptomyces* spp. *Aspergillus oryzae* Some animal sources such as calf stomach	Enzymatic hydrolysis of proteins widely used in detergents and in brewing, baking and meat tenderization
Lipases	*Mucor* spp. *Myriococcum* spp. Animal pancreas	Enzymatic hydrolysis of lipids. Used in dairy industry for flavour development in foods and also used in detergents

into the surrounding medium. Various species of *Aspergillus* are used as producer organisms as they share many of the positive attributes exhibited by *Bacillus* species. A few industrially important enzymes may be obtained from yeast, mainly from species of *Saccharomyces*.

Candidate producer organisms are usually identified by screening programmes. Many wild-type organisms isolated by such screening systems initially produce small quantities of the enzyme of interest. Traditionally, yields were increased by methods of strain improvement such as mutagenesis followed by identification of hyper-producing microbial strains. More recently, recombinant DNA technology has been used to increase product yields.

Most industrially significant enzymes are secreted into the surrounding medium by the producer microorganism. Extracellular production of enzymes simplifies subsequent product recovery and purification (see Chapter 3).

As previously outlined, most bulk enzyme products are relatively crude preparations containing many contaminant activities. All production methods must conform to the principles of good manufacturing practice as applied to bulk enzyme production. The whole process is also subject to numerous quality assurance checks. Production of bulk enzymes should not be considered to be technically straightforward.

Thermophilic microorganisms as enzyme producers

Methods of enzyme stabilization have been outlined in Chapter 3. However, in recent years, the use of thermophilic organisms as sources of more stable enzymes has generated considerable interest. Many thermophiles have been isolated from environments such as hot springs. Most enzymes produced by such organisms are catalytically active at temperatures in excess of 80 °C, temperatures that would quickly inactivate similar enzymes produced by mesophiles, which grow at much lower temperatures. The introduction of such thermostable enzymes would allow many enzyme-catalysed industrial processes to be operated at elevated temperatures. This would be desirable due to the increased reaction rates attainable. Most thermostable enzymes also exhibit enhanced general stability and are more robust than their counterparts from mesophilic organisms.

The majority of thermophiles have not been extensively characterized and hence are not GRAS listed. In addition, many of these organisms are difficult to culture, even on a small scale. Such drawbacks have so far limited the application of such thermophiles in industry; however, a few exceptions exist. Some *Bacillus* species for example, are thermophilic. Many enzymes produced by such organisms have attracted considerable industrial and academic interest. Recombinant DNA technology facilitates the production of thermostable enzymes in GRAS organisms.

SALES VALUE OF BULK ENZYMES

Annual worldwide sales of bulk enzymes are considered to be in the region of US $600 million (Table 8.3). Various proteolytic preparations account for almost two-thirds of such sales. Isomerases, most notably glucose isomerase, command annual sales in the region of $50 million. Lipases account for only a fraction of bulk enzyme sales, but this market is set to grow.

Table 8.3. Estimated value of annual worldwide sales of major enzymes

Enzyme type	Market value (US $ million)
Proteases	330
Carbohydrases	150
Isomerases	50
Lipases	20

INDUSTRIAL APPLICATIONS OF BULK ENZYMES

Bulk enzymes are used in a great many biotechnological processes. Many technologies, such as brewing, wine-making and cheese manufacture, may be traced to the dawn of history. In such instances, individuals unknowingly used microorganisms as a source of the enzymes required to transform initial substrates into products such as ethanol or cheese. A greater understanding of the molecular basis of these conversions facilitated the subsequent utilization of isolated enzymes for specific industrial purposes. Such enzyme preparations are now used to supplement or replace the natural enzymatic complement.

Enzymatic preparations currently find application in the brewing, bread-making and cheese-making industries. The availability of such enzyme preparations has also facilitated the development of numerous additional biotechnological approaches which produce a wide range of industrially important commodities. Enzymes are also used in the production of sweeteners and in modification of the flavour, texture and appearance of many foodstuffs. Enzymatic preparations are used to tenderize meat, in the clarification of beers, wines and other drinks, and are also included in many detergent preparations. Specific examples of the industrial usage of a variety of enzymes are discussed later in this chapter.

CARBOHYDRASES

Polysaccharide-degrading enzymes represent one of the most significant groups of industrially important bulk enzymes. Such enzymes include amylases, pectinases and cellulases. In addition, several other carbohydrate transforming enzymes such as glucose isomerase, invertase and lactase also enjoy significant commercial niche markets. A list of some industrially important carbohydrates is presented in Table 8.4.

Table 8.4. Some industrially important carbohydrates

Monosaccharides
    Glucose
    Fructose
    Galactose
    Arabinose

Disaccharides
    Sucrose
    Lactose
    Maltose
    Cellobiose

Polysaccharides
    Starch
    Glycogen
    Cellulose
    Hemicellulose
    Pectin

Amylases

Enzymes that participate in the hydrolytic degradation of starch are referred to as amylolytic enzymes or amylases. Specific enzymes classified within this group include α-amylase, β-amylase, glucoamylase (also known as amyloglucosidase), pullulanase and isoamylase. Enzymatic degradation of starch yields glucose, maltose and other low molecular weight sugars. Furthermore, ezymatically mediated isomeration of glucose yields high-fructose syrups.

Abundant supplies of starch may be obtained from seeds and tubers, such as corn, wheat, rice, tapioca and potato. The widespread availability of starch from such inexpensive sources, coupled with large-scale production of amylolytic enzymes, facilitates production of syrups containing glucose, fructose or maltose, which are of considerable importance in the food and confectionery industry. Furthermore, they may be produced quite competitively when compared with the production of sucrose, which is obtained directly from traditional sources such as sugar-beet or sugar-cane.

The starch substrate. Starch represents the most abundant storage form of polysaccharides in plants. As previously mentioned, it is especially abundant in seeds such as corn and in a variety of tubers. It is stored in granular form in the plant cell.

The starch polymer consists exclusively of glucose units. Two forms exist, namely α-amylose and amylopectin (Figure 8.1). Alpha-amylose is a long, linear polymer, in which successive D-glucose molecules are linked by an $\alpha 1 \rightarrow 4$ glycosidic bond. Individual α-amylose chains may vary in length and hence in molecular weight. The larger chains have molecular weights in the region of 500 000 Da.

Amylopectin, in contrast, is a highly branched molecule. Successive glucose residues are linked via $\alpha 1 \rightarrow 4$ glycosidic linkages, as in amylose, along the linear portion of the molecule, with branch points consisting of $\alpha 1 \rightarrow 6$ glycosidic linkages. Such branch points generally occur every 25 to 30 glucose residues. Starch isolated from most plants consists of 70–80% amylopectin. In some cases, such as rice, the starch granule consists exclusively of amylopectin.

Starch may be hydrolysed by chemical or enzymatic means. Chemical hydrolysis was used formerly and involves heating in the presence of acid. However, enzymatic hydrolysis generates fewer byproducts and produces higher yields of end product as compared with the chemical method.

Alpha-amylase. The initial step in starch hydrolysis entails disruption of the starch granule. Solubilization of the granules, the process of "gelatinization", facilitates subsequent catalytic degradation. Gelatinization is normally achieved by heating the starch to temperatures often in excess of 100 °C for several minutes. Alpha-amylase (α-amylase) may be added immediately prior to the heating step, in order to render more efficient the process of granule disruption. Once the granules have been disrupted, additional α-amylase is added in order to liquefy the starch slurry. This process reduces the viscosity of the starch solution.

α-Amylase activity is widely distributed in nature. The enzyme may be isolated from microbial sources and from animal and plant tissues. α-Amylase is an endo-acting enzyme, catalysing the random hydrolysis of internal $\alpha 1 \rightarrow 4$ glycosidic linkages present in

Figure 8.1. Structures of (a) α-D-glucose; (b) segment of amylose chain; (c) section of amylopectin. Glycosidic bonds between successive glucose residues link carbon atom no. 1 of one glucose residue to carbon atom no. 4 of the adjacent glucose residue. The bonds are in the α conformation and are termed $\alpha 1 \rightarrow 4$ glycosidic bonds. At branch points found in amylopectin (c), carbon atom no. 6 of the glucose residue in the main chain is linked to carbon no. 1 of the first glucose residue in the branch. This bond is termed an $\alpha 1 \rightarrow 6$ glycosidic bond $\longrightarrow$

the starch substrate. These enzymes are incapable of hydrolysing $\alpha1 \rightarrow 6$ glycosidic linkages present at branch points of amylopectin chains. One exception to this is the α-amylase produced by *Thermoactinomyces vulgaris*, which can hydrolyse both $\alpha1 \rightarrow 6$ and $\alpha1 \rightarrow 4$ glycosidic linkages. All α-amylases characterized to date are metalloproteins. These enzymes can also generally catalyse

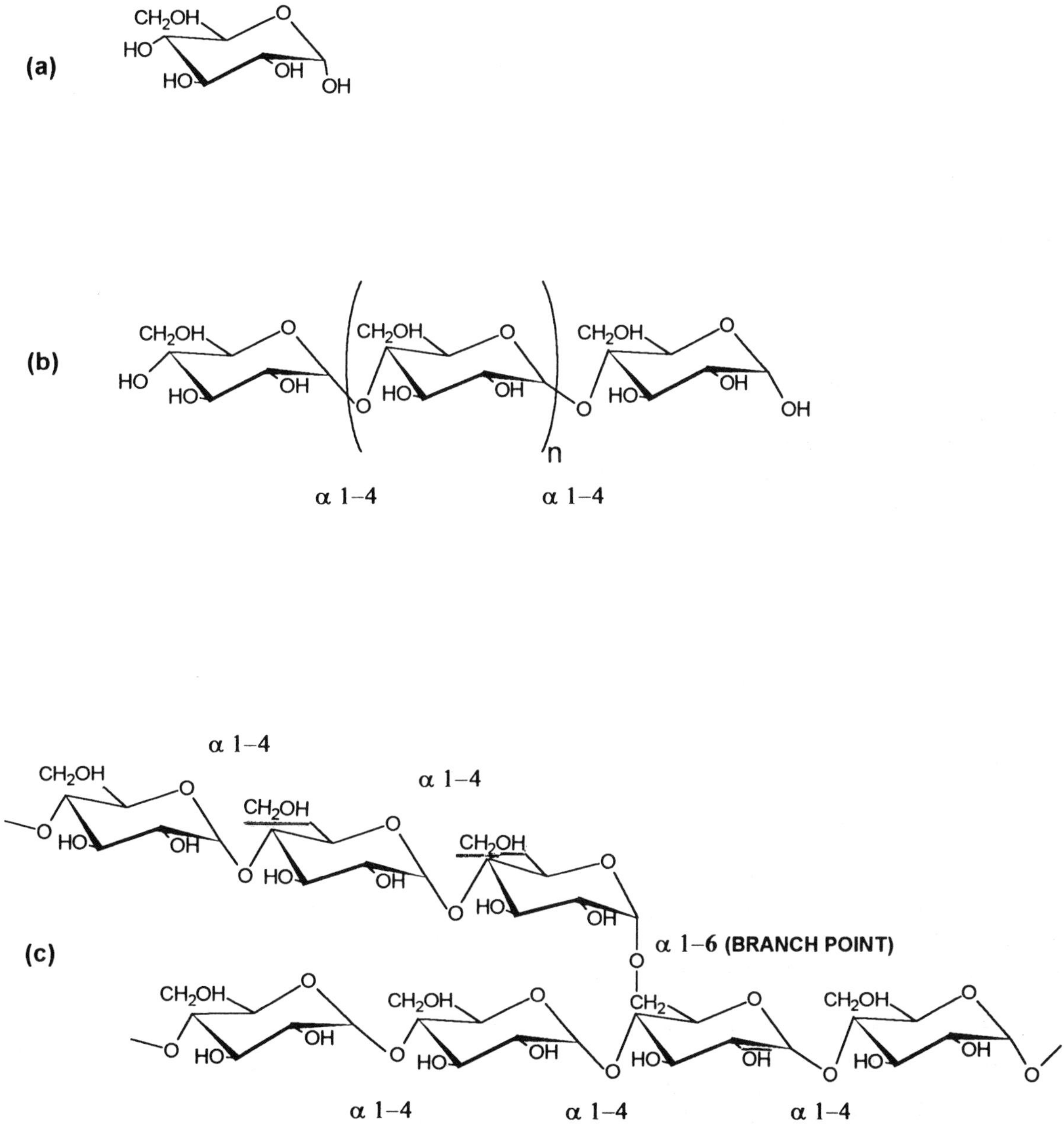

(a)

(b)

$\alpha\ 1-4$ $\alpha\ 1-4$

(c)

$\alpha\ 1-4$ $\alpha\ 1-4$ $\alpha\ 1-4$ $\alpha\ 1-6$ (BRANCH POINT) $\alpha\ 1-4$ $\alpha\ 1-4$ $\alpha\ 1-4$

the cleavage of internal $\alpha 1 \rightarrow 4$ glycosidic bonds in glycogen and a variety of additional oligosaccharides.

Two of the more commonly used α-amylases are those isolated from *Bacillus amyloliquefaciens* and *Bacillus licheniformis*. *Bacillus* amylases exhibit a pH optimum close to neutrality, and are stabilized by the presence of calcium ions. α-Amylase produced by *Bacillus licheniformis* is particularly suited to industrial applications because of its thermal stability. This enzyme consists of 483 amino acids and has a molecular weight of 55 200 Da. Its pH optimum is 6.0 and its temperature optimum is 90 °C. Most other α-amylases, including those produced by *B. amyloliquefaciens*, are rapidly inactivated at temperatures in excess of 40 °C.

The advent of recombinant DNA technology has facilitated the cloning and expression of genes coding for various α-amylases in a variety of recombinant organisms. Human, wheat and bacterial α-amylase have, for example, been expressed in *Saccharomyces cerevisiae*. More recently, the gene coding for *B. licheniformis* α-amylase has been expressed in transgenic tobacco plants. This was among the first examples of the production of bulk industrial enzymes in a plant. The molecular weight of the recombinant protein was found to be 64 000 Da, compared with 55 200 Da in the native *Bacillus* species. This discrepancy was shown to be due to extensive post-translational glycosylation of the molecule. Direct application of the α-amylase-containing transgenic seeds in starch liquefaction studies was found to be highly effective.

α-Amylase activities are also produced by a variety of fungi. Fungal α-amylases most commonly used at an industrial level are produced by species of *Aspergillus*, most notably *A. oryzae* and *A. niger*.

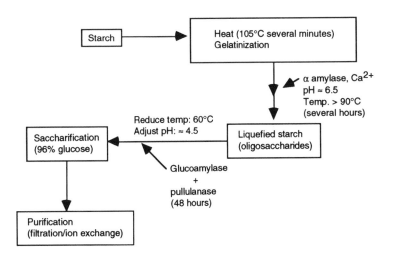

Figure 8.2. Controlled hydrolysis of starch, yielding up to a 96% glucose syrup. Pullulanase may be added in addition to glucoamylase during the saccharification process: this enzyme hydrolyses $\alpha 1 \rightarrow 6$ glycosidic bonds

Glucoamylase. Glucoamylase, also known as amyloglucosidase, is produced as an extracellular enzyme by a variety of fungal species, most notably members of the *Aspergillus* and *Rhizopus* family. The enzyme produced by *Aspergillus niger* is the most widely used on an industrial scale.

Glucoamylases catalyse the sequential hydrolysis of $\alpha 1 \rightarrow 4$ glycosidic bonds from the non-reducing end of the starch molecule. The enzyme also catalyses the hydrolysis of $\alpha 1 \rightarrow 6$ glycosidic bonds present at branch points in amylopectin, although at a much slower rate. Glucoamylase thus catalyses the hydrolysis of starch, yielding glucose. It is normally used industrially after the liquefaction of the starch by bacterial α-amylase in order to produce glucose syrup. This process is described as the saccharification of starch.

Amyloglucosidase is a relatively thermolabile enzyme and is unstable at temperatures in excess of 60 °C; thus the temperature of liquefied starch must be adjusted downwards before addition of the saccharifying enzyme. Adjustment of pH to more acidic values is also required to ensure optimal enzyme activity. The overall scheme of production of glucose syrups from starch is summarized in Figure 8.2.

The glucose syrup produced by this method is often used directly by the food and related industries, in addition to the production of crystalline glucose. A further application entails conversion of some of the glucose to fructose by the enzyme glucose isomerase, thus producing a high-fructose syrup.

Beta-amylase. In contrast to α-amylase, β-amylases are exo-acting enzymes catalysing the sequential hydrolysis of alternate $\alpha 1 \rightarrow 4$ glycosidic linkages present in starch, from its non-reducing end. This reaction produces maltose units with inversion to the β form (Figure 8.3). As in the case of α-amylases, β-amylases are incapable of hydrolysing $\alpha 1 \rightarrow 6$ linkages. Conversion of amylopectin to maltose by β-amylase is therefore limited by branch points. The amylose molecule, on the other hand, being devoid of such branch points, may be fully degraded by β-amylases. Overall, hydrolysis of most starches by β-amylase yields a product mix of maltose and larger oligosaccharides, often termed β limit dextrans.

β-Amylase activity is present in many higher plants. For example, the enzyme found in barley may be used in the production of maltose from starch as happens in the malting process. β-Amylases have also been obtained from a variety of bacterial sources, most notably from *Bacillus circulans*.

Alpha$1 \rightarrow 6$ *glucosidases.* $\alpha 1 \rightarrow 6$ glucosidases are amylolytic enzymes capable of hydrolysing the $\alpha 1 \rightarrow 6$ linkages present at the branch points in amylopectin. Amyloglucosidase is one such

$\alpha 1 \rightarrow 6$ glucosidase. These debranching enzymes play a central role in the complete degradation of starch as neither α- nor β-amylases possess the catalytic ability to hydrolyse the $\alpha 1 \rightarrow 6$ bonds of amylopectin. Several additional $\alpha 1 \rightarrow 6$ glycosidases hydrolyse branch-point linkages much more efficiently and rapidly than does amyloglucosidase. The most important of these enzymes are pullulanase and isoamylase. These enzymes are often used to aid the saccharification process outlined in Figure 8.4.

Although both pullulanase and isoamylase cleave the $\alpha 1 \rightarrow 6$ linkages of amylopectin, they may be differentiated by their ability to degrade the polysaccharide pullulan: pullulanase degrades pullulan, whereas isoamylase does not. Pullulan is a linear polysaccharide consisting of up to 1500 glucose molecules. The basic recurring structure consists of three glucose residues linked via two $\alpha 1 \rightarrow 4$ glycosidic linkages. Each such maltotriosyl unit is linked to the next via an $\alpha 1 \rightarrow 6$ bond, as shown in Figure 8.5.

Pullulanase was first discovered in a species of *Aerobacter* in the early 1960s. It is produced by a variety of bacteria, including some bacilli and streptococci. Isoamylase, produced by a number of microbial species, was initially isolated from yeast. This enzyme is also produced by a variety of bacteria, including some bacilli. Extracellular isoamylase produced in large quantities by a particular mutant strain of *Pseudomonas amyloderamosa* enjoys widespread industrial use.

More recently, a number of pullulanases exhibiting novel activities have been isolated from several thermophilic organisms. Such producer microorganisms include a variety of clostridia and a number of strains of *Thermoanaerobium*. Some of these enzyme have been produced as heterologous protein products in recombinant systems such as *E. coli* and *Bacillus subtilis*. Most exhibit excellent thermal stability and remain active for prolonged periods at temperatures in excess of 90 °C. Perhaps the most striking attribute of many such novel pullulanases, sometimes termed amylopullulanases, is their ability to hydrolyse $\alpha 1 \rightarrow 6$ linkages in some carbohydrates such as pullulan and $\alpha 1 \rightarrow 4$ linkages in others such as starch. Conventional pullulanase fails to hydrolyse the $\alpha 1 \rightarrow 4$ glycosidic linkages of either pullulan or amylose. Purified pullulanase from *T. brockii*, for example, hydrolysed only $\alpha 1 \rightarrow 6$ glycosidic bonds in pullulan but exhibited an almost exclusive preference for $\alpha 1 \rightarrow 4$ bonds in starch.

Glucose isomerase

The enzymatic hydrolysis of large quantities of inexpensive, readily available starch facilitates the economical production of large quantities of glucose syrup. Although glucose syrups may be used directly, many are first converted into high-fructose syrups. The

Figure 8.3. Hydrolysis of starch, in this example amylopectin, by α-amylase and β-amylase. Alpha-amylase catalyses the random hydrolysis of internal $\alpha 1 \rightarrow 4$ glycosidic linkages; it is incapable of cleaving $\alpha 1 \rightarrow 6$ linkages. Beta-amylase catalyses the sequential removal of maltose units from the non-reducing end of the starch molecule. It too fails to hydrolyse $\alpha 1 \rightarrow 6$ glycosidic linkages found at branch points $\longrightarrow$

conversion of glucose into fructose is desirable in so far as fructose is almost twice as sweet as glucose (Table 8.5), and is therefore more attractive industrially when used as a sweetener in the confectionery, ice-cream and soft drinks industries.

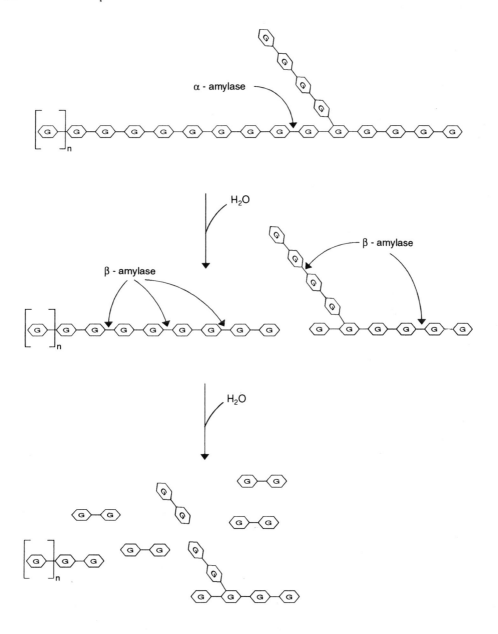

Key: $\langle G \rangle$ = Glucose

$\langle G \rangle - \langle G \rangle$ = Maltose

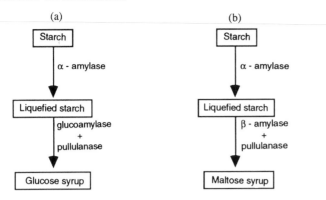

Figure 8.4. Hydrolytic degradation of starch, yielding industrially important end products. In (a) a combination of α-amylase, glucoamylase and pullulanase yields a glucose-rich syrup, whereas in (b) a combination of α-amylase, β-amylase and pullulanase yields a maltose syrup. Process parameters such as adjustments of pH and temperature, and addition of stabilizers have been omitted for clarity

Glucose may be converted into fructose by chemical or enzymatic isomerization (Figure 8.6). Chemical conversion relies upon the use of alkali; low yields and undesirable side-reactions limit the applicability of this particular method. The enzyme glucose isomerase catalyses the isomerization reaction at ambient temperatures and at near neutral pH values, yielding a syrup containing 42% fructose. Recently developed process refinements now allow production of syrups of even higher fructose content.

Glucose isomerase obtained from a wide variety of microbial species has found industrial application in the production of high-fructose syrups. Bacterial species from which this enzyme may be obtained include a variety of *Aerobacter* and of *Bacillus*, and organisms such as *Streptomyces albus*, *Lactobacillus brevis* and *Actinoplanes missouriensis*. Glucose isomerase obtained from plant sources also has found application in the conversion of glucose to fructose.

Glucose isomerase as produced by most microorganisms is an intracellular enzyme. Isolation of the enzyme thus requires disruption of the producer cells, followed by an appropriate purification scheme. The intracellular location renders the production of purified glucose isomerase more technically and economically demanding than production and isolation of extracellular enzymes.

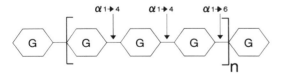

Figure 8.5. Structure of pullulan

Table 8.5. The relative sweetness of a number of commonly occurring sugars of industrial importance. For comparative purposes, sucrose has been assigned a relative sweetness value of 100%

Sugar	Relative sweetness (%)
Sucrose	100
Glucose	70
Fructose	130
Maltose	40

Many of the initial studies designed to investigate the industrial potential of glucose isomerization used soluble enzymes. However, most industrial-scale isomerization systems now use an immobilized form of the enzyme. Economically this is quite significant, as it facilitates reuse of the enzyme. A variety of techniques have been developed in the immobilization of glucose isomerase and are discussed in Chapter 9.

Industrial importance of starch conversion

End-products of starch hydrolysis are incorporated into a wide variety of foodstuffs and soft drinks. Some are used simply as food sweeteners, while others are incorporated to impart a particular texture to the food product.

Starch-degrading enzymes are also utilized in the production of alcoholic beverages and in bread-making. The production of alcoholic beverages by brewing relies upon the ability of yeast to ferment carbohydrates present in the malted barley and other added sugars.

Yeast cells do not possess the enzymatic ability to degrade starch and other complex carbohydrates associated with grains, as they utilize only glucose or other monosaccharides and disaccharides as substrates. Germination of grain is therefore promoted prior to

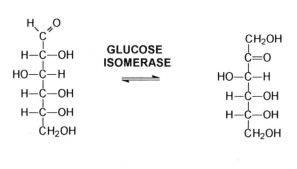

GLUCOSE   **FRUCTOSE**

Figure 8.6. The isomerization of D-glucose, an aldohexose, forming D-fructose, a ketohexose

the fermentation step. The germinating seeds produce endogenous enzymes capable of hydrolysing not only the stored starch but also cellulose and other structural polysaccharide components of the seed. Germination is subsequently arrested by controlled heat, to prevent further seed growth. The seed then contains enzymes such as α- and β-amylases which are capable of hydrolysing stored starch, thus producing glucose and other sugars which the yeast cells are capable of fermenting. This process is termed *malting*.

This traditional process, by which the seeds are induced to produce amylolytic enzymes, may now be supplemented or replaced by the addition of exogenous amylolytic enzymes obtained from microbial sources.

The level of β-amylase present in cereals used for bread-making remains relatively constant but the content of α-amylase is low prior to germination. Milled flour therefore often contains low concentrations of α-amylase. Supplementation of such flour with fungal α-amylase results in a more effective degradation of flour starch, rendering the dough easier to work and allowing yeast fermentation to proceed. This in turn promotes leavening, which increases loaf volume and enhances bread texture.

Additional sugar-degrading enzymes

A number of other sugar-degrading enzymes have also found important, if limited, industrial application. Lactase (β-galactosidase, Figure 8.7), for example, catalyses the hydrolytic cleavage of lactose, the sugar of milk, yielding glucose and galactose. Hydrolysis of lactose is of interest for a number of reasons. Lactose present in ingested milk must be hydrolysed to monosaccharides prior to being absorbed from the small intestine. Although high levels of intestinal lactase activity are present in infants, some adult human

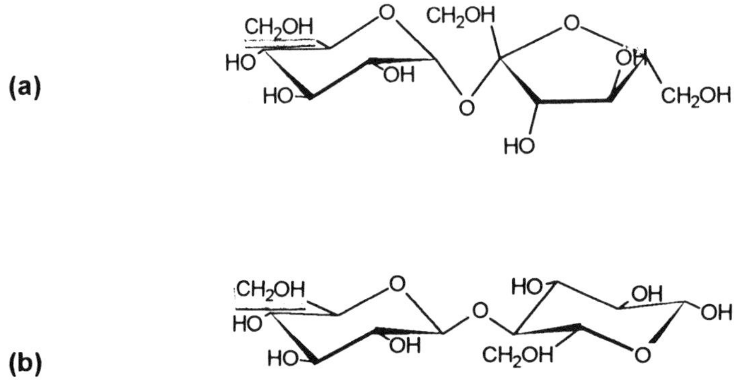

Figure 8.7. Structure of the disaccharides, sucrose (a) and lactose (b)

populations are virtually devoid of this enzyme. Adults of northern European origin, however, do produce significant levels of intestinal lactase throughout their lives. Those adults exhibiting little or no lactase activity are lactose intolerant, and cannot digest dietary lactose. In such instances the ingested lactose causes stomach upsets and promotes diarrhoea. This phenomenon severely restricts the use of milk as a nutrient source by many adult populations. Prior enzymatic hydrolysis of the milk lactose overcomes this difficulty, and immobilized preparations of lactase from species of *Aspergillus* and *Saccharomyces* have been successfully used in this regard. Furthermore, lactase-containing tablets have been formulated which may be taken orally by lactose-intolerant individuals prior to ingestion of milk or milk products.

The cheese-making industry produces large quantities of a by-product termed whey, which is difficult to dispose of. Whey, not surprisingly, has a high lactose content. Hydrolysis of the lactose yielding glucose and galactose could potentially render this by-product a useful food or feed supplement.

Sucrase or invertase is also used industrially to promote hydrolysis of sucrose (Figure 8.7), forming glucose and fructose. This enzyme is often utilized in the candy industry to form the semisolid filling present in some soft-centred sweets. It has also been used in the manufacture of ham, artificial honey and marzipan. Invertase used on an industrial scale is usually produced by fungi such as *Aspergillus* or by some species of yeast.

Lignocellulose-degrading enzymes

The most abundant natural carbohydrate reserves available on the planet are present in plant biomass. Biomass has been defined as everything that grows. Plant biomass consists largely of three polymeric substances: cellulose (40%), hemicellulose (33%) and lignin (23%). Perpetual renewal of plant biomass via the process of photosynthesis ensures an inexhaustible supply of such material. It has been estimated that approximately 4×10^{10} tonnes of cellulose are synthesized annually by higher plants. In practice, this means that in excess of 70 kg of cellulose is synthesized per person per day. Enzymes capable of degrading cellulose therefore are of obvious industrial interest.

Figure 8.8. Structure of a portion of the cellulose backbone. Individual glucose molecules are linked via $\beta 1 \rightarrow 4$ glycosidic bonds. Successive glucose residues are rotated at an angle of 180° relative to the preceding glucose residue

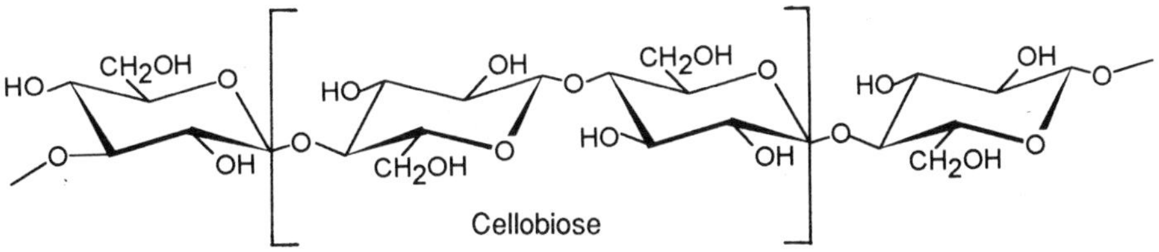

Cellobiose

The cellulose substrate. Cellulose is a linear unbranched homo-polysaccharide consisting of glucose subunits linked via $\beta 1 \rightarrow 4$ glycosidic linkages. Individual cellulose molecules vary widely with regard to polymer length, with some molecules containing as few as 3000–4000 glucose residues, whereas others may contain as many as 20 000 units. The majority of cellulose molecules consist of between 8000–12 000 glucose molecules. Each glucose molecule present in cellulose is rotated at an angle of 180 ° with respect to its immediately adjacent residues. The actual repeating structural subunit, therefore, is the disaccharide cellobiose (Figure 8.8).

The underlying molecular structure confers upon cellulose properties that are very different to those of amylose, which also consists exclusively of glucose residues linked by $1 \rightarrow 4$ glycosidic linkages. Due to the β form of the linkage and the spatial arrangement of alternate glucose molecules, cellulose adopts an extended conformational structure. Furthermore, individual cellulose molecules are usually arranged in bundles or fibrils consisting of several parallel cellulose molecules, held in place by an extensive network of intermolecular hydrogen bonding. Glucose residues within the cellulose molecule also seem to engage in intramolecular hydrogen bonding which contributes to the overall rigidity of the molecule. Within cellulose fibrils, there are extended areas exhibiting a completely ordered structure (crystalline areas) and areas that are less well ordered (amorphous areas).

Vertebrates are devoid of endogenous enzymatic activities capable of hydrolysing the cellulose $\beta 1 \rightarrow 4$ linkages, and therefore are incapable of digesting cellulose and utilizing it as a nutrient source. Some microorganisms, in particular various fungi, do synthesize cellulase enzymes and hence are capable of utilizing cellulose as an energy source. Ruminant animals, by virtue of the fact that their digestive system contains such microorganisms, can utilize the glucose molecules in cellulose.

Hemicelluloses and lignin. While cellulose is the principal con-stituent of the plant cell wall, it is rarely found in pure form as it is in intimate association with other polymeric substances termed *hemicelluloses* and *lignin*.

Hemicelluloses are generally lower molecular weight polysacchar-ides. They consist predominantly of D-xylose, D-mannose, D-glucose, D-galactose, L-arabinose and 4-O-methyl-D-glucuronic acid. The most abundant hemicellulose types present in the cell walls include glucans, mannans and xylans. Xylan, the single most abundant hemicellulose, is a homopolymer, consisting of $\beta 1 \rightarrow 4$ linked D-xylose residues.

Lignin is a complex aromatic polymer found in higher plants, predominantly located within the plant cell wall, interspersed with hemicellulose. This mixture forms a cement-like matrix in

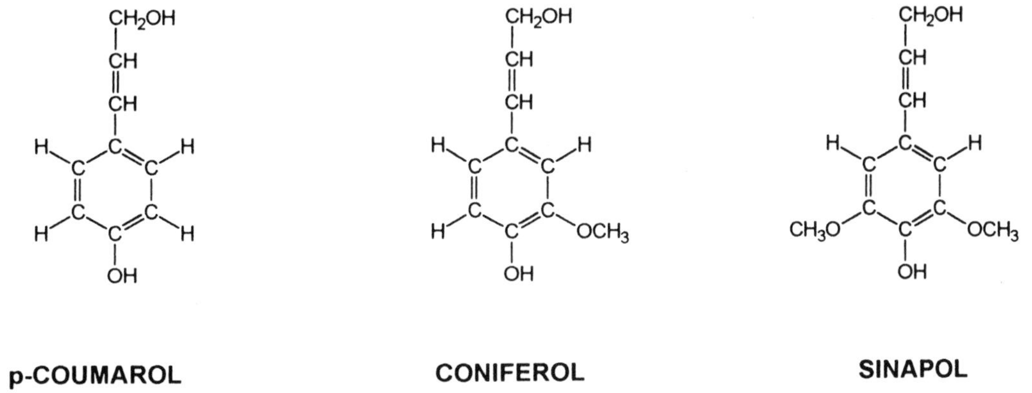

p-COUMAROL   **CONIFEROL**   **SINAPOL**

Figure 8.9. Structure of the alcohol molecules found in lignin

which the ordered cellulose fibrils are embedded. The presence of hemicellulose and lignin serves to increase the overall structural strength of the cell wall. The woody portion of tree trunks consists of over 20% lignin. In such woody tissues, lignin is also present in the intercellular spaces, the middle lamella, where it serves as an adhesive, holding adjacent cells together. Unlike cellulose or hemicellulose, lignin is not a polysaccharide. It is a polymeric molecule composed mainly of three alcohol molecules: coumaryl alcohol, coniferyl alcohol and sinapyl alcohol (Figure 8.9). Seemingly random cross-linking of such alcohols yields highly dispersed lignin molecules. Angiosperm lignins generally contain equal quantities of coniferyl alcohol and sinapyl alcohol monomers. Gymnosperm lignins, on the other hand, consist largely of coniferyl units alone, whereas grass lignins contain all three alcohols.

Cellulases. Agricultural and household waste also contain appreciable quantities of cellulose, originally derived from plant material. Wood and wood pulp contain over 40% cellulose, straw and bagasse contain 30–50% cellulose, while newspapers can contain up to 80% cellulose. Cotton is almost pure cellulose. The complete hydrolysis of cellulose yields glucose. Any process that could efficiently and economically achieve conversion of cellulolytic material to glucose would be of immense industrial significance. Enzymes capable of hydrolysing cellulose are termed *cellulases*.

Cellulose is not degraded by a single enzyme, but by a combination of enzymatic activities which function in a concerted manner. Degradation of cellulolytic material occurs slowly in nature for a number of reasons. Very few microorganisms produce complete cellulase systems capable of total and systematic degradation of cellulose to glucose molecules. The ordered crystalline structure of individual cellulose molecules present in cellulose fibres renders enzymatic attack very difficult. Amorphous, non-structured areas of cellulose, on the other hand, are degraded more rapidly.

Furthermore, the close natural association of cellulose with hemi-cellulose, lignin and sometimes pectin, further reduces the accessibility of cellulases to their substrates. Cellulolytic materials may be subjected to various pretreatments in order to render the cellulose molecules more accessible to enzymatic attack. Adoption of such treatments on an industrial scale would render glucose production from cellulose uneconomical.

Cellulolytic enzymes are synthesized by a number of micro-organisms, most notably fungi (Table 8.6). Some bacterial species also exhibit cellulose-degrading ability. Many cellulases produced by bacteria appear to be bound to the cell wall and are unable to hydrolyse native lignocellulose preparations to any significant extent. Many fungi capable of degrading cellulose synthesize large quantities of extracellular cellulases which are more efficient in depolymerizing the cellulose substrate. It is not surprising, there-fore, that fungal cellulases have received most attention. Some fungal species, most notably *Trichoderma* species such as *T. viride*, *T. reesei* and *T. koningii*, as well as *Penicillium funiculosum*, produce cellulases capable of degrading—at least in part—crystalline regions of native cellulose.

Table 8.6. Selected microbial sources of cellulases

Fungal sources	Bacterial sources
Trichoderma viride; *T. reesei*; *T. koningii*	Selected species of *Aeromonas*
Fusarium solani	Selected species of *Bacillus*
Sporotrichum pulverulentum; *Sporotrichum cellulophilum*	Selected thermophilic *Actinomyces* spp.
T. terrestris	*Clostridium thermocellum*
Talaromyces emersonii	
Penicillium verruculosum; *P. funiculosum*	

Most cellulolytic enzymes produced by fungi may be classified as one of three major types: (a) endocellulases (endo-$\beta 1 \rightarrow$ 4-D-glucanases), (b) cellobiohydrolases and (c) β-glucosidases. Any one fungal species capable of degrading cellulose may produce multiple forms of each of these three enzymatic activities. Some such multiple forms are genetically distinct, whereas others may result from partial proteolysis or differential glycosylation of a single protein. All these differing cellulolytic activities act synergistically to solubilize the cellulose substrate.

The endocellulases (endoglucanases) catalyse the random internal hydrolytic cleavage of the cellulose molecule. As many as six endocellulase activities may be associated with some fungi. Endo-cellulases appear to hydrolyse cellulose chains primarily within amorphous regions and display low hydrolytic activity towards crystalline cellulose.

Cellobiohydrolases generally catalyse the sequential removal of cellobiose units from the non-reducing end of the cellulose molecule. The β-glucosidases, on the other hand, hydrolyse both short-chain oligosaccharides derived from cellulose in addition to cellobiose, yielding glucose monomers. Many cellobiohydrolases exhibit product inhibition, as their activity is decreased in the presence of increasing concentrations of cellobiose. Beta-glucosi-dase activity thus prevents product inhibition of cellobiohydrolases.

This summary of cellulose hydrolysis is somewhat simplistic. The enzymatic degradation of this polymer is complex, and controversy still exists regarding the number of enzymes required in this process and the exact role that each enzyme plays in the overall degradative process. Cellulase systems produced by bacteria are less well under-stood than their fungal counterparts.

Industrial application of cellulose hydrolysis. The potential industrial applications of the cellulases and related enzymes are enormous. Glucose produced from the cellulose substrate could be used directly in animal or human nutrition. Alternatively, the glucose product could be used as a substrate for subsequent fermentations or other processes which could yield valuable end products such as ethanol, methanol, butanol, methane, amino acids, organic acids, and single cell protein. Cellulolytic enzymes could also be used directly to increase the digestibility of food having a high-fibre content, and to enhance food flavour, texture or other qualities.

Despite the abundance of such potential applications, cellulases have as yet been used in few industrial processes. This is due to both technical difficulties and economic factors. As previously outlined, cellulose-degrading organisms synthesize a very complex complement of molecules required for cellulose breakdown. In most cases, the exact mechanism by which the molecule is sequen-tially degraded remains poorly understood. The crystalline arrangement of cellulose molecules into fibrils, and the associa-tion of other polymeric substances such as lignin, pectin and hemicellulose with such fibrils, greatly retards the process of its enzymatic degradation. Efficient hydrolysis necessitates expensive pretreatments which are economically unattractive and the current high costs associated with production of cellulolytic enzymes further decreases its economic feasibility. Cellulases are produced in small quantities by most wild-type producer strains and the enzymes generally exhibit disappointing specific activities. While mutational approaches have yielded hyperproducing strains, their specific activity still remains low. Many cellulases are strongly inhibited by the products they form. Such product inhibition can further retard the degradation rate. In recent years, increasing attention has been focused on the study of genes encoding cellulases. Such investigations, coupled with ongoing biochemical research, may facilitate the production of more efficient cellulase digesting systems in the future.

Despite the technical and economic barriers, some cellulolytic preparations have found limited industrial applications, most notably as digestive aids. Such preparations have been fed to animals and used to enhance the extraction process of various commercially significant materials from plant sources. Molecular biology has also facilitated application of cellulases—or more correctly, cellulose binding domains of cellulases—for a number of novel purposes. Many cellulases contain discrete cellulose binding domains adjacent to the catalytic domains. Nucleotide sequences coding for such cellulose binding domains have been synthesized and fused to the genes coding for other proteins. The resultant recombinant product retains the biological activity of the original protein, but also binds cellulose tightly via its newly acquired cellulose binding domain. This may be used for affinity purification of the recombinant product on a cellulose column. Alternatively, it may be used to immobilize the protein on such cellulose columns.

Pectin and pectic enzymes

Pectin is yet another structural carbohydrate found in higher plants. It is located primarily in the cell wall and in the middle lamella, where it serves to bind adjacent cells. Various enzymatic activities capable of degrading pectin may be isolated from both plant and microbial sources. Degradation of pectin plays an important role in the growth of plant cells and the ripening of fruit. Microbial pectic enzymes, mainly produced by fungi, are utilized in many large-scale industrial processes. Advances in recombinant DNA technology have facilitated the detection, cloning and sequencing of genes coding for many such enzymes, and enabled their expression in heterologous systems.

The pectic substrate. Pectic substances are a relatively diverse group of polysaccharides, which vary in both their composition and molecular weight. Galacturonic acid is the major molecule present, constituting up to 60–80% of some pectic preparations. Other sugars often present in pectic preparations include rhamnose, arabinose, galactose and xylose (Figure 8.10).

The pectin molecule, present in intact immature plant tissue, is often referred to as protopectin. Protopectin is insoluble, which seems to be due to its polymer size and its association with calcium

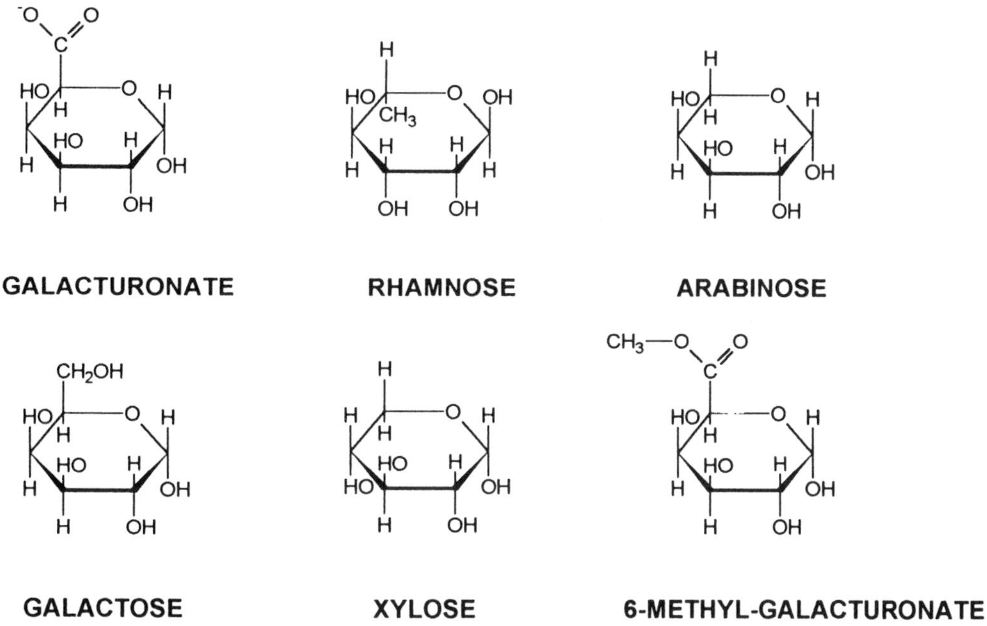

GALACTURONATE

RHAMNOSE

ARABINOSE

GALACTOSE

XYLOSE

6-METHYL-GALACTURONATE

and other divalent cations. All other pectic substances, which are soluble, are derived from protopectin by hydrolysis. Pectins may thus be described as polysaccharides composed mainly of galacturonic acid, at least 75% of which is esterified with methanol. Enzymatic de-esterification of such galacturonide yields a polymeric substance termed pectic acid. Pectic substances are often classified as galacturonans, rhamnogalacturonans, arabinans, galactans and arabinogalactans.

Rhamnogalacturonans represent the major constituent of pectic substances. The polysaccharide backbone of rhamnogalacturonans consists mainly of α-D-galacturonate units linked via $\alpha 1 \rightarrow 4$ bonds. Molecules of L-rhamnose are interspersed in the backbone, occurring on average every 25–30 galacturonate units. The rhamnose units are linked via $\beta 1 \rightarrow 2$ and $\beta 1 \rightarrow 4$ bonds to the D-galacturonate residues. Side chains of variable length, often consisting of galacturonans, galactans, arabinans or arabinogalactans, branch off from the main chain (Figure 8.11).

Pectic enzymes. There are two broad classes of pectic enzymes. Enzymes of the first group are termed pectin esterases, also known as pectin methyl esterases. The second group are termed depolymerases. Pectin methyl esterases remove methoxy groups from methylated galacturonides. Pectin esterase activity is present in all higher plants and is particularly abundant in citrus fruits and vegetables. Esterase activities are also found with a variety of microorganisms, most notably fungi.

Figure 8.10. Structural formulae of the more common monomers found in pectin. In native pectin, 75% or more of the galacturonic acid units are esterified with methanol, thus forming methyl galacturonides

The depolymerases catalyse the cleavage of glycosidic bonds via hydrolysis (hydrolases) or via β-elimination (lyases). In many instances, pectin esterase must firstly remove methoxyl groups from the galacturonide before depolymerase activity can commence.

Polygalacturonases catalyse the hydrolysis of $\alpha 1 \rightarrow 4$ linkages between β-galacturonic acid residues. A number of distinct polygalacturonase activities have been recognized, both in higher plants and in microbes. Endopolygalacturonases catalyse the hydrolysis of internal $\alpha 1 \rightarrow 4$ glycosidic linkages in stretches of polygalacturonic acid. As cleavage sites are chosen more or less at random, a number of lower molecular weight oligosaccharides are produced. Cleavage by endopolygalacturonases requires the prior removal of methoxy groups by pectin methylesterase.

Exopolygalacturonases catalyse the sequential removal of galacturonic acid residues from the non-reducing end of polygalacturonate.

Lyases are a group of pectin-degrading depolymerases produced almost exclusively by microorganisms. These enzymes catalyse the cleavage of $\alpha 1 \rightarrow 4$ glycosidic bonds, which link galacturonic acid residues, by β-elimination. Endopectate lyases catalyse the cleavage of internal glycosidic bonds in regions of polygalacturonic acids devoid of methoxy groups. Such enzymes have been isolated from a number of microbial plant pathogens. They have high pH optima and require the presence of calcium ions to maintain activity.

Exopectate lyases are found mainly in bacteria. Most such enzymes cleave the penultimate glycosidic bond of galacturonans, thus releasing dimeric molecules composed of galacturonic acid. Some such lyases, however, split the terminal galacturonan glycosidic bond, yielding single galacturonic acid moieties. As in the case of endopectate lyases, most exopectate lyases preferentially attack polygalacturonic acid which is essentially free of methyl ester groups. Pectin lyases display a preference for highly esterified polygalacturonic acid sequences (galacturonans) as substrates. These enzymes are produced mostly by fungi and usually catalyse the random internal cleavage of glycosidic bonds, thus producing esterified oligogalacturonates.

Industrial significance of pectin and pectin-degrading enzymes. Pectin enjoys widespread industrial application, as solutions of pectin are viscous, and readily gel when heated in the presence of sugar under acidic conditions. This renders pectin useful as a gelling agent, emulsifier or thickener in the production of a number of foodstuffs. Pectin used for such purposes has been classified as high-methoxy or low-methoxy pectin. As the name suggests, the majority of galacturonic acid residues present are methoxylated in high-methoxy pectin. In low-methoxy pectin, many such methoxy groups have been removed. While gelation of high-methoxy

Figure 8.11. Structure of a segment of rhamnogalacturonans. Regions containing a high density of side chains are termed "hairy regions"; they are normally separated by extended sequences devoid of side chains, the smooth regions $\longrightarrow$

pectin is dependent upon the presence of significant quantities of sugar, low-methoxy pectin may be induced to gel by the addition of certain metal ions, in the absence of sugar. Use of the latter pectin substrate thus makes possible the manufacture of jams and jellies of low sugar content. Commercial pectin is normally produced by extraction from the rind of citrus fruits or from sugar-beet pulp, both of which are rich sources of this substance.

Pectic enzymes are also used in a number of industrial processes. Such enzymes have found particular favour within the fruit juice extraction and clarification industries. Fruit juices are normally manufactured by the mechanical pressing of the relevant fruits. In many instances, the physical characteristics of the fruit hinder maximal juice extraction. The addition of commercial preparations of pectic enzymes generally facilitates a greatly enhanced yield of juice.

Enzymatic degradation of pectin by fungal pectinase preparations is routinely used to maximize juice yields from grapes and apples. Juice extraction from most soft fruits, such as raspberries, strawberries and blackberries, is also increased by supplementation with exogenous pectinolytic enzymes.

Commercial pectinase preparations are routinely used to clarify "sparkling" fruit juice preparations, such as apple juice and pear juice. When freshly pressed, most such juices contain relatively high levels of soluble pectins which contribute to the characteristic viscosity and haziness associated with such products. Haze formation often reflects a decreased solubility of one or more of the components present in the juice with which pectins associate. Partial degradation of the pectins is achieved by addition of commercial pectinase preparations, which results in a significant drop in product viscosity. Furthermore, partial degradation of the pectin destabilizes the haze particles, resulting in their coagulation and precipitation from solution. Subsequent removal of the precipitate is easily achieved by centrifugation or filtration, yielding the sparkling, clear juice. The decrease in solution viscosity observed upon treatment with pectic enzymes permits the production of concentrated juice extracts.

Pectic enzyme preparations are also used in the maceration of fruits and vegetables. Maceration usually entails the conversion of fruit or vegetable tissue into a suspension. This may be achieved by selectively hydrolysing the pectin present in the middle lamella, which binds plant cells. Maceration is normally achieved by treatment with pectic enzyme preparations exhibiting high levels of polygalacturonase activity. This enzyme has also been directly associated with the process of fruit softening during ripening. Enzymatic maceration may be used in the production of fruit nectars, "pulpy" drinks, and in the preparation of some baby foods. Pectic enzymes are also used to aid extraction of various citrus oils and pigments from orange and lemon peel.

Most commonly available pectinase preparations are obtained from fungal sources such as various species of *Aspergillus* or *Penicillium*. A source of pectin (such as apple pomace, citrus peel or dried sugar beet pulp) is normally included in the fermentation medium used to culture pectinase-producing fungi. This enhances not only pectinase production, but also promotes increased secretion of these enzymes from the mycelium. The enzymes produced are usually concentrated and partially purified by techniques such as precipitation. Stabilizers, preservatives and other additives are then incorporated into the final enzymatic preparation, which may be marketed in liquid or powdered form. Commercial pectinase preparations usually contain a variety of pectinolytic activities. Most also contain appreciable quantities of additional enzymes such as cellulases and hemicellulases.

PROTEOLYTIC ENZYMES

Enzymes exhibiting proteolytic activities constitute the single most important group of proteins produced in bulk quantities. Such proteases find application in a multitude of industrial applications, some of which are listed in Table 8.7. Proteolytic enzymes have long been used, often unknowingly, in processes

Table 8.7. Some important proteolytic enzymes

Protease	Application	Source
Microbial proteases	Detergent manufacture Leather bating Brewing Cheese-making	Bacteria, fungi
Tissue plasminogen activator	Thrombolytic agent	Recombinant, mainly from mammalian cells
Urokinase	Thrombolytic agent	Human urine
Trypsin	Debriding agent Digestive aid	Animal
Papain	Debriding agent Meat tenderizer Brewing Digestive aid	Plant (papaya plant)
Collagenase	Debriding agent	Bacterial
Bromelains	Anti-inflammatory agent Digestive aid	Plant (pineapple)
Pepsin	Digestive aid	Animal
Rennin	Cheese-making	Animal (stomach of unweaned calves); also recombinant

such as brewing, baking and cheese-making. Such enzymes have also found important application in the tanning industry and in medicine.

Many of the proteolytic enzymes traditionally used in industrial processes were obtained from plant or animal sources. The widespread incorporation of microbial proteases in detergent powders now renders microorganisms the major producers of these enzymes. Microbial proteases used industrially are generally extracellular in nature and are produced on a large scale by either semisolid or submerged fermentation. Incorporation into detergent preparations represents by far the single largest application of such "industrial" proteinases.

Recombinant DNA technology has also made possible the production of important proteolytic enzymes in novel host organisms. Specific examples include the production of tissue plasminogen activator in recombinant animal cell lines and the production of chymosin in recombinant microbial systems. Genetic engineering allows the production of unlimited quantities of proteolytic enzymes which are synthesized in low quantities by the natural producer organisms. This technology also facilitates the production in GRAS-listed organisms of certain novel or desirable proteases which are produced naturally by microbial species whose safety is questionable.

Proteolytic enzymes of medical significance

A variety of proteolytic enzymes have found important medical applications (Table 8.7). In industrial terms, most such enzymes are produced in small to moderate quantities and are subject to significant downstream processing. Most, therefore, could not truly be described in the traditional sense as bulk industrial enzymes. The properties and applications of such enzymes have already been described in Chapters 5 and 6.

Protease enzymes used in the brewing and baking industries

The cooling of beer after brewing often promotes haze formation. The haze is composed largely of protein, carbohydrate and polyphenolic compounds. Haze formation can be arrested by addition to the beer of proteolytic enzymes. Although various microbial enzymes have been assessed, plant-derived proteases such as papain and bromelain are most commonly used for such purposes.

Fungal proteases enjoy limited application in the baking industry. Such enzymes, generally from various species of *Aspergillus,* are used in order to modify the protein components of flour, and thus alter the texture of the dough. Gluten represents one major protein of flour. This protein aids the retention in dough of carbon dioxide produced by fermentation which, in turn, determines the pore structure of leavened bread.

Proteases used in the meat and leather industries

Papain is often used as a meat tenderizing agent (Chapter 2). The proteolytic preparation may be applied directly to the meat or it may be injected into the animal immediately prior to slaughter. This practice facilitates even distribution of the enzyme. The proteolytic activity renders meat more tender by enzymatically degrading connective tissue collagen and elastin. It is primarily these components that render meat tough. This process may also be promoted naturally by storing fresh carcases in cold rooms for several days after slaughter. During this time various degradative enzymes are released as the integrity of cells is disrupted. This process is known as "ageing".

Proteolytic enzymes are used extensively in the dehairing and bating of leather. Hair may be removed from animal hides by treatment with a combination of lime and sodium sulfide. Such chemicals, however, are unpleasant to work with and often give rise to problems of waste disposal. Proteolytic enzymes are often used as alternatives, either alone or (more usually) in combination with reduced concentrations of lime. The alkaline conditions generated by the lime renders it essential to employ alkaline proteases, such as subtilisin, in such a process.

Various proteolytic enzymes are also used in the bating of leather. The bating process renders the leather soft and pliable. Leather employed in the manufacture of gloves is highly bated, whereas shoe sole leather is not bated. Trypsin has traditionally been used in the bating process. More recently, microbial enzymes produced by various species of *Aspergillus* and *Bacillus* have become more popular in this regard.

Proteases incorporated into detergent products

The major quantity of proteases produced in industrial quantities are incorporated into detergents. Enzymes were first introduced into detergent preparations at the beginning of the twentieth century. Few such products were successful, as the enzymes chosen, generally from animal sources, were invariably inactivated by other components present in the detergent mix or by the ensuing washing process.

Most clothing becomes soiled by substances such as dyes, biological molecules, soil and miscellaneous particulate matter. Biological "dirt" includes protein, lipid and carbohydrate-based materials. Such dirt components may be derived directly from humans or animals, such as shedding of skin and blood, or may

Table 8.8. Principal ingredients of modern detergent preparations

Detergent component	Function
Soap and various surfactants	Removal of dirt particles especially hydrophobic molecules from fabrics
Sodium perborate	Removal of dyes and stains from fabrics
Sodium tripolyphosphate	Used to soften water
Enzymes	Removal of biological dirt which is mainly protein
Sodium carbonate and silicate	Maintenance of an alkaline pH
Polycarboxylates	Helps disperse dirt particles in water
Silicones	Controls foaming
Perfume	Imparts an appropriate scent to fabrics

be derived from other sources such as foodstuffs or grass. Modern detergent formulations contain a range of ingredients capable of efficiently removing both biological and non-biological dirt. Typical ingredients are listed in Table 8.8.

Soaps and surfactants remove the majority of dirt particles from fabrics. Perborates exhibit a bleaching action and thus help remove dyes and stains, such as those from tea, coffee and wine. Proteolytic enzymes function to degrade protein dirt such as blood, egg and gravy. Proteins are often denatured and aggregated by the washing process. This renders them even more difficult to remove from clothing fibres. The addition of proteolytic enzymes active under washing conditions greatly facilitates the degradation and subsequent removal of such stubborn stains. Phosphates, such as sodium tripolyphosphate, function to soften the water thus ensuring maximal detergent efficiency. Hard water contains appreciable quantities of calcium ions. Calcium and other divalent ions such as Mg^{2+} or Fe^{2+} will react with soap, forming a precipitate. Calcium-containing water may be softened by a number of means. One popular method entails the addition of sequestering agents such as sodium polyphosphates. The sequestering agent complexes or sequesters the divalent cations in solution and thus prevents precipitate formation. One consequence of using such sequestering agents is that divalent cation-dependent proteases may not be used. The presence of sodium carbonate in the detergent formulation ensures the maintenance of an alkaline environment, required for maximal cleaning efficiency.

Proteolytic enzymes found widespread application in detergent preparations during the 1960s. This fact was reflected in an enormous increase in worldwide protease sales during this period. By the late 1960s, however, the presence of proteolytic enzymes in detergents had become the subject of considerable controversy. This was due mainly to the not uncommon occurrence of allergic responses against powder constituents. At that time crude protease products were incorporated into the detergent in fine powder form, which encouraged significant dust formation. Inhalation of enzymes and other detergent constituents often resulted in allergic reactions. Production personnel handling and packing such enzyme-containing products were particularly prone to such adverse allergic reactions. Many end-users also suffered from responses ranging from respiratory problems to dermatitis.

By the early 1970s, enzymes had been removed from most detergent preparations. This was reflected in an immediate slump in world bulk proteinase sales. However, subsequent scientific investigations concluded that the inclusion of enzymes *per se* was a safe industrial practice. Granulation technologies were developed which minimized dust formation by dry enzyme preparations. Initially, enzyme powders were mixed with a liquefied wax and subsequently spray cooled. This process, known as prilling, generated granules in which both the wax and enzymes were homogeneously distributed. An alternative process involved granulation of a pure enzyme-containing paste; these granules were coated with wax, which ensured retention of integrity and thus minimized dust formation. Granulation technology allowed the widespread reincorporation of enzymes in detergents while eliminating the health risks associated with dust formation.

Proteolytic enzymes incorporated into detergent formulations must exhibit satisfactory catalytic activities in the presence of other detergent components, and under standard washing conditions. Such enzymes must, therefore, be stable at alkaline pH at relatively high temperatures and in the presence of sequestering agents and surfactants.

One of the first proteolytic enzymes successfully used in detergent products was subtilisin Carlsberg. Subtilisin Carlsberg is a single-chain serine protease produced by *Bacillus licheniformis*. It consists of 274 amino acids and has a molecular weight of 27 500 Da. It is a single domain popypeptide whose active site is formed by serine 221, aspartate 32 and histidine 64. The enzyme exhibits typical Michaelis–Menten kinetics and broad product specificity. The protein is maximally active at pH values of 8–10 and is stable at temperatures in excess of 50 °C. The subtilisin gene has been cloned and expressed in heterologous production systems. Its relatively straightforward structure and kinetic mechanism, along with its almost unparalleled industrial importance, has rendered it subject to extensive investigation.

Alkaline proteases synthesized by a variety of additional bacteria have been studied in order to assess their potential application in detergent preparations. Activities produced by a variety of alkalophilic species of *Bacillus* have gained a particular importance in this regard. Like subtilisin, these enzymes are single-chain serine proteases. Many exhibit overall characteristics which render them as suitable as subtilisin Carlsberg in terms of applications in detergents.

Proteases used in cheese manufacture

Rennin, also termed chymosin, represents another proteolytic enzyme subject to considerable industrial demand. This protease finds application in the cheese manufacturing process. The initial step in cheese-making involves the enzymatic coagulation of milk. Rennin catalyses limited proteolytic cleavage of milk kappa casein. This destabilizes casein micelles and promotes their precipitation, thus forming curds. The remaining liquid or whey is removed and the curd is further processed, yielding cheese or other dairy products.

Rennin is obtained from the fourth stomach of suckling calves. Extraction often involves prolonged treatment of dried strips of the calf stomach with a salt solution containing boric acid. The enzymatic extract obtained, rennet, contains a variety of proteolytic and other enzymatic activities. Rennin accounts for no more than 2–3% of such preparations.

Rennin (chymosin) is an aspartic proteinase and is synthesized as preprorennin. The hydrophobic leader sequence which directs the newly synthesized molecule through the endoplasmic reticulum, is cleaved yielding prorennin, the inactive rennin zymogen. Prorennin undergoes autocatalytic activation under the acidic conditions associated with the stomach, thus yielding the active enzyme. Rennin has a molecular weight of 35 000 Da and consists of 323 amino acids. Two isoforms of the enzyme occur naturally, termed chymosin A and chymosin B. Both exhibit acid pH optima in the region of pH 4.0.

Calf rennin catalyses the proteolytic cleavage of a single peptide bond, the Phe^{105}-Met^{106} bond in the κ-casein molecule, thus inducing protein precipitation and curd formation. The extreme specificity of rennin towards its casein substrate renders this enzyme ideally suited to cheese-making operations. Further proteolytic cleavage of casein would result in the production of inferior quality, unpalatable cheeses.

The rate of slaughter of young calves reflects market demand for veal. Fluctuation of this demand affects the availability and price of rennet. Many alternative sources have thus been screened in an effort to identify a suitable replacement protease. Pepsin preparations, obtained from a variety of slaughterhouse animals, have had limited success in this regard, although they are sometimes used in combination with rennet preparations. Of all the microbial enzymes thus far screened, few induce satisfactory curd formation. Several members of the *Mucor* genus of thermophilic fungi were found to produce acceptable alternative activities to calf rennin. The *Mucor* enzymes may be produced economically and in satisfactory quantities by fermentation technology. Like rennin, these enzymes catalyse limited cleavage of the casein molecule and function adequately under the conditions of the cheese-making process. Most such *Mucor* enzymes, however, are thermostable, and high temperatures are required to bring about their inactivation.

An alternative approach involves the expression of the rennin cDNA in recombinant microbial species. The chymosin cDNA has been expressed in a variety of systems including *E. coli*, *Aspergillus nidulans*, *Saccharomyces cerevisiae* and *Trichoderma reesei*. Recombinant rennin produced as a heterologous protein product in *E. coli* K-12, was the first food ingredient produced by recombinant DNA technology approved for use in the USA by the FDA in 1990.

The recombinant rennet forms inclusion bodies in *E. coli*, and recovery of the inclusion bodies, followed by solubilization and subsequent renaturation, yields active rennin. The recombinant product displays biological properties identical to those of native rennin. Furthermore, this preparation is in excess of 60% pure, as opposed to a typical value of 2–3% for native rennet preparations. Alternative recombinant systems capable of synthesizing large quantities of chymosin produced as an extracellular protein would successfully compete on a commercial basis with the recombinant *E. coli* production systems.

Proteolytic activity is also thought to contribute to flavour development in cheese. A number of peptides are known to impart a characteristic flavour to foodstuffs. So-called "bitter" peptides have been isolated from a number of cheese types. Most bitter peptides exhibit a highly hydrophobic amino acid content, and their presence in cheese is generally considered to be undesirable. The existence of peptides that impart a desirable flavour to cheese products still remains to be confirmed.

LIPASES

Enzymes that catalyse the degradation of lipids are termed *lipases*. In contrast to catalysts that effect the hydrolytic cleavage of other biopolymers, these enzymes are produced industrially on a relatively small scale, reflecting a moderate industrial demand, although they are used for a number of specialized functions.

These enzymes are often used to promote flavour development in foodstuffs. Suitable lipases are produced by a selected number of microorganisms, including some species of *Bacillus*, *Aspergillus* and *Mucor*.

Lipase activities are often incorporated into milk and milk products in order to impart particular flavour characteristics to dairy products such as cheese. The lipolytic enzymes hydrolyse triglycerides, producing free fatty acids. Some of these fatty acids in turn are further metabolized yielding a range of metabolites such as ketones, which together with the shorter-chain fatty acids, contribute to the flavour characteristics of the food.

Although the current world lipase market is modest, future demand for such enzymes is set to increase steadily. Lipases may be used in a variety of additional processes such as in the production of modified fats and oils, or in organic chemical synthesis. The aspiration to render detergent preparations more environmentally friendly may prompt the incorporation of lipases in detergents. A variety of other potential industrial applications of such enzymes are under active investigation at laboratory level.

SAFETY ASPECTS

The occurrence of allergic reactions to proteases included in detergent preparations raises a question of the safety of industrial enzyme preparations. The significance of this question can be more easily appreciated if one considers the central role played by enzymes in the food processing industry.

Extensive research and literature studies commissioned by many responsible organizations such as the FDA, the World Health Organization and the Food and Agricultural Organization, all support the thesis that enzymes used in the food processing industry are inherently non-toxic and are safe to consume. In many (if not most) instances, enzymes introduced in order to mediate a desired effect during food processing are inactivated or destroyed at a subsequent stage of the manufacturing process.

It is essential that enzymes utilized in the food industry or for medical purposes are obtained from non-toxic sources, otherwise the possibility exists that such enzymatic preparations could harbour toxic contaminants. Microbial sources must thus be GRAS listed (Chapter 2). Furthermore, all ingredients utilized in the formulation of the microbial fermentation media must be non-toxic. The manufacturing process used must ensure that potentially toxic or otherwise harmful substances are not introduced into the product during downstream processing. In the same way enzymes obtained from plants or animals must be obtained from edible, non-toxic sources.

FURTHER READING

Books

Beynon, R.J. & Bond, J.S. (eds) (1989). *Proteolytic Enzymes, A Practical Approach*. IRL Press, Oxford.

Coughlan, M.P. & Hazelwood, G.P. (eds) (1993). *Hemicellulose and Hemicellulases*. Portland Press, Colchester.

Fogarty, W. (ed.) (1983). *Microbial Enzymes and Biotechnology*. Applied Science.

Price, N.C. & Stevens, L. (1982). *Fundamentals of Enzymology*. Oxford University Press.

Rose, A.H. (ed.) (1980). *Economic Microbiology*, Vol. 5, *Microbial Enzymes and Bioconversions*. Academic Press, London.

Russell, G.E. (ed.) (1984). *Biotechnology and Genetic Engineering Reviews*, Vol. 1. *Applications of Biocatalysts to Biotechnology*. Intercept.

Wiseman, A. (ed.) (1985). *Handbook of Enzyme Biotechnology*, 2nd ed. Ellis Horwood, Chichester.

Articles

Arbige, M. & Pitcher, W. (1989). Industrial enzymology: a look towards the future. *Trends in Biotechnol.*, **7**, 330–335.

Barker, P. & Joshi, K. (1991). The recovery of fructose from inverted sugar beet molasses using continuous chromatography. *J. Chem. Tech. Biotechnol.*, **52**, 93–108.

Beguin, P. (1990). Molecular biology of cellulose degradation. *Ann. Rev. Microbiol.*, **44**, 219–248.

Bjorkling, F. *et al.* (1991). The future impact of industrial lipases. *Trends in Biotechnol.*, **9**, 360–363.

Brown, D. (1983). Lignocellulose hydrolysis. *Phil. Trans. R. Soc.* (London), **B300**, 305–322.

Cheetham, P. (1987). Screening for novel biocatalysts. *Enzyme Microb. Technol.*, **9**, 194–213.

Coughlan, M. *et al.* (1985). Colloquium. Cellulases: production properties and applications. *Biochem. Soc. Trans.*, **13**, 405–417.

Cullen, D. *et al.* (1987). Controlled expression and secretion of bovine chymosin in *Aspergillus nidulans*. *Bio/Technology*, **5**, 369–375.

Cygler, M. *et al.* (1992). Advances in structural understanding of lipases. In Tombs, M.P. (ed.), *Biotechnology and Genetic Engineering Reviews*, Vol. 10, pp. 143–184. Intercept.

Dubos, R. (1971). Toxic factors in enzymes used in laundry products. *Science*, **173**, 259–260.

Flamm, E. (1991). How the FDA approved chymosin: a case history. *Bio/Technology*, **9**, 349–351.

Harwood, C. (1992). *Bacillus subtilis* and its relatives: molecular biological and industrial workhorses. *Trends in Biotechnol.*, **10**, 247–256.

Kirk, T. (1987). Enzymatic "combustion": The microbial degradation of lignin. *Ann. Rev. Microbiol.*, **41**, 465–505.

Kovaleva, I. *et al.* (1989). Synthesis and secretion of bacterial α amylase by the yeast, *Saccharomyces cerevisiae*. *FEBS Lett.*, **251**, 183–186.

Kramer, M. *et al.* (1989). Progress towards the genetic engineering of tomato fruit softening. *Trends in Biotechnol.*, **7**, 191–194.

Kristjansson, J. (1989). Thermophilic organisms as sources of thermostable enzymes. *Trends in Biotechnol.*, **7**, 349–353.

Lambert, P. & Meers, J. (1983). The production of industrial enzymes. *Phil. Trans. R. Soc.* (London), **B300**, 263–282.

Lee, S. *et al.* (1991). Crystallization and a preliminary X-ray crystallographic study of α-amylase from *Bacillus licheniformis*. *Arch. Biochem. Biophys.*, **291**, 2, 255–257.

Lützen, N. *et al.* (1983). Cellulases and their application in the conversion of lignocellulose to fermentable sugars. *Phil. Trans. R. Soc.* (London), **B300**, 283–291.

Mulholland, F. (1991). Flavour peptides: the potential role of lactococcal peptidases in their production. *Biochem. Soc. Trans.*, **19**, 3, 685–690.

Ong, E. *et al.* (1989). The cellulose-binding domains of cellulases: tools for biotechnology. *Trends in Biotechnol.*, **7**, 239–243.

Palmer, J. & Evans, C. (1983). The enzymatic degradation of lignin by white-rot fungi. *Phil. Trans. R. Soc.* (London), **B300**, 293–303.

Pariza, M. & Foster, E. (1983). Determining the safety of enzymes used in food processing. *J. Food Protection*, **46**, 5, 453–468.

Pen, J. *et al.* (1992). Production of active *Bacillus licheniformis* alpha amylase in tobacco and its applications in starch liquefaction. *Bio/Technology*, **10**, 292–296.

Pitts, J. *et al.* (1991). Protein engineering of chymosin and expression in *Trichoderma reesei*. *Biochem. Soc. Trans.*, **19** (3), 663–666.

Saha, B. & Zeikus, J. (1989). Novel highly thermostable pullulanase from thermophiles. *Trends in Biotechnol.*, **7**, 234–239.

Schwardt, E. (1990). Production and use of enzymes degrading starch and some other polysaccharides. *Food Biotechnol.*, **4**, 1, 337–351.

Shanley, N. *et al.* (1993). Physicochemical and catalytic properties of three endopolygalacturonases from *Penicillium pinophilum*. *J. Biotechnol.*, **28**, 199–218.

Wells, J. & Estell, D. (1988). Subtilisin—an enzyme designed to be engineered. *Trends Biochem. Sci.*, **13**, 291–297.

Whitaker, J. (1990). Microbial pectolytic enzymes. In: Fogarty, W. & Kelly, C. (eds.), *Microbial Enzymes and Biotechnololgy*, 2nd edn, pp. 133–175. Elsevier. Amsterdam.

Whitaker, J. (1990). New and future uses of enzymes in food processing. *Food Biotechnol.*, **4**, 2, 669–697.

Wong, K. *et al.* (1988). Multiplicity of β-1,4-xylanase in microorganisms: functions and applications. *Microbiol. Rev.*, **52**, 3, 305–317.

Chapter 9

Other proteins of industrial significance

- Cyclodextrins and cyclodextrin glycosyltransferase.
- Penicillin acylase.
- Additional proteins used in the food industry.
    - Alteration of food texture and flavour development.
    - Single-cell protein.
    - Sweet proteins.
- Enzymes and animal nutrition.
    - Removal of antinutritional factors.
    - Factors affecting enzyme efficacy and stability.
    - Enzymes and animal nutrition: future trends.
    - Enzymatic conversion of biological waste into a nutritional source.
    - Dietary amino acid balance and amino acid production.
- Restriction endonucleases.
- Reverse enzymatic activities for synthesis.
- Immobilized enzymes.
- Further reading.

The previous four chapters have reviewed a variety of proteins which are of considerable industrial significance. A number of other proteins also command industrial attention, some of which are considered below. Many are used in the health-care industry, although not always as end products. Some are used in the food, beverage and allied industries. Both academic and industrial research endeavours have created a market for proteins such as restriction endonucleases. The increased research and commercial emphasis placed on biotechnology ensures that the range of economically significant proteins will continue to increase.

CYCLODEXTRINS AND CYCLODEXTRIN GLYCOSYLTRANSFERASE

Cyclodextrins are cyclic oligosaccharides enzymatically derived from starch. Three main cyclodextrin types have been identified

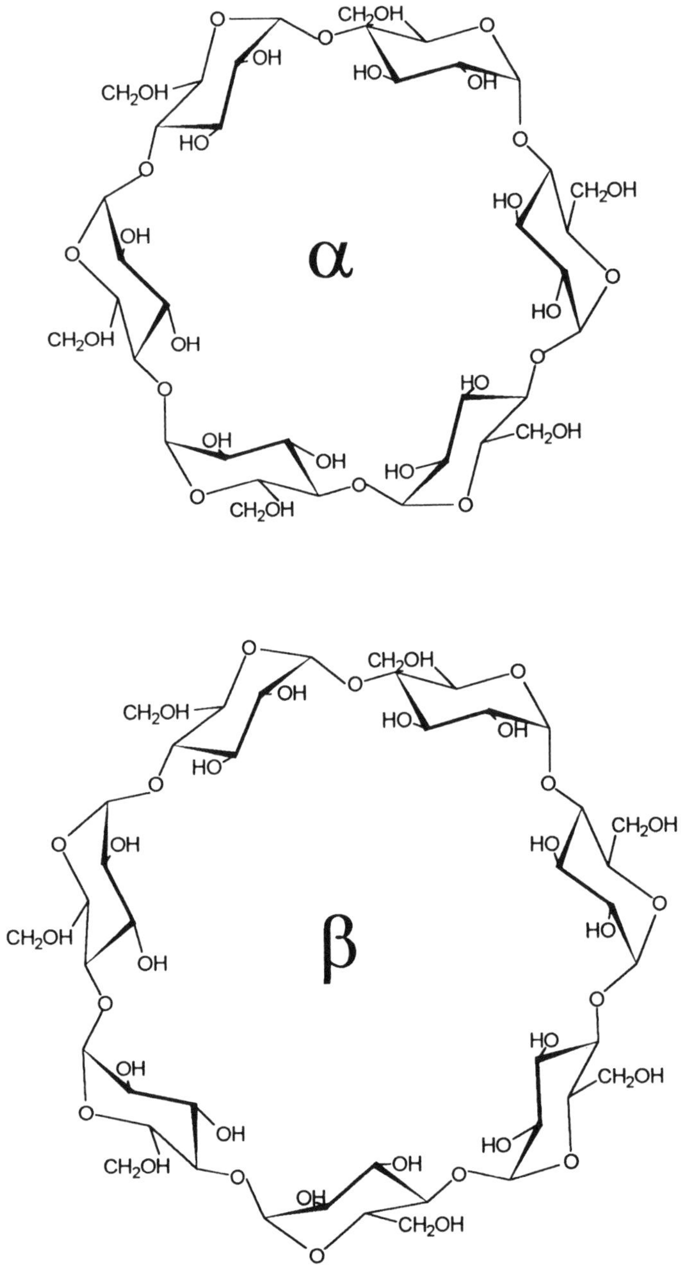

Figure 9.1. Structures of α-, β- and γ-cyclodextrins

(Figure 9.1): α-cyclodextrins are composed of six glucose molecules, β-cyclodextrins are composed of seven glucose molecules, while γ-cyclodextrins consist of eight glucose molecules. In each case, individual glucopyranose units are linked via $\alpha 1 \rightarrow 4$ glycosidic linkages characteristic of linear starch molecules.

Cyclodextrins were first studied over a hundred years ago, when their presence was noted in bacterial digests of starch. It is only recently, however, that these substances have begun to enjoy widespread industrial demand. They are now utilized in the pharmaceutical industry as well as in the food, cosmetic and allied industries.

Cyclodextrins are doughnut-shaped molecules, the outer surface of which is hydrophilic in nature, the internal cavity is apolar. When cyclodextrins are dissolved in aqueous media, the internal cavity is occupied by water molecules. The polar nature of water renders this energetically unfavourable. Added substances which are less polar than water, and which are of appropriate molecular dimensions, will replace the water as "guest molecules" within the cavity of the molecule. Such guest molecules are retained within the cavity solely by non-covalent interactions. The overall guest molecule–cyclodextrin structure is termed an *inclusion complex.*

Only molecules of an appropriate size will form stable inclusion complexes. The internal cavity of α-cyclodextrin is the smallest of the three types, and will only accommodate molecules of low molecular weight, such as chlorine, bromine or iodine. Much larger molecules such as steroids or antibiotics may be accommodated in the cavity of γ-cyclodextrin molecules. Specific side groups present in macromolecules may also interact with the cyclodextrin cavity and, in this way, form a complex.

Of the three cyclodextrins, the β form is the most commonly used on an industrial scale. The internal dimensions of the β-cyclodextrin cavity makes it ideally suited for a variety of applications. The β-cyclodextrins are also economically attractive as they are the least expensive to produce. Native β-cyclodextrins, however, exhibit poor solubility characteristics. A 14% solution of α-cyclodextrin and a 23% solution of γ-cyclodextrin are readily achievable in aqueous media. However, the maximum solubility of β-cyclodextrin in water is 1.8 g per 100 ml, i.e. 1.8%. Such poor solubility characteristics limit the industrial potential of the native β-cyclodextrin molecule.

This limitation has been overcome by the introduction to the marketplace of modified β-cyclodextrins. Substitution of the cyclodextrin hydroxyl groups with a variety of alkyl, ester or other residues dramatically improves solubility while having no effect on its complex-forming ability. Molecules such as hydroxypropyl β-cyclodextrin and hydroxyethylated β-cyclodextrin are now subject to increasing industrial demand.

Interaction of suitable substances with cyclodextrins effectively results in their molecular encapsulation. Encapsulation may be advantageous for a number of reasons. The guest molecule is protected from a wide variety of chemical and other reactants which might otherwise lead to its destruction. Such protected molecules are less likely to undergo polymerization or autocatalytic reactions. Complex formation may also mask undesirable tastes or odours normally associated with the guest molecule. Volatile compounds may be effectively stabilized by complexation with appropriate cyclodextrins.

Crystallization of cyclodextrin complexes can effectively convert guest molecules from liquid form into powder form. This is often of value in the pharmaceutical industry. Another significant consequence of complex formation is the solubilization of hydrophobic substances in aqueous solution. A variety of medically important drugs are hydrophobic and are known to form stable inclusion complexes. Such hydrophobic drugs can be safely transported through the blood stream and gastrointestinal tract in such a format. The hydrophobic drug is then released at the cell surface due to the hydrophobic nature of the plasma membrane's lipid bilayer. Cyclodextrins are therefore used medically to improve the bioavailability of a range of poorly soluble medicinal substances. This can often allow administration of lower doses with consequent economic, therapeutic and other benefits.

Cyclodextrins may also be used to stabilize certain proteins. The macromolecular structure of the smallest polypeptide precludes complete inclusion complex formation. However, virtually all proteins contain amino acid subunits containing non-polar or hydrophobic side chains. Cyclodextrins may freely interact with such side chains and in this way become intimately associated with the protein molecule. In some instances, and at high concentrations, cyclodextrins may actually destabilize or denature proteins. Under most circumstances, protein–cyclodextrin interactions serve to enhance protein stability. This and other beneficial effects render attractive the inclusion of cyclodextrin in many biopharmaceutical preparations. Cyclodextrins have been shown to reduce loss of enzymatic activity due to chemical or physical influences such as heating, freeze-drying, storage or the presence of oxidizing agents.

Upon standing in aqueous solutions, many proteins undergo limited aggregation. Powdered protein preparations, such as freeze-dried products, also sometimes form aggregates upon reconstitution in an aqueous media. Aggregate formation prevents intravenous administration of any preparation. Formulation of biopharmaceutical products in order to minimize or eliminate protein–protein interactions does not always yield satisfactory results. Many important biopharmaceutical products, such as growth hormone, urokinase and interleukin 2, still exhibit a marked tendency towards aggregate formation upon reconstitution. In many instances, inclusion of a suitable cyclodextrin preparation will eliminate such undesirable intermolecular interactions.

The ever-increasing industrial demand for cyclodextrins is reflected in an increased demand for cyclodextrin glycosyltransferase preparations. Cyclodextrin glycosyltransferase (CGTase) is the enzyme used in the synthesis of cyclodextrins from starch. Cyclodextrins may also be synthesized chemically, but their enzymatic production is technically and economically more attractive.

Three major types of cyclodextrin glycosyltransferases have been identified: α, β and γ. As the name suggests, α-CGTase predominantly yields αcyclodextrins, whereas β- and γ-CGTases yield β- and γ-cyclodextrins respectively. In all cases, however, prolonged reaction times result in the formation of a mixture of all three cyclodextrin types with β-cyclodextrin representing the predominant reaction product.

Most cyclodextrin glycosyltransferases identified to date are derived from various species of *Bacillus* (Table 9.1). The genes coding for many of these enzymes have been identified, sequenced and expressed in a variety of recombinant systems. Significant amounts of heterologous CGTase has been produced in host

Table 9.1. Some bacterial producers of cyclodextrin glycosyltransferase activity

---

Bacillus stearothermophilus

Bacillus megaterium

Bacillus subtilis

Bacillus circulans

E. coli (recombinant)

---

species such as *E. coli* and *Bacillus subtilis*. Increased production capacity of CGTases should help reduce the overall cost of cyclodextrin preparations, which in turn is likely to promote increased utilization of cyclodextrins in a variety of industrial applications.

Current annual production of cyclodextrins is in excess of 100 tonnes. Production of modified β-cyclodextrin is also increasing rapidly. Production levels secure future large-scale demand for cyclodextrin glycosyltransferases, thus these enzymes may now be classified as "bulk" industrial enzymes. The commercial significance of these enzymes within the industry is set to increase steadily over the coming years.

PENICILLIN ACYLASE

The discovery of penicillin heralded a revolutionary advance in the control of bacterial infections. Subsequently, a variety of other compounds exhibiting antimicrobial activity were introduced into the clinical arena. The penicillins and the structurally related cephalosporins remain the most popular antibiotic compounds prescribed by the medical community. The passage of time has witnessed an increasing number of bacterial populations which have become resistant to the antimicrobial action of penicillin. Resistance has been counteracted in part by the development of semisynthetic penicillins, to which many such resistant bacteria remain sensitive. The enzyme penicillin acylase plays an important role in the production of such semisynthetic penicillins, and hence is the subject of significant industrial demand.

The chemical structures of some naturally occurring and semi-synthetic penicillins are illustrated in Figure 9.2. All contain an identical core ring structure termed 6-aminopenicillanic acid. Different penicillin types differ in their attached side chains.

Semisynthetic penicillins may be produced by the enzymatic removal of the side chain of native penicillins, with subsequent attachment of a novel side chain to the resultant 6-aminopenicillanic acid core. The enzymatic removal of the side chains from penicillin G is illustrated in Figure 9.3.

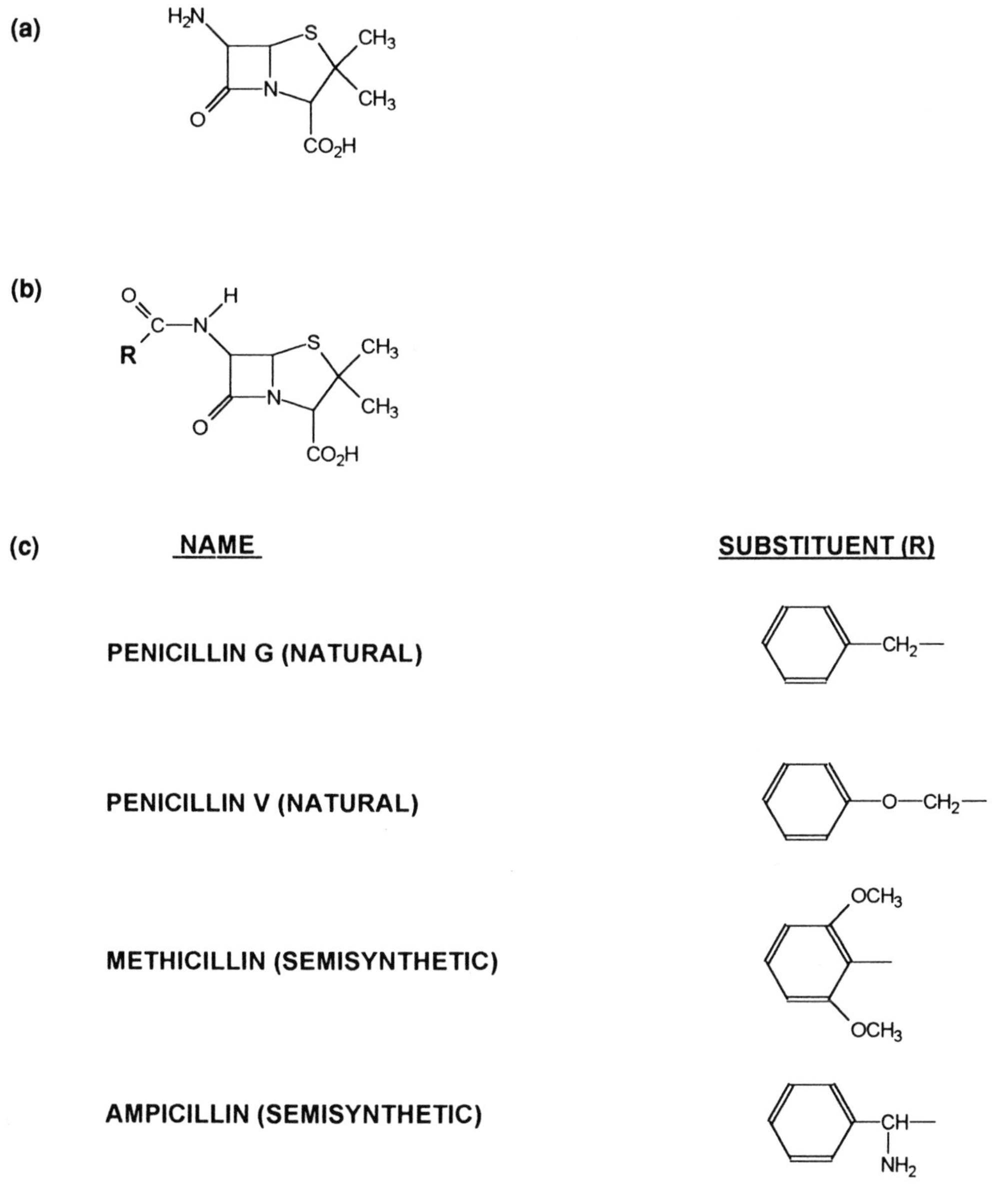

(a)

(b)

(c)

NAME	SUBSTITUENT (R)
PENICILLIN G (NATURAL)	
PENICILLIN V (NATURAL)	
METHICILLIN (SEMISYNTHETIC)	
AMPICILLIN (SEMISYNTHETIC)	

Quite a number of microorganisms produce penicillin acylase (Table 9.2). Among these, the enzyme produced by *E. coli* has received most attention. Although it is possible to utilize the free enzyme, an immobilized form is generally used in the production of semisynthetic penicillins as this is more attractive economically, facilitating the reuse of the enzyme over many production runs.

The increasing incidence of bacterial resistance to natural penicillins renders crucial the continued production of semisynthetic

Figure 9.2. Structure of 6-aminopenicillanic acid (a); generalized penicillin structure (b); and side groups present in two natural penicillins and two semisynthetic penicillins (c)

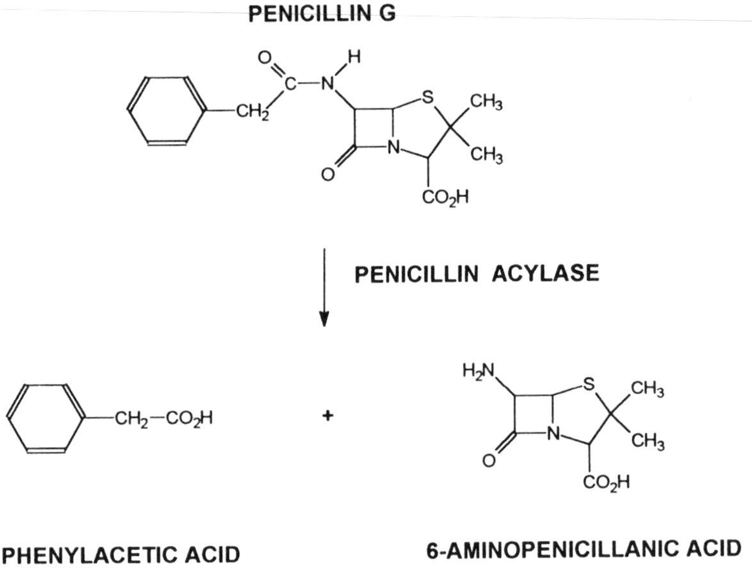

PENICILLIN G

PENICILLIN ACYLASE

PHENYLACETIC ACID   **6-AMINOPENICILLANIC ACID**

Figure 9.3. Action of penicillin acylase on penicillin G, a natural penicillin produced in large quantities by fermentation. The reaction proceeds in the direction indicated under alkaline conditions. As one of the reaction products is an acid, the pH must be continually adjusted to maintain alkaline values. Upon completion of the conversion, the pH value may be adjusted downwards to a value of 4.2–4.3. At this pH, 6-aminopenicillanic acid precipitates from solution and thus may be harvested. Some penicillin acylase preparations are also capable of catalysing the reverse reaction under mildly acidic conditions. Such enzyme preparations may thus be employed in the synthesis of semisynthetic penicillin from 6-aminopenicillanic acid and the relevant side chain group

penicillins. In addition to overcoming the problem of resistance, several semisynthetic penicillins exhibit improved clinical properties. Many inhibit a greater variety of bacterial pathogens than does penicillin G. Others are more acid-stable and hence are particularly suited for oral administration.

The advent of recombinant DNA technology has facilitated the detailed molecular elucidation of biosynthetic pathways for β-lactam antibiotics, such as penicillin and the cephalosporins. A number of genes coding for specific enzymes involved in β-lactam biosynthesis have now been cloned and characterized. Introduction

Table 9.2. Various microorganisms producing penicillin acylase

Escherichia coli

Pseudomonas spp.

Proteus rettgeri, P. morganii

Brevibacterium spp.

Bacillus spp.

Flavobacterium spp.

Streptomyces spp.

Aerobacter spp.

Kluyvera spp.

of additional copies of such genes into producing microorganisms facilitates strain improvement. Such genetic manipulations have already led to enhanced production of certain industrial strains of antibiotic producing fungi. An increased understanding of β-lactam biosynthetic pathways could also make possible alteration of such pathways at the molecular level in order to produce novel antibiotic molecules.

ADDITIONAL PROTEINS USED IN THE FOOD INDUSTRY

Proteins are widely used in the food industry, as a wide range of enzymatic activities play pivotal roles in several food processing systems. Examples include the use of carbohydrases in the production of glucose and fructose syrups from starch, the use of pectinases and related enzymes in the fruit drinks industry and the use of proteases in processes such as cheese-making. Enzymes such as these are used industrially on a large scale. Many such processes have been slowly developed and perfected over decades and, in some cases, hundreds of years. A variety of other proteins are also used in the food industry, specific examples of which are considered below.

Alteration of food texture and flavour development

Some food biotechnology companies have come to specialize in developing specific protein preparations which, when added to certain foodstuffs, enhance a particular physical characteristic of that food. For example, certain protein-based extracts may be added to enhance "whipping" performance; such proteins promote efficient aeration and enhanced mouthfeel of products such as whipped desserts, milk-shakes and mousses. Most such proteins are included in the food at levels up to 1% and may be used in addition to or as a replacement for more variable materials such as egg albumen or gelatin. Owing to intense competition, few such companies care to discuss the exact composition of their protein additives.

Other proteins are used in the food industry in order to enhance food flavour. Extensive research efforts are investigating flavour development in commodities such as cheese and chocolate. As discussed in Chapter 8, both protease and lipase activities have a significant effect on cheese texture and taste. Chocolate is probably one of the most widely used flavouring agents in the food industry. Chocolate flavour development is a complex process which remains largely undefined. Its development is dependent upon fermentation of cocoa beans, with subsequent drying and roasting steps. Chocolate manufactured from fresh, unfermented cocoa beans is devoid of its characteristic flavour. The fermentation process promotes hydrolysis of cocoa storage proteins in addition to carbohydrates and other components, yielding a mixture of peptides, amino acids and sugars. Upon roasting, this mixture yields a myriad of volatile molecules which impart the characteristic flavour to chocolate.

Proteolysis of two major cocoa storage proteins appears to be particularly critical to development of full chocolate flavour. A greater understanding of this process and the nature of the resultant protein hydrosylate would greatly assist logical biotechnological intervention in chocolate flavour development.

Monosodium glutamate is the sodium salt of the amino acid glutamate. This amino acid salt is extensively used to enhance flavour characteristics of many food products.

Single-cell protein

Microorganisms can constitute a valuable source of dietary protein. Indeed, the production of single-cell protein has received attention since the beginning of the twentieth century. Production of microbial biomass as a source of protein is attractive for a number of reasons. Most microbial populations have short generation times compared with traditional animal or plant protein sources. Fermentation technology required to produce single-cell protein is well established. This process may be carried out independently of seasonal or climatic conditions, and can utilize raw materials that are readily available and inexpensive. In addition, the protein content of most microbial species is high, and genetic intervention may be employed to enhance the quality and quantity of the protein produced.

Any microorganism used in the production of single-cell protein must be GRAS listed. While such products may be consumed directly by humans, they may also be utilized in the animal feed industry. A steadily increasing world population is likely to render the future microbial production of protein and other nutrients more significant.

Sweet proteins

The production of high-fructose syrups and their use as sweeteners was considered in Chapter 8. Sweetness, however, is not a monopoly of sugar-based molecules. Several proteins and peptides have been identified that, if tasted, are perceived as being intensely sweet. The sweetest natural substances thus far discovered are two plant proteins termed thaumatin and monellin; both are isolated from the fruit of plants native to Africa. These proteins are 100 000 times as sweet as sucrose on a molar basis. Solutions containing as little thaumatin as 1×10^{-8} moles per litre are perceived as being sweet.

Monellin is a dimeric protein of relatively low molecular weight—its larger polypeptide chain consists of 50 amino acids, whereas the other chain contains 44 amino acids. In contrast, thaumatin is a single-chain polypeptide consisting of approximately 207 amino acids. Very little amino acid sequence homology is shared between these two proteins; however, polyclonal antibodies raised against one protein cross-react with the other. Several different thaumatins have been identified, all of which are products of distinct thaumatin genes. The two predominant forms are termed thaumatin I and II respectively; they differ in amino acid sequence by only 5 amino acids and have molecular weights in excess of 22 kDa.

The cDNA coding for thaumatin has been cloned and expressed in *Bacillus subtilis*, *E. coli*, and in yeasts such as *Saccharomyces cerevisiae*. Levels of expression were disappointingly low. The thaumatin gene has also been expressed in transgenic plants. Extracts from such plants exhibited a sweet taste.

Production of recombinant thaumatin is a prerequisite for its widespread industrial use. The plant in which it is naturally produced will not bear fruit when cultivated in climates removed from its native Africa. Such proteins could enjoy widespread industrial usage as non-nutritive sweeteners, and the facts that they do not promote tooth decay and that they can be safely consumed by diabetic patients heightens this potential. Other sweet proteins have also been identified, among them are miraculin and curculin, which are unusual in that they specifically confer a sweet taste on sour-tasting substances.

Aspartame is yet another low-energy sweetener. It is a dipeptide, L-aspartyl-L-phenylalanine methyl ester, which was initially synthesized by chemical means. Its enzymatic synthesis has subsequently gained favour. Several methods of production have been developed, one of which involves direct coupling of the amino acids by the enzyme thermolysin. Aspartame is approximately 160 times sweeter than sucrose. It has been approved as a food additive and enjoys significant use. Increased levels of production of this dipeptide is reflected in an increased requirement for both L-phenylalanine and L-aspartic acid.

ENZYMES AND ANIMAL NUTRITION

Higher organisms have developed sophisticated digestive systems by which they procure nutrients from ingested matter. Degradation of polymeric nutrients, such as proteins, carbohydrates and lipids, is a prerequisite of efficient nutrient assimilation. Most such degradative events are mediated by specific digestive enzymes. Evolutionary pressure has ensured the development of an efficient digestive system; modern intensive livestock production practices place added pressure on this digestive process. Biotechnological intervention can serve to redress any digestive imbalance caused by modern production methods.

Weaning of young animals such as piglets serves as a good example. Modern production practices transform weaning from a gradual process to an abrupt event. The sudden alteration in dietary composition from a milk-based feed to a more complex nutritional intake frequently causes digestive upsets in young animals. Young piglets often display a physiological deficiency in gastric acid production, resulting in stomach pH values above those required for optimal digestive function. In addition, the overall digestive capability of such animals may not have fully developed at the time of weaning. Poorly digested food results in suboptimal nutrient assimilation by the weaned animal, and incompletely digested matter promotes vigorous growth of many microorganisms in the large intestine. Such factors contribute to digestive upsets and an increased incidence of post-weaning diarrhoea.

Digestive difficulties associated with weaning may be averted by feeding specially formulated, readily digestible rations, and by inclusion of acidifiers such as citric acid in the diet to generate low stomach pH. The addition of exogenous enzymes, capable of hydrolysing more complex components of the weaned animal's rations, also promote increased feed digestibility. Proteolytic enzymes, in addition to cellulases and hemicellulases, have been used as digestive aids for weanlings. Not surprisingly, the beneficial effects of exogenous enzyme addition are most noticeable when animals are fed rations containing more complex dietary components.

Removal of antinutritional factors

Various enzyme preparations have been used to bring about the removal of specific antinutritional factors from animal feeds. The addition of β-glucanase to barley-containing poultry feed is perhaps the most noted example. The inclusion of pentosanase in wheat-based poultry diets serves as an additional example. Incorporation of phytase in cereal-based animal feedstuffs represents a particularly exciting development, as it not only removes a major antinutritional factor but may also considerably reduce the pollutive effect of animal wastes.

β-Glucan is a non-starch polysaccharide associated with barley consisting of glucose units linked via $\beta 1 \rightarrow 3$ and $\beta 1 \rightarrow 4$ linkages. The level of β-glucan present in barley can vary considerably and is influenced by factors such as soil type, growing conditions and time of harvesting. Although β-glucan may be present at levels below 2–3%, values in excess of 10% of grain have been recorded. When ingested, the β-glucan present in barley becomes solubilized in the gut. Animals are devoid of endogenously-produced digestive enzymes capable of hydrolysing the β-glucan molecule. Soluble β-glucans form solutions of high viscosity (Figure 9.4), thus their presence in the diet promotes formation of highly viscous intestinal contents. This has a particularly negative effect on the digestive function of poultry. Maximum nutrient utilization is impeded and the animals develop difficulty in passing faeces. Due to its viscous nature, the faeces adhere to the birds' feathers, to their bedding and to the eggs of laying hens. The multiple negative effects associated with ingested β-glucan have traditionally limited the level of incorporation of barley in poultry feed, although it is an economically attractive foodstuff.

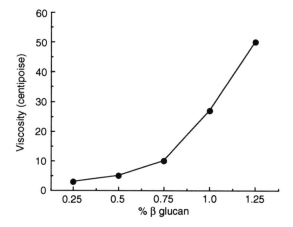

Figure 9.4. The effect of increasing β-glucan concentrations on the viscosity of an aqueous solution

Some microbial populations produce enzymes termed β-glucanases, which are capable of hydrolysing β-glucans; incorporation of β-glucanase preparations in poultry diets now facilitates the use of barley in such diets. Upon ingestion, the microbially derived β-glucanase enzyme degrades the β-glucans present, thus destroying their antinutritive properties and providing sugars for energy (Table 9.3).

Wheat contains an antinutritive factor termed pentosan. Pentosan is a non-starch polysaccharide consisting predominantly of sugars such as pentose, xylose and raffinose. The digestive complement of animals is incapable of hydrolysing this substance and—as in the case of β-glucan in barley—solubilization of pentosans in the gastrointestinal tract results in increased viscosity of digesta. Addition of microbial pentosanase preparations to wheat-based diets will destroy these antinutritional molecules. The presence of pentosans does not, however, present such acute dietary difficulties as does the presence of β-glucans.

Table 9.3. Effect of supplementing barley based diets with β-glucanase on performance of poultry. The wheat-based diet serves as a control. Bird performance may be assessed by the ratio of feed to weight gain: the lower this value, the more efficiently the bird has utilized the feed provided. Enzyme was included in the barley-based diet at a rate of 1 kg per tonne of feed

Dietary treatment	Feed consumption per bird per day (g)	Weight gain per bird per day (g)	Feed : gain ratio
Wheat-based diet	102.3	49.9	2.05
Barley-based diet	91.5	41.5	2.2
Barley and β-glucanase	96.3	50.2	1.92

Phytin. More than 50% of the total phosphorus present in cereals is in the form of phytin (*myo*-inositol hexaphosphate, Figure 9.5). The phosphate groups present in phytic acid are biologically unavailable to non-ruminant animals, who lack the enzymatic activity required to release the phosphate groups from the core ring structure. In ruminant species, phytin is degraded by microbial populations within the rumen. Lack of phosphorus availability in monogastric animals, such as pigs and poultry, renders likely the possibility of dietary phosphorus deficiency. This is generally avoided by the addition of inorganic phosphorus in the form of dicalcium phosphate to such animal feeds.

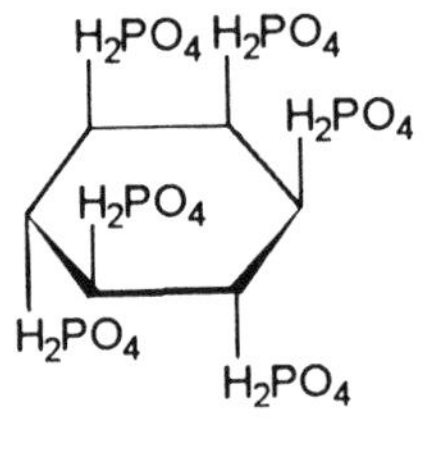

Figure 9.5. Structure of *myo*-inositol hexaphosphate (phytin)

Phytin is also an antinutritional factor because it binds a range of essential minerals such as calcium, zinc, iron, magnesium and manganese within the digestive tract, rendering them unavailable for absorption.

The digestive complement of animals is devoid of an enzymatic activity capable of degrading phytin. Several such "phytase' enzymes are produced in plants and in a variety of micro-organisms. Addition of microbial phytase to dietary rations has been shown to promote degradation of phytin within the digestive tract. Such enzyme-mediated degradation has several beneficial effects. Degradation of phytate destroys its antinutritive effect,

Table 9.4. Effect of phytase on phosphorus balance in pigs. The phytase diet was similar to the control diet but was supplemented with phytase. Both phosphorus intake and absorption were monitored. Inclusion of phytase resulted in increases in absorbed and retained phosphorus. All differences recorded were found to be statistically significant

Parameter	Phytase diet	Control diet
Phosphorus intake (g/day)	8.3	7.3
Phosphorus absorbed (g/day)	5.4	3.5
Percentage absorption	65	48
Percentage retained	60	47

From Pointillard, A. *et al.* (1987). Importance of cereal phytase activity for phytate phosphorus utilization by growing pigs fed diets containing triticale or corn. *J. Nutr.*, **117**, 907.

thus promoting increased mineral absorption. The phosphate groups released from the *myo*-inositol ring are biologically available to the animal (Table 9.4), reducing the amount of supplemental inorganic phosphorus that must be added to the diet. Reduction in the level of such supplementation is desirable on both economic and environmental grounds. Phosphate pollution derived from the animal production sector has long been a source of concern. It is estimated that over 100 million tonnes of animal manure is generated in the USA each year. Such quantities contain in excess of 1 million tonnes of phosphorus. Although animals fail to digest phytic acid, many microorganisms present in faeces display appreciable levels of phytase activity. This form of phosphorus may therefore greatly contribute to the pollutive effect of animal slurry.

Although the efficacy with which microbial phytases degrade dietary phytate is beyond doubt, the enzyme is not yet widely used within the animal feed industry. Difficulties still remain, most notably with regard to the identification of a specific phytase which is highly active in the digestive tract. Microbial strains producing elevated levels of this activity must also be developed if the enzyme is to be utilized on a large scale. Some fungi, such as certain strains of *Aspergillus*, are capable of producing moderate levels of this enzyme. Recombinant DNA technology, however, offers the most effective way of developing hyperproducing strains of microorganisms.

A phytase gene obtained from *Aspergillus niger* has been expressed in transgenic tobacco seeds. The recombinant enzyme was expressed as 1% of the soluble protein in the mature seed. Direct application of these seeds to animal feed promoted enhanced phytic acid-phosphorus utilization by recipient animals.

Factors affecting enzyme efficacy and stability

The addition of specific enzymes to animal feedstuffs in order to achieve a particular effect is one of the most exciting areas of modern agricultural biotechnology. Several issues must however, be addressed in order to ensure that maximal beneficial effects are obtained from such enzymatic supplementation. One such issue relates to the ability of enzymes to function under the conditions encountered within the digestive tract. The majority of nutrients are absorbed within the small intestine. Supplemental enzymes should ideally function within the stomach and/or the initial portion of the small intestine, prior to the sites of nutrient absorbtion.

Issues concerning enzyme stability must also be addressed. The stability and catalytic capability of enzymes when exposed to the acidic conditions within the stomach must be assessed. In addition, the stability of the enzyme during the pelleting process must be determined. Most animal feeds are subjected to pelleting during their manufacturing process. This involves exposing the feed to high temperatures under elevated pressure within specialized pelleting equipment. The elevated temperatures associated with the pelleting process (typically 70–90 °C) give rise to concern regarding possible detrimental effects on heat-labile feed constituents, such as vitamins and enzymes. This concern is heightened by the trend to increase the pelleting temperature still further to minimize the possibility of pathogen transmission via contaminated feed rations. However, the feed matrix probably serves to protect the incorporated enzymatic activity from the full effects of the heat applied. Several post-pelleting enzyme application systems have been developed. Such systems normally entail controlled spraying of the enzyme preparation onto the external surface of the pellet immediately subsequent to pellet formation. As the pellet cools, the enzyme is effectively drawn into its interior. A wax or other protectant may then be applied to the pellet surface in order to minimize loss of enzyme during handling.

Enzymes and animal nutrition; future trends

It is now generally accepted that incorporation of selected enzyme preparations in animal feed can promote positive production responses. More recent research has focused upon taking this concept one step further, i.e. the generation of transgenic animals capable of synthesizing novel digestive enzymes. A cellulase gene obtained from *Clostridium thermocellum* has been successfully expressed in the pancreas of transgenic mice. The enzyme was secreted into the small intestine where it was found to be resistant to proteolytic inactivation by endogenous proteases. Such research demonstrates the feasibility of generating monogastric animals with endogenous capacity to digest plant structural polysaccharides.

Enzymatic conversion of biological waste into a nutritional source

The food production and processing industries generate large quantities of waste protein. Such waste includes dead animals and inedible portions of animals such as heads, feet, guts and feathers. Such materials are at best of marginal economic value and often are regarded as a waste disposal problem. Proteinaceous waste may also be viewed as a rich source of certain nutrients, in particular amino acids. Some waste material may be effectively recycled as food by rendering facilities, which process or render dead animals and animal offal, usually by cooking at high pressure. This yields a valuable source of protein which may subsequently be incorporated into livestock feed. The process also effectively sterilizes the rendered material, thus preventing potential transmission of disease from any infected starting material.

The rendering process requires dedicated, well-equipped facilities. High energy inputs are also required in the cooking process. An alternative method of converting such biological waste into a valuable nutrient source involves the use of degradative enzymes. Such a biological process would require less sophisticated processing facilities and should function with lower energy costs. Enzymatic conversion of waste is also more flexible than traditional rendering processes, and may expand the range of convertible waste products. Enzymatic digestion of poultry feathers represents a good example.

It is estimated that in excess of 400 million chickens are killed each week on a worldwide basis. Typically, each bird has up to 125 g of feathers. The weekly worldwide production of feather waste would thus be 5000 tonnes. Feathers are poorly digested by animals, mainly because of the poor ability of the latter to hydrolyse disulphide bonds characteristic of α-keratins. An enzymatic process has been developed which facilitates comprehensive digestion of such feather waste. The process is summarized in Figure 9.6. Briefly, it entails blending a slurry of ground feathers with sodium sulphite and an enzyme cocktail consisting predominantly

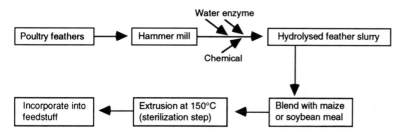

Figure 9.6. Conversion of poultry feathers from waste to valuable protein source

of proteolytic enzymes. The chemical environment generated promotes dissolution of the disulphide bonds. The enzyme activities hydrolyse the ground feather protein (α-keratin), yielding a peptide-rich, digestible product. The feather digest may be formulated with soybean meal in order to upgrade the overall nutritional value of the product. The resultant mixture can then be sterilized by heat in order to prevent potential transmission of pathogens.

Feather digest products are already used in the animal feed industry and are routinely fed to both poultry and pigs. It has also been included in the diets of fish and domestic pets. The enzymatic degradation of feather waste has thus proved to be both economically and technically feasible.

Animal hair and wool consist almost entirely of α-keratins. Thousands of tonnes of such material are deposited in landfill sites annually. Such waste products also represent potential substrates for enzymatic digestion. A process similar in design to that utilized in feather digestion could be used to convert such substrates into a valuable nutrient source.

Large volumes of blood generally accumulate as a byproduct in most slaughterhouse facilities. Whole blood may be centrifuged in order to collect red blood cells. This red blood cell fraction contains 70% of total blood protein. Incubation of the collected red cell fraction with alkaline proteases results in their degradation. Enzymatic activity may be arrested simply by reducing the reaction pH to an acidic value. Filtration of the hydrosylate removes the colour. Evaporation is then used to concentrate the protein hydrosylate to 40% solids. This slurry is finally dried by drum or spray-drying. The final product is almost 100% digestible and is often incorporated into the rations of young animals.

The fish processing industry also generates large quantities of waste byproducts which are proteinaceous in nature. Such products include not only inedible portions of fish, but also undersized or damaged whole fish. Such waste products may be enzymatically converted into protein hydrosylates of high nutritional value, and fish hydrosylates are frequently incorporated into the diets of animals such as mink.

Dietary amino acid balance and amino acid production

Unlike many microorganisms, most animals cannot synthesize all the amino acids required for protein synthesis. Those essential amino acids that animals are incapable of producing must be procured from dietary sources. Nutritionists have long recognized that protein dietary constituents must contain a well-balanced amino acid composition, in terms of essential amino acids, in order to ensure optimal animal growth.

Both lysine and threonine are essential amino acids in the mammalian diet. They are also generally the first and second limiting amino acid for pigs and poultry fed a cereal-based diet. Maximal utilization of cereal-derived protein requires the addition of almost 4 kg of crystalline lysine and almost 2 kg of threonine per tonne of feed. Other essential amino acids which may become limiting, in particular if animals are fed diets high in soybean, include tryptophan and the sulphur-containing amino acids cysteine and methionine (Figure 9.7).

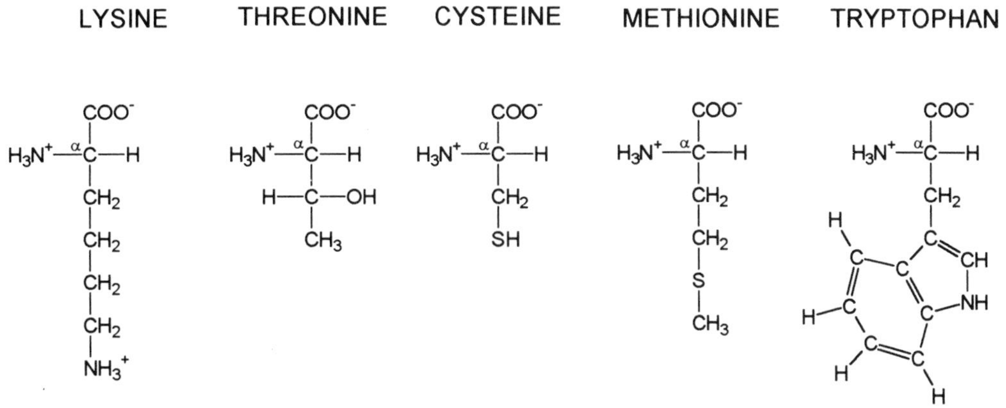

LYSINE THREONINE CYSTEINE METHIONINE TRYPTOPHAN

Figure 9.7. Structure of amino acids most likely to become limiting when monogastric animals are fed cereal-based diets. In contrast to such, animals ruminants procure a significant level of all naturally occurring amino acids from ruminal microorganisms, as these microorganisms can synthesize such essential amino acids *de novo*

The problem of attaining a well-balanced dietary amino acid complement could be overcome by generating transgenic animals capable of synthesizing their own essential amino acids. Such a project, however, would be long-term in nature. In the meantime nutritionists must ensure that animals are fed rations containing an optimal amino acid balance. This often necessitates supplementation of the diet with specific amino acids.

An alternative strategy currently pursued entails the generation of transgenic plants capable of producing proteins of altered amino acid composition. The major proteins present in soybean are deficient in methionine and cysteine. A number of alternative seed proteins rich in methionine have been identified. Production of these proteins in transgenic soybeans could negate the necessity to supplement soybean-based rations with methionine, and as cysteine can be synthesized from methionine, it also would not be required.

Several essential amino acids are required in bulk quantities by the animal feed industry. A number of other amino acids are produced in bulk for a variety of additional industrial processes. Glutamic acid, for example, is widely used in the food industry. Inclusion of its sodium salt, monosodium glutamate, enhances the natural flavour associated with many food types. Hundreds of thousands of tonnes of this amino acid are now produced annually.

The development of the sweetener aspartame has, as discussed earlier, substantially increased the industrial demand for its two amino acids, phenylalanine and aspartic acid. Other amino acids are also required in moderate quantities for medical, research and other specialist purposes.

Amino acids may be produced either chemically or enzymatically. Chemical production has traditionally been the method of choice, mainly on economic grounds. Chemical production, however, yields a racemic mixture consisting of the D and L isoforms of the amino acids. As the L form is biologically utilizable by most species, separation of L from D isoforms is required subsequent to chemical synthesis. This is most readily achievable by using the enzyme L-aminoacylase. The racemic amino acid mixture is firstly chemically acetylated. The mixture is then passed over a bed of immobilized L-aminoacylase. This enzyme will deacylate only the L-form of the acylated amino acids. The chemical properties of acylated amino acids differ from those of native amino acids, thus subsequent isolation of the L-amino acids is straightforward. Large-scale amino acid production has generated a very substantial demand for L-aminoacylase. The enzyme normally used is derived from fungal sources such as *Aspergillus*.

RESTRICTION ENDONUCLEASES

Most microorganisms have developed a variety of mechanisms by which they protect themselves from invading viral and other pathogens. One such method relies upon specific microbial enzymes termed *restriction endonucleases*. These enzymes are capable of cleaving foreign double-stranded DNA and, hence, preventing its replication. Host microbial DNA may be modified, usually by methylation of specific DNA sequences, and in this way is protected from destruction by endogenous restriction endonucleases.

Some 600 restriction endonucleases have been identified. There are three types: types I and III restriction enzymes are complex, consisting of a number of subunit types and requiring a variety of cofactors to maintain activity. These enzymes cleave DNA at sites removed from the DNA sequence which the enzyme recognizes. Type II restriction endonucleases are less complex, generally consisting of a single subunit, require only Mg^{2+} for activity, and cleave DNA at the sequence that the enzyme recognizes and binds. Type II restriction endonucleases have thus found wide application in both pure and applied molecular biology.

Restriction endonucleases recognize and bind DNA sequences of a defined base sequence (Table 9.5). These sequences normally exhibit a bilateral symmetry around a specific point and are usually 4, 6 or 8 base pairs in length. Such sequences are often termed

Table 9.5. Some restriction endonucleases, their DNA recognition sites and cleavage points. *Mac* I recognizes a tetrameric site and generates staggered ends. *Bsu* I, on the other hand, while recognizing a tetrameric site, generates flush or blunt ends. *Eco*RI and *Hind*II both recognize hexameric sites; the former generates staggered (sticky) ends, while the latter generates flush ends. *Not* I cuts an octameric recognition site and will generate sticky ends. For a more comprehensive list of commercially available restriction enzymes, refer to *Methods in Enzymology*, Vol. 152, cited at the end of this chapter

Restriction enzyme	Enzyme DNA recognition sequence
Mac I	...C↓TAG... ...GAT↑C...
Eco RI	...G↓AATTC... ...CTTAA↑G...
Not I	GC↓GGCCGC CGCCGG↑CG
Bsu I	GG↓CC CC↑GG
*Hind*II	...CTPy↓PuAC... ...CAPu↑PyTG...

G, guanine; C, cytosine; A, adenine; T, thymine; Pu, any purine; Py, any pyrimidine. Arrow indicates the exact site of cleavage.

"palindromes". In general, the larger the recognization sequence, the fewer such sequences will be present in a given DNA molecule, hence, the smaller the number of DNA fragments that will be generated. Depending upon the specific restriction endonuclease used, DNA cleavage may yield blunt ends or staggered ends—the latter are often referred to as "sticky" ends.

Each microorganism expresses one or more restriction endonucleases of defined recognition sequence. Several hundreds of these enzymes have now been described and characterized in detail. Many are available commercially. Restriction endonucleases used either alone or in combination find a ready market in research and industrial applications relying on DNA manipulation. Many restriction endonucleases are biologically labile and must be transported and stored at temperatures below −20 °C. More recent studies have shown that in many cases, no loss of activity occurs if the enzymes are dried at room temperature in the presence of a glucose-containing disaccharide termed trehalose (see Figure 5.7). Furthermore, the dried enzymatic preparation was found to have an extended shelf-life, even when stored at elevated temperatures. Exposure of such dried preparations to temperatures as high as 70 °C had little effect on the enzymatic activity of the reconstituted product. Many microorganisms subject to periodic desiccation are known to contain high levels of trehalose. The trehalose may well serve to protect the enzymatic complement and other structural elements during the drying process, and facilitate complete recovery of cellular function upon rehydration.

REVERSE ENZYMATIC ACTIVITIES FOR SYNTHESIS

While emphasis on the use of enzymes in biotechnological processes is based upon their hydrolytic activity in the degradation of biopolymers, sometime enzymes can be used to synthesize compounds. Some hydrolytic enzymes can function in synthesis through reverse enzymatic reactions. This approach is receiving considerable attention and can be illustrated by reference to the synthesis of an ester from an organic acid and an alcohol.

$$R_1 - \overset{\overset{\textstyle O}{\|}}{C} - OH \ + \ R_2 - OH \ \underset{\text{esterase}}{\rightleftharpoons} \ R_1 - \overset{\overset{\textstyle O}{\|}}{C} - O - R_2 + H_2O$$

carboxylic acid   alcohol   ester

In this reaction a molecule of water is produced. Thus, to achieve this synthesis, a low concentration of water is essential. This is promoted by performing the reaction in non-aqueous solvents, for example hexane. It is essential that the substrates are soluble in the non-aqueous solvents and that the enzyme is stable and active in such solvents. Such reverse enzymatic reactions have some advantages over chemical synthesis in the production of certain synthetic compounds, as the specificity of the enzymic reaction ensures that

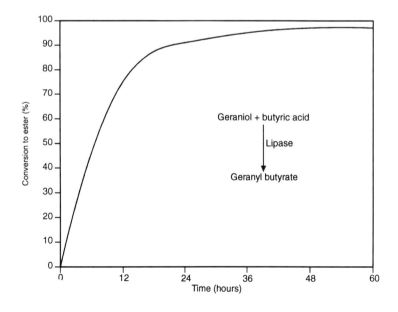

Figure 9.8. Production of geranyl butyrate by reverse enzymatic synthesis. Levels approaching 100% conversion of substrates can be achieved

only one particular isomer is formed. This removes the necessity to separate isomers which are formed in equal quantities by chemical reactions. As with all enzymatic reactions, the conditions required to promote the reaction are mild and its associated specificity reduces or eliminates undesirable side reactions. Such reverse enzymatic reactions have been used to produce flavouring compounds such as geranyl butyrate (Figure 9.8).

To achieve ester synthesis, the concentration of water in the system must be low—below 2%. In many cases, the reacting alcohol can serve as the solvent system. The level of enzyme use is generally 1–2% v/v based upon total volume. The reaction is carried out at 30 °C, at which temperature the enzyme is quite stable. As is evident from Figure 9.8, the rate of reaction is relatively slow—generally taking days to achieve the conversion. Nonetheless, for the reasons outlined above, there are advantages to using reverse enzymatic synthesis over chemical synthesis for many compounds including glycerides.

IMMOBILIZED ENZYMES

The vast majority of enzymes used in industry are not recovered following completion of their catalytic conversions. In some instances however enzymes are used in an immobilized form such that they may be recovered at the end of the catalytic process. This facilitates recycling of the enzyme preparation, which is of obvious economic benefit. The decision to use immobilized or free enzyme in any given situation depends upon economical, technical and practical considerations. The more expensive the enzyme preparation, the greater the impetus to use it in an immobilized form. However, immobilization must not adversely affect enzyme stability, or kinetic or other relevant properties. In many instances (enzymes used therapeutically, or processes employing bulk quantities of crude enzymes), subsequent recovery of an immobilized enzyme is rendered impractical. Some of the more notable industrial processes which do use immobilized enzymes are listed in Table 9.6.

A variety of methods may be used to immobilize any enzyme: these generally involve either trapping the enzyme within a confined matrix or binding the enzyme to an insoluble support matrix (Figure 9.9).

Enzymes may be entrapped within the matrix of a polymeric gel. This is achieved by incubating the enzyme together with the gel monomers and then promoting gel polymerization. Enzymes entrapped within polyacrylamide or polymethacrylate gels serve as two such examples. The enzyme-containing gels may be further processed in order to produce smaller gel particles of appropriate size and shape. In order to function successfully, the gel pore size

Table 9.6. Some biotechnological processes utilizing immobilized enzyme preparations

Enzyme	Process	Reaction catalysed
Glucose isomerase	Production of high-fructose corn syrup	Conversion of glucose to fructose
Amino acid acylase	Amino acid production	Deacetylation of L-acetyl amino acids
Penicillin acylase	Production of semisynthetic penicillins	Removal of side chains from naturally produced penicillin, yielding 6-aminopenicillanic acid
Lactase	Hydrolysis of lactose	Hydrolysis of lactose, yielding glucose and galactose

generated must be small enough to retain the entrapped enzyme but must be large enough to allow free passage of enzyme reactants and products. Enzymes may also be entrapped within molecular-sized pockets formed during spinning of industrial fibres such as cellulose acetate.

Encapsulation involves entrapping enzymes within a spherical semipermeable membrane. Membrane pore diameter is generally about 100 μm. Smaller molecules, such as enzyme substrates, cofactors and products, must be able to pass freely through such pores, while enzymes or other macromolecules are retarded. Cellulose nitrate and nylon-based membranes have been extensively used in this regard.

Enzymes may also be immobilized by promoting their binding to an insoluble matrix. A variety of matrices and methods of attachment have been developed. Perhaps the simplest such method involves physical adsorption of the enzyme to a suitable carrier substance. This may often be achieved by directly mixing the enzyme and support, under incubation conditions (ionic strength, pH and temperature) at which maximal adsorption is observed. These parameters are normally determined empirically. Leakage of enzymes from the support is often a problem as the forces of attraction retaining the enzyme on its surface are relatively weak. Supports such as aluminium hydroxide are most often used, and a variety of enzymes have been immobilized by this procedure.

Enzymes may also be immobilized by promoting ionic interactions with a suitably charged matrix. Ion-exchange resins have found particular favour in this regard. Anion exchange resins such

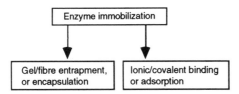

Figure 9.9. The most popular methods of enzyme immobilization

as DEAE-cellulose or DEAE-Sephadex have been used in the immobilization of negatively-charged enzymes, while cation exchange media such as CM-cellulose may be used to immobilize positively-charged enzymes. This method of enzyme immobilization is technically undemanding and economically attractive. Enzyme leakage normally does not present a major problem providing the column is operated under appropriate conditions. Such systems may also be regenerated by passing through a solution of soluble enzyme.

Perhaps the best-known industrial-scale enzyme, used in immobilized form on DEAE-Sephadex, is amino acylase, used in the production of synthetic L-amino acids. Glucose isomerase may also be immobilized by this method. Glucose isomerase is an intracellular enzyme, and immobilized forms generally consist of dead microbial cells which have been cross-linked by glutaraldehyde and subsequently pelleted. The enzyme is then effectively immobilized within the producing microbial cells. The pelleted cellular preparations may be poured into a reactor column, through which the glucose syrup is passed.

The most widely used method of enzyme immobilization involves covalent attachment of the enzyme to a suitable insoluble support matrix. Covalent attachment is technically more complex than most other methods of immobilization. The attachment procedures are undertaken by highly specialized technical personnel, and require a variety of often expensive chemical reagents. Immobilization procedures are also often time-consuming. The covalent nature of the binding renders the immobilized enzyme preparations very stable, and leaching of the enzyme from the column is minimal.

A wide variety of different immobilization procedures have been developed. All promote the formation of a covalent bond between a

Table 9.7. Some amino acids normally involved in the formation of covalent linkages in enzyme immobilization

Amino acid	Functional group
Lysine	$-NH_2$
Arginine	$-NH_2$
Tyrosine	—⟨ ⟩—OH
Aspartic acid	$-COOH$
Glutamic acid	$-COOH$
Cysteine	$-SH$
Methionine	$-S-CH_3$

suitable reactive group present on the surface of the insoluble matrix and a suitable group present on the surface of the protein. Hydroxyl groups present in carbohydrate-based matrices such as cellulose, dextrans or agarose often participate in covalent bond formation. Amino, carboxyl and sulphydryl groups present in various amino acids are also generally involved in covalent bond formation (Table 9.7).

Further Reading

Books

Berger, S. & Kimmel, A. (eds), (1987). *Methods in Enzymology, Vol. 152, Guide to Molecular Cloning Techniques*. Academic Press, London.

Haresign, P. & Cole, D. (eds) (1992). *Recent Advances in Animal Nutrition*. Garnsworthy, Butterworth-Heinemann, Oxford.

Lyons, T.P. (ed.) (1992). *Biotechnology in the Feed Industry*. Alltech technical publications, Nicholasville, Ky.

Wiseman, A. (ed.) (1987). *Handbook of Enzyme Biotechnology*, 2nd edn. Halsted Press, New York.

Woodward, J. (ed.) (1985). *Immobilized Cells and Enzymes, A Practical Approach*. IRL Press, Oxford.

Articles

Altenbach, S. & Simpson, R. (1990). Manipulation of methionine rich protein genes in plant seeds. *Trends in Biotechnol.*, **8**, 156–160.

Bar, R. (1989). Cyclodextrin-aided bioconversions and fermentations. *Trends in Biotechnol.*, **7**, 2–4.

Beck, C. & Ulrich, T. (1993). Biotechnology in the food industry. *Bio/Technology*, **11**, 895–902.

Bell, G. *et al.* (1978). Ester and glyceride synthesis by *Rhizopus arrhizus* mycelia. *FEMS Microbiol. Lett.*, **3**, 223–225.

Colaco, C. *et al.* (1992). Extraordinary stability of enzymes dried in trehalose: simplified molecular biology. *Bio/Technology*, **10**, 1007–1011.

Cromwell, G. *et al.* (1993). Efficacy of phytase in improving the bioavailability of phosphorus in soybean meal and corn-soybean meal for pigs. *J. Anim. Sci.*, **71**, 1831–1840.

Hall, J. *et al.* (1993). Manipulation of the repertoire of digestive enzymes secreted into the gastrointestinal tract of transgenic mice. *Bio/Technology*, **11**, 376–379.

Kim, S. *et al.* (1988). Crystal structures of two intensely sweet proteins. *Trends Biochem. Sci.*, **13**, 1, 13–15.

Knorr, D. & Sinskey, A. (1985). Biotechnology in food production and processing. *Science*, **229**, 1224–1229.

Lambrechts, C. *et al.* (1992). Utilization of phytate by some yeasts. *Biotechnol. Lett.*, **14**, 1, 61–66.

Lyons, T.P. (1992). Enzymes—a world of solutions. *Feed Comp.*, June/July, 22–27.

MacDonald, H. (1993). Flavour development from cocoa bean to chocolate bar. *Biochemist*, **15**, 3–5.

Mazid, M. *et al.* (1993). Biocatalysis and immobilized enzyme/cell bioreactors. *Bio/Technology*, **11**, 690–695.

Pen, J. *et al.* (1993). Phytase-containing transgenic seeds as a novel feed additive for improved phosphorus utilization. *Bio/Technology*, **11**, 811–814.

Pointillard, A. *et al.* (1987). Importance of cereal phytase activity for phytate phosphorus utilization by growing pigs fed diets containing triticle or corn. *J. Nutr.*, **117**, 907.

Rees, W.D. *et al.* (1990). A molecular biological approach to reducing dietary amino acid needs. *Bio/Technology*, **8**, 629–632.

Schmid, G. (1989). Cyclodextrin glycosyltransferase production: yield enhancement by overexpression of cloned genes. *Trends in Biotechnol.*, **7**, 244–248.

Simons, P. *et al.* (1990). Improvement of phosphorus availability by microbial phytase in broilers and pigs. *Br. J. Nutr.*, **64**, 525–540.

Skatrud, P. (1992). Genetic engineering of β-lactam antibiotic biosynthetic pathways in filamentous fungi. *Trends in Biotechnol.*, **10**, 324–329.

Szejtli, J. (1991). Cyclodextrins in drug formulations: Part 1. *Pharm. Technol. Int.*, **3**, 2, 15–22.

Szejtli, J. (1991). Cyclodextrins in drug formulations: Part 2. *Pharm. Technol. Int.*, **3**, 3, 16–22.

Walsh, G.A. *et al.* (1993). Enzymes in the animal-feed industry. *Trends in Biotechnol.*, **11**, 424–430.

Witty, M. (1990). Thaumatin II—a palatability protein. *Trends in Biotechnol.*, **8**, 113–116.

Index

Note: Page references in *italics* refer to Figures; those in **bold** refer to Tables

Index compiled by Annette Musker